GENERA ORCHIDACEARUM

VOLUME 1

General Introduction, Apostasioideae, Cypripedioideae

To our dear friend Robert L. Dressler,
for a lifetime of devotion to the
biology and classification of Orchidaceae
and for inspiring the next generations of orchid scientists

GENERA ORCHIDACEARUM

VOLUME 1

General Introduction, Apostasioideae, Cypripedioideae

Edited by

ALEC M. PRIDGEON, PHILLIP J. CRIBB, MARK W. CHASE,
Royal Botanic Gardens, Kew

and

FINN N. RASMUSSEN
Botanical Laboratory, University of Copenhagen

OXFORD
UNIVERSITY PRESS

OXFORD
UNIVERSITY PRESS

Great Clarendon Street, Oxford OX2 6DP

Oxford University Press is a department of the University of Oxford.
It furthers the University's objective of excellence in research, scholarship,
and education by publishing worldwide in
Oxford New York

Athens Auckland Bangkok Bogotá Buenos Aires Calcutta
Cape Town Chennai Dar es Salaam Delhi Florence Hong Kong Istanbul
Karachi Kuala Lumpur Madrid Melbourne Mexico City Mumbai
Nairobi Paris São Paolo Singapore Taipei Tokyo Toronto Warsaw
with associated companies in Berlin Ibadan

Oxford is a registered trade mark of Oxford University Press
in the UK and in certain other countries

Published in the United States
by Oxford University Press Inc., New York

A catalogue record for this book is available from the British Library

Library of Congress Cataloging in Publication Data
(Data available)
ISBN 0 19 850513 2

Typeset by EXPO Holdings, Malaysia

Printed in Great Britain
on acid-free paper by
The Bath Press, Avon

FOREWORD

by *Robert L. Dressler*

In the early years of orchid classification, botanists were primarily concerned with naming the unending flood of new species from the tropics and had little time for subfamilies, tribes or subtribes. Indeed plant classification has evolved quite a bit in the last two centuries. In Linnaeus' time there was no real theory of classification. Even the introduction of the concept of natural selection by Darwin and Wallace had little effect on the methods of taxonomists. They continued to classify by intuition or by their own unique theories for many years. Early classification laboured under some very unbiological concepts inherited from the Greek philosophers, such as the ideas of defined categories and idealized 'types'. Unfortunately, when biologists thought to select name-bearing specimens for new species, they chose the term 'type', which is very close to the philosophers' *'Typus'*. However unbiological, such ideas have not died easily, and even now the type specimen is sometimes misunderstood. A type is not necessarily typical. Ideally a student should carefully study all other materials of a study group to decide what the species are and then finally consult the type specimens to determine the correct name for each species.

Keys often emphasize single features, but natural groups do not differ from each other only by single features. One may use single features to identify, but classification should be based on as many features as possible. This is especially important in the orchids, where there is a great deal of parallelism and convergence, both in vegetative and floral features. The occasionally contentious and dogmatic development of cladistics has been concerned more with method than with theory, but behind the method there must be theory, and we now have a much closer approach to a theory of classification than we had 50 years ago. Cladistics not only promotes the analysis of all adequately sampled features but makes a clear distinction between ancestral and derived features at each level.

The early attempts to classify the orchids remind one of the old tale of the elephant and the blind men. Lindley based his classification primarily on the features of the pollinia. The pollinia show very clear polarity and are certainly important, yet any one-character classification is suspect. The number of states in the pollinia is limited, and there is a good deal of parallelism. Perhaps the most glaring case of bad taxonomy based on pollinia is the supposed distinction between *Laelia*, with eight pollinia, and *Cattleya*, with four pollinia. The Mexican laelias (including the type of the genus) are not closely related to the Brazilian 'laelias', and *Laelia* section *Cattleyodes* especially is very closely related to the Brazilian cattleyas, with natural hybridization wherever they coexist. *'Laeliocattleya' dormaniana*, however, is a distinct species of *Cattleya* with four additional rudimentary pollinia. Some self-pollinating orchids have been misplaced because their pollinia germinate within the anther and appear soft (as in *'Crybe'* = *Bletia*) or because they seem to lack stipe or viscidium, both more obvious in their outcrossing relatives (as in *'Neourbania'* = *Maxillaria*).

Pfitzer rejected Lindley's system, partly because pollinia may be lacking in some specimens, and chose leaf vernation as a major feature. Leaf development may be difficult to determine in a specimen without leaves or with only mature leaves, and one occasionally finds a *Zygopetalum* with conduplicate development in one shoot and convolute development in another at the same time. Both pollinia and leaf development are useful features, but to use either without considering other features may lead to classifications as meaningful as the blind men's hypotheses about the elephant.

The last 50 years have seen an increased interest in orchid phylogeny. There were inconsistencies in the Schlechterian system, and we tried to modify that system to reflect biological reality. This was fairly easy at the generic level, but at higher levels it is difficult to make sense of the complex patterns of variation. The analysis of DNA is a powerful tool for dealing with these complexities. Factors such as pollination syndromes may cause convergence and lead taxonomists to create artificial groups apparently united by several derived features; such groups may be difficult to distinguish by analysis of structural features. Whereas there may be parallelism in single base pairs of DNA, there is no reason to think that any sort of selection pressure can cause convergence in many base pairs in separate regions of the genome. Further, the number of base pairs varying in a well-chosen region of DNA may be much greater than the number of useful morphological features in a given group, so that molecular data may be treated statistically with more confidence and little worry about convergence. The comparison of DNA will often falsify our earlier hypotheses, but in most cases the new arrangement makes sense and seems better than the old. Many of the results are about what one would expect from Hennigian theory. Groups united by primitive features do not hold up well. The union of the Apostasioideae and the Cypripedioideae, under whatever name, cannot be supported. The old Polychondreae, or Neottieae, or Neottioideae, are a grade of several distantly related groups, sharing only soft pollen masses. At the same time, molecular systematics favours the union of the Cranichideae, Diurideae, and Orchideae into a single, broader subfamily Orchidoideae, with some realignments in Cranichideae and Diurideae. Such a subfamily is easily recognized and gives the lie to the old dogma that floral features are more 'important' than vegetative features. At the same time, molecular systematics indicates that the derivation of the Diseae from the Orchideae is as upside-down as the anther of *Satyrium*. Rather, the Orchideae were derived from a (presumably African) *Disa* grade. How to classify them is not yet clear, but this relationship may explain the fixation of the 'basitonic' character of the Orchideae. A group with the anther upside-down had to have the viscidia basal to achieve pollination. Thus basitony became fixed in the *Disa* complex and has remained so in the derivative Orchideae.

Within the Epidendroideae similar patterns are evident. Relatively primitive groups such as Arethuseae, Bletiinae, Calypsoeae, and Cymbidieae may break up under careful analysis, though we need more thorough sampling to decide what to do with some of these polyphyletic or paraphyletic grades. Some groups, such as the Epidendreae, neither primitive nor highly derived, have been difficult to delimit, at least without much more structural evidence than we have. Molecular studies, however, promise to bring order into this chaos, separating the Epidendreae into two major New World groups, the Epidendreae and the Pleurothallideae, each with a few Old World members, and associating most Old World Epidendreae with other groups, including the fragments of the old Arethuseae or Bletiinae. More derived groups are likely to hold up or to be expanded in response to molecular analysis. The majority of the 'vandoid' New World orchids make up a natural Maxillarieae, whereas the Vandeae themselves appear to be derived from epidendroid ancestors (but not along the lines I suggested in 1993). This is a bad time to offer hypotheses about orchid phylogeny based only on morphology.

PREFACE

Orchidaceae are the largest monocot family and perhaps the largest flowering plant family as well in terms of number of species (*c.* 19 000; Atwood 1986), but for a variety of reasons, such as lack of a fossil record, relative scarcity of scientific research until recent years, size and geographical distribution of the family, and the largely tropical distribution of most which hinders access, it remains one of the least understood.

The early classifications and generic circumscriptions of Lindley (1830–40), Bentham and Hooker (1883), and Schlechter (1927) relied entirely on floral morphology. The problem with this at higher levels of classification is that floral morphology is extremely plastic in evolutionary terms, so that unrelated species may have developed similar structures in response to similar selection, particularly pollinator pressures. Such parallelisms are rife in Orchidaceae and largely responsible for dramatically shifting taxonomic concepts over the last 250 years, so it is necessary to apply more objective criteria and character choices in trying to unravel the complexities of the family's history. Dressler's (1993) classification introduced characters from pollen, seeds, and anatomy to supplement traditional characters and identified many synapomorphies, making it the most natural system at that time.

The recent surge in molecular approaches, combined with several other types of new data and phylogenetic analysis, has already provided support for the recircumscription of many taxa from the rank of genus to even subfamily. We have found that the best systematic approach to handling such a species-rich group as Orchidaceae is first to work out higher-level relationships with plastid DNA. Using this information to assign outgroups, we can then focus more closely on lower taxa with both plastid and nuclear DNA and other types of data—morphological, anatomical, embryological, etc.—to produce the first truly comprehensive phylogenetic treatment of this large family. For example, we have adopted the concept of Vanilloideae (*Vanilla*, *Pogonia*, and their relatives, subtribes Vanillinae and Pogoniinae *sensu* Dressler 1993) on the basis of *rbcL* studies (Cameron *et al.* 1999) demonstrating that this group is clearly distinct from all other monandrous orchids but (quite distantly) related to them. Following the same methods and principles we have abandoned Spiranthoideae (*sensu* Dressler 1993) in light of several recent molecular studies (Chase *et al.* 1994; Kores *et al.* 1997; Cameron *et al.* 1999) showing them embedded within Orchidoideae as the close relatives of Chloraeinae and other Diurideae, classified as Orchidoideae by Dressler (1993).

The aim of *Genera Orchidacearum* is therefore to produce a more robust and natural account of the orchids at the generic level and above, incorporate the wealth of new molecular data in a truly phylogenetic classification, and identify those disciplines and taxa needing additional work by students of the future. We and our contributors apply the same phylogenetic principles and methods to systematics of Orchidaceae that have been successfully applied to many other groups of the world's organisms in the knowledge that those evolutionary processes governing other plant families and indeed other phyla have not been idiosyncratically repealed for orchids.

Although the project is based at the Royal Botanic Gardens, Kew, where there are currently several staff members, students, short-term and long-term visitors working exclusively on the morphology, anatomy, and molecular phylogenetics of orchids, more than 50 collaborators from around the world will help to make *Genera Orchidacearum* the authoritative orchid classification for years to come and ensure that there are minimal gaps in coverage.

We anticipate publication in five volumes over a six-year period as follows:

Volume 1: Introduction; subfamilies Apostasioideae and Cypripedioideae

Volume 2: Subfamily Orchidoideae (Orchideae, Diseae, Diurideae)

Volume 3: Subfamily Orchidoideae (remaining taxa, formerly in Spiranthoideae)

Volume 4: Subfamilies Vanilloideae and Epidendroideae (part 1)

Volume 5: Subfamily Epidendroideae (part 2); summary and final classification

For each subfamily, tribe, and subtribe there will be a description with keys to respective lower taxa. These will be followed by generic descriptions with complete nomenclature and type citation; distribution (with maps); treatments of anatomy, palynology, cytogenetics, phytochemistry, phylogenetics, ecology, pollination, economic uses, and cultivation; and concluding with taxonomic notes and select bibliography. Endmatter in each volume will include literature cited and indices of subjects and scientific names. Authority names follow the usage recommended by Brummitt and Powell (1992).

Opening this volume are several chapters by different contributors briefly reviewing the taxonomic history of Orchidaceae and our state of knowledge in several disciplines as they relate to modern phylogenetics, especially with respect to synapomorphies (shared derived features), methods of analysis, and measures of support. *Genera Orchidacearum* is principally a systematic rather than a broadly biological treatise, and therefore reviews of orchid physiology, tissue culture, and mycorrhiza, which have been summarized so well in numerous other books, are not included here.

If we have learned anything about orchid systematics in the last 250 years, it is that classification and nomenclatural changes are frustrating but necessarily protean as new technologies evolve to offer new or refreshed insights. For example, development of the first electron microscopes in the 1930s eventually helped researchers half a century later to discover new characters never imagined by Lindley or to refine previous observations using the light microscope. The relatively recent advent of molecular techniques has done nothing less than revolutionize the way we view orchid phylogeny in that it focuses directly on the genotype instead of the phenotype. This is not to say that molecular data are any more important than other types of data, but they do afford a more easily objective way of viewing the natural world around us.

Kew   AMP
Copenhagen   PJC
                                                                                                        MWC
                                                                                                        FNR

ACKNOWLEDGEMENTS

First and foremost we wish to thank all other contributors whose extensive knowledge of Orchidaceae, particularly Apostasioideae and Cypripedioideae, made this volume as authoritative and up-to-date as possible: John Atwood, Peter Brandham, Mark Clements, Tony Cox, Renée Grayer, Harold Koopowitz, Mia Molvray, Holger Perner, William L. Stern, Nigel Veitch, and Jeffrey Wood. Reviewers of submitted text included Carol Furness, Hanne Rasmussen, Paula Rudall, and Judy West. John Freudenstein offered valuable suggestions as well. We acknowledge all those who supplied photographs and the numerous individuals worldwide who helped with travel and logistics for photography and fieldwork, as well as Judi Stone, Eleanor Catherine, and Valerie Price who prepared excellent diagnostic drawings and Jeff Eden who produced the distribution maps. We thank the Botanical Society of America, the Linnean Society of London, and Dr. Wendy Zomlefer for permission to reprint original photographs or line drawings.

We are most grateful to Professor Sir Ghillean Prance, Director of the Royal Botanic Gardens, Kew, and the Trustees of the Royal Botanic Garden, Kew, for supporting this international project from the outset and providing laboratories and computer support, without which the molecular research and editing would have been impossible. We also wish to thank in this respect Professor Gren Lucas, former Keeper of the Herbarium, Royal Botanic Gardens, Kew, for initially approving the project as well as Professor Simon J. Owens, present Keeper of the Herbarium, for continuing that support. Professor Michael D. Bennett, Keeper, and Dr. David F. Cutler of the Jodrell Laboratory also offered unwavering encouragement. Anette de Bruijn, Jeffrey Joseph, and Martyn Powell assisted in extractions and sequencing in the Molecular Systematics Laboratory of the Jodrell, and Sarah Thomas assisted with distribution maps. The Living Collections Division of the Royal Botanic Gardens, Kew, made available many species for study.

PC would like to thank Peter Taylor and Mark Wood for stimulating his interest in slipper orchids; Sir Robert and Lady Sainsbury for their support in the genesis of this project and their continuing enthusiasm for it; and the many colleagues and friends who have helped him either with access to materials or photographs or with fieldwork on slipper orchids in both the Old and New Worlds.

AP wishes to thank Sir Robert and Lady Sainsbury for their generous patronage through the years; the Australian Orchid Foundation, Hermon Slade, Gerald McCraith, Lindsay Pryor and the Australia and Pacific Science Foundation, and Mark Clements, David Jones, Susan Walker, and Corinna Broers for making possible fieldwork and laboratory studies in Australia; and especially his wife Sherry for her tolerance and loving endurance.

ALEC M. PRIDGEON
PHILLIP J. CRIBB
MARK W. CHASE
FINN N. RASMUSSEN

CONTENTS

List of contributors   xiii

List of colour plates   xv

HISTORY, ELEMENTS, AND SCIENCE OF ORCHID CLASSIFICATION

I. The development of orchid classification   3
Finn N. Rasmussen

II. Morphology   13
Phillip J. Cribb

III. Anatomy   24
Alec M. Pridgeon

IV. Palynology   33
Alec M. Pridgeon

V. Embryology   38
Mark A. Clements

VI. Seed morphology   59
Mia Molvray and Mark W. Chase

VII. Cytogenetics   67
Peter Brandham

VIII. Molecular systematics, parsimony, and orchid classification   81
Mark W. Chase

CLASSIFICATION

ORCHIDACEAE   91

  A. APOSTASIOIDEAE   94

    1. *Neuwiedia*   96

    2. *Apostasia*   100

  B. CYPRIPEDIOIDEAE   105

    3. *Cypripedium*   114

    4. *Selenipedium*   132

    5. *Paphiopedilum*   137

    6. *Phragmipedium*   153

    7. *Mexipedium*   161

Glossary	165
References	174
Index to scientific names	191
Subject index	195

Contributors

Abbreviations used in text for authorship credit

DR JOHN ATWOOD (JA)
Marie Selby Botanical Gardens, 811 S. Palm Avenue, Sarasota, FL 34236 USA

DR PETER BRANDHAM (PB)
Royal Botanic Gardens, Kew, Richmond, Surrey TW9 3AB UK

DR MARK W. CHASE (MWC)
Royal Botanic Gardens, Kew, Richmond, Surrey TW9 3AB UK

DR MARK A. CLEMENTS (MAC)
Centre for Plant Biodiversity Research, Australian National Herbarium, CSIRO Plant Industry, GPO Box 1600, Canberra, ACT 2601 Australia

DR ANTONY COX (AC)
The Sanger Centre, Wellcome Trust Genome Campus, Hinxton, Cambs CB10 1SA UK

DR PHILLIP J. CRIBB (PC)
Royal Botanic Gardens, Kew, Richmond, Surrey TW9 3AB UK

DR ROBERT L. DRESSLER (RD)
Missouri Botanical Garden, PO Box 299, St Louis, MO 63166-0299 USA

DR RENÉE GRAYER (RG)
Royal Botanic Gardens, Kew, Richmond, Surrey TW9 3AB UK

DR HAROLD KOOPOWITZ (HK)
Ecology and Evolutionary Biology, University of California, Irvine, CA 92697-2525 USA

DR MIA MOLVRAY (MM)
G. L. Cross Hall, Department of Botany and Microbiology, University of Okalhoma, 70 Van Vleet Oval, Norman, OK 73019-0245 USA

DR HOLGER PERNER (HP)
GKSS-Forschungszentrum Geesthacht GmbH, Max-Planck-Strasse, D-21502 Gesthacht, Postfach 11 60, D-21494 Geesthacht, Germany

DR ALEC M. PRIDGEON (AP)
Royal Botanic Gardens, Kew, Richmond, Surrey TW9 3AB UK

DR FINN N. RASMUSSEN (FR)
Botanical Institute, University of Copenhagen, Gothersgade 140, DK-1123 Copenhagen K, Denmark

PROF. WILLIAM LOUIS STERN (WS)
Department of Botany, University of Florida, Gainesville, FL 32611 USA

DR NIGEL VEITCH (NV)
Royal Botanic Gardens, Kew, Richmond, Surrey TW9 3AB UK

JEFFREY WOOD (JW)
Royal Botanic Gardens, Kew, Richmond, Surrey TW9 3AB UK

Colour Plates

The plates section is between pp. 88 and 89.

Plate 1. Epiphytic, lithophytic, and terrestrial habits of orchids.

Plate 2. Terrestrial, climbing, and subterranean habits of orchids.

Plate 3. Growth forms of orchids and plicate leaves.

Plate 4. Conduplicate leaves and common inflorescence types.

Plate 5. Ovary and uncommon inflorescence types in orchids.

Plate 6. Floral morphology of orchids.

Plate 7. Column of *Aspasia principissa* showing anther, stigmatic cavity, pollinarium, pollinia, 'stipe' (tegula), and viscidium.

Plate 8. Pollinarium, dimorphic flowers, capsule, and protocorms.

Plate 9. Representative species of *Neuwiedia*, *Apostasia*, and *Cypripedium*.

Plate 10. Representative species of *Cypripedium*.

Plate 11. Representative species of *Cypripedium*.

Plate 12. Representative species of *Cypripedium and Selenipedium*.

Plate 13. Representative species of *Paphiopedilum*.

Plate 14. Representative species of *Paphiopedilum*.

Plate 15. Representative species of *Paphiopedilum and Phragmipedium*.

Plate 16. Representative species of *Phragmipedium* and *Mexipedium xerophyticum*.

History, Elements, and Science of Orchid Classification

The Development of Orchid Classification

The first orchid systems—Swartz and the explorers

The classification of orchids has developed by accumulation of discoveries of new species and a steadily growing knowledge of the life history, morphology, and anatomy of the species in this large family of plants. This growing body of information has also been accompanied by changing and controversial views on methods and objectives of biological classification.

The extraordinary diversity of the orchids has only been fully realized during the last 150 years, yet orchids are mentioned in the earliest botanical literature as sources of wonderment or medicinal power. The well-known belief in the powers of the testicle-like tubers of *Orchis* was described by Dioscorides in his *De Materia Medica* (c. 70 AD), applying a name already used by Theophrastos. Several recognizable species were described and depicted in medieval herbals; see Garay (1976) and Jacquet (1994) for thorough reviews of orchids in early European botanical literature. Through works like Rheede (1678–1693) and Rumphius (1741–1750; see de Wit 1977) European naturalists started to learn about the astonishing tropical orchid flora. Linnaeus, however, recognized only eight genera in his class XX 'Gynandria Diandria' for *Species Plantarum* (1753), the basis for botanical nomenclature. The generic names thus established by Linnaeus are *Orchis, Satyrium, Ophrys, Serapias, Limodorum, Cypripedium,* and *Epidendrum*. He mentioned 69 species, of which 21 were placed in *Orchis*, 17 in *Ophrys*, and 14 in *Epidendrum*. Less than half of these species are represented by material in his herbarium, but synonyms in older literature were often cited. Many of the Linnean species are now considered mixtures of more than one taxon. The second edition of *Species Plantarum* (1762–63) lists 30 species of *Epidendrum*; in present-day classifications they are distributed in at least 15 genera. The family name Orchidaceae is ascribed to de Jussieu, who recognized orchids as one of the 'Ordines naturales' in his *Genera Plantarum* (1789).

Olof Swartz, who studied under the younger Linnaeus in Uppsala, became the first specialist in orchid taxonomy when he published a comprehensive critical review of orchid literature, a morphological analysis, and a classification with a key to the 25 genera that he recognized, of which ten were his own creations (1800a). He defended this remarkable inflation by referring to the insight he had gained during field studies in the West Indies and with access to new material collected by his friends Thunberg, Sparrman, and Afzelius. He was convinced that many more genera would have to be accepted but adopted the commendable position of treating only taxa for which he had studied plant material. Swartz was the first to realize that most orchids have only one stamen whereas slipper orchids have two, and he introduced the subdivisions Monandrae and Diandrum (the latter with only one genus, *Cypripedium*). Later the same year a monograph of the family followed (1800b), which included diagnostic drawings of 16 genera. Swartz subdivided the monandrous orchids into three groups without formal names based on the position of the anther on the column and other characters. These groups correspond essentially to what would be called orchidoids, neottioids, and epidendroids in modern orchid jargon (Table I.1).

Much of the special descriptive terminology used in orchid literature stems from a paper by L. C. Richard (1817) on European orchids. Richard emphasized the structure of the pollinia and details of the gynostemium in his key and depicted details of the critical characters on a plate with the heading '*Genitalia Orchidearum*'. At the beginning of the nineteenth century the knowledge of tropical orchids grew quickly with the works of botanists such as Brown (1810), du Petit-Thouars (1809, 1822), and Blume (1825a,b), who all had opportunities to study orchids in the tropics. These early orchidologists amended Swartz's classification and added characters as new genera were discovered.

Table I.1. A tabular overview of major orchid classifications. The sequence of the groups in some systems has been changed to facilitate comparison. Groups may differ considerably in content, making a linear comparison impossible.

Swartz 1800*b*	Lindley 1830–40	Bentham 1881	Pfitzer 1887
		Cypripedieae trib.	DIANDRAE subf. *(later as Pleonandrae)* [2 TRIBES]
(apostasioids not known)	*(apostasioids excluded)*		
Diandrum	Cypripedieae trib.		
			MONANDRAE subf.
Monandrae — Anthera subterminali	Ophreae trib.	Ophrydeae trib.	Basitonae [1 TRIBE]
Anthera erecta	Neottieae trib.	Neottieae trib.	Acrotonae — Acranthae — Convolutae — Continentes [1 TRIBE] / Articulatae [4 TRIBES]
Anthera operculari	Arethuseae trib.	Epidendreae trib.	Duplicatae [6 tribes] — Pleuranthae — Convolutae — Homoblastae [3 TRIBES] / Heteroblastae [3 TRIBES]
	Malaxeae trib.		
	Epidendreae trib.		Duplicatae — Sympodiales [9 TRIBES]
	Vandeae trib.	Vandeae trib.	Monopodiales [2 TRIBES]

Classic orchid systematics—Lindley and the evolutionists

John Lindley is a legendary figure in the history of orchid classification and also a very important figure in general plant systematics. He was a prolific writer who published books on many aspects of systematics. Lindley was, as were most of his contemporaries, an ardent pursuer of the 'natural' system, to be understood as a system that as closely as possible reflected the great plan in nature. This could only be achieved by studying patterns of characters and abandoning the artificial Linnean arrangements, which Lindley dismissed as 'once popular, but superficial and useless' (1847, p. 1). Although not 'evolutionary', Lindley's discussion of relations between taxa did not differ very much from later discussions by post-Darwinian systematists before the appearance of cladistics in the 1960s.

Lindley was the first to use a category between family and tribe in his *Orchidearum Sceletos* (1826). In this classification scheme he divided the orchids into four groups, Neottieae, Orchideae, Epidendreae, and Cypripedieae, which are now recognized as subfamilies and cited with the subfamily suffix -*oideae*. Lindley changed his primary classification of orchids several times. In *Nixus Plantarum* (1833) he listed three families in 'cohors Gynandrae': Orchideae, Cypripedieae, and Apostasieae, a view later revived by Vermeulen (1966). In the second edition of *A Natural System of Botany* (1836) he recognized a separate family, Vanillaceae, comprising *Vanilla* and *Epistephium* because of their succulent fruits and unusual seeds. This position was abandoned in later works, in which he also simplified the system by omitting the subfamilies and reducing the number of tribes from eight to seven.

Lindley probably examined more orchid material than any botanist before him when he started publishing *The Genera and Species of Orchidaceous Plants*

Table I.1. A tabular overview of major orchid classifications. The sequence of the groups in some systems has been changed to facilitate comparison. Groups may differ considerably in content, making a linear comparison impossible (continued).

SCHLECHTER 1926	MANSFELD 1937	DRESSLER AND DODSON 1960	GARAY 1960 AND 1972
		CYPRIPEDIOIDEAE subf.	APOSTASIOIDEAE subf.
(apostasioids excluded)	(diandrous orchids not treated)		
			CYPRIPEDIOIDEAE subf.
DIANDRAE subf.			
CYPRIPEDILOIDEAE trib.			
MONANDRAE subf.	MONANDRAE subf.	ORCHIDOIDEAE subf.	ORCHIDOIDEAE subf.
Basitonae	Thraurosphaereae	ORCHIDEAE trib.	ORCHIDEAE trib.
OPHRYDOIDEAE trib.	OPHRYDEAE trib.		DISEAE trib.
			DISPERIDEAE trib.
Acrotonae	NEOTTIEAE trib.		NEOTTIOIDEAE subf.
POLYCHONDREAE trib.		NEOTTIEAE trib.	EPIPOGONEAE trib.
			NEOTTIEAE trib.
			CRANICHIDEAE trib.
KEROSPHAEREAE trib.	Kerosphaereae	EPIDENDREAE trib.	EPIDENDROIDEAE subf.
Acranthae	EPIDENDREAE trib.		EPIDENDREAE trib.
Pleuranthae			
Sympodiales			
Monopodiales	VANDEAE trib.		VANDEAE trib.

(1830–1840) (Table I.1). Lindley stated in the preface (written in 1840) that he had treated 1980 species, for 75% of which he had analysed the 'organs of fructification'. He acknowledged plant material from many correspondents and encouraged his readers to send him more for identification but made clear that he could not return such material and would incorporate it with his own herbarium. Lindley's herbarium contained about 7000 specimens when it was purchased by the Royal Botanic Gardens, Kew, after Lindley's death in 1865 (Cribb 1998). Lindley presented the primary divisions of his most thorough orchid system in the 'tabular view of the tribes of Orchidaceae', which is also a key to the seven tribes:

I. Anther one only
  A. Pollen masses waxy.
    a. No caudicula or separable stigmatic gland   **Tribe I.** Malaxeae (or Malaxideae)
    b. A distinct caudicula, but no separable stigmatic gland   **Tribe II.** Epidendreae
    c. A distinct caudicula, united to a deciduous stigmatic gland   **Tribe III.** Vandeae
  B. Pollen powdery, granular or sectile
    a. Anther terminal, erect   **Tribe IV.** Ophreae (or Ophrydeae)
    b. Anther terminal, opercular   **Tribe V.** Arethuseae
    c. Anther dorsal   **Tribe VI.** Neottieae
II. Anthers two   **Tribe VII.** Cypripedieae

The numbering of the tribes was not meant to express a 'natural sequence' as in the systems of some later authors. The Malaxidiae were subdivided

Table I.1. A tabular overview of major orchid classifications. The sequence of the groups in some systems has been changed to facilitate comparison. Groups may differ considerably in content, making a linear comparison impossible (continued).

Vermeulen 1966	Dressler 1981	Burns-Balogh and Funk 1986
Apostasiaceae fam.	APOSTASIOIDEAE subf.	NEUWIEDIOIDEAE subf.
		APOSTASIOIDEAE subf.
Cypripediaceae fam.	CYPRIPEDIOIDEAE subf.	CYPRIPEDIOIDEAE subf.
Orchidaceae fam.	ORCHIDOIDEAE subf.	ORCHIDOIDEAE subf.
ORCHIDOIDEAE subf.	Orchideae trib.	[3 tribes]
	Neottieae trib.	
	Diurideae trib.	
EPIDENDROIDEAE subf.		NEOTTIOIDEAE subf.
Neottianthae	SPIRANTHOIDEAE subf.	[4 tribes]
Neottieae trib.	Erythrodeae trib.	
	Cranichideae trib.	SPIRANTHOIDEAE subf.
Arethuseae trib.		[3 tribes]
Vanilleae trib.	EPIDENDROIDEAE subf.	
	Epipogieae trib.	EPIDENDROIDEAE subf.
	Vanilleae trib.	[10 tribes]
Epidendranthae	Gastrodieae trib	
Epidendreae trib.	Arethuseae trib.	
	Malaxideae trib.	
	Coelogyneae trib.	
	Calypsoeae trib.	
	Cryptarrheneae trib.	
	Epidendreae trib.	
Vandeae trib.	VANDOIDEAE subf.	
	Cymbidieae trib.	
	Maxillarieae trib.	
	Polystachyeae trib.	
	Vandeae trib.	

into 'sections', the Arethuseae and Neottieae into 'divisions'. In the second edition of *The Vegetable Kingdom* (1847), Lindley mentioned three families (still referred to as 'natural orders'), Burmanniaceae, Orchidaceae, and Apostasiaceae under Alliance (now order) Orchidales. He counted 394 genera and estimated the number of species to be at least 3000. *Folia Orchidacea*, begun in 1852, was intended to be a series of monographs of all orchid genera, the last part being published in 1859. No taxonomic rearrangements at the tribe or family level were suggested in this work. Lindley also contributed to orchidology with a number of beautifully illustrated articles in *Edwards's Botanical Register* and in folio books such as *Illustrations of Orchidaceous Plants* (Bauer and Lindley 1830–38) and *Sertum Orchidaceum* (1837–41).

In 1881 George Bentham published a new orchid classification (Table I.1) as a precursor to the treatment of Orchidaceae in his and J. D. Hooker's great systematic work, *Genera Plantarum*. Bentham's system was based on Lindley's, but the Malaxideae were included in the Epidendreae (an action already suggested by Lindley) and the Arethuseae in Neottieae. Each of the five remaining tribes except the Cypripedieae were subdivided into subtribes, a level of classification that since then has been fundamental in orchid systematics. Bentham was aware that the 'caudicles' and other characters used in orchid classification at the tribal level were insufficiently known and introduced the term *stipes* for the pollinium stalk in the Vandeae. He challenged the illustrious German orchid specialist H. G. Reichenbach to publish an orchid classification or at least reveal his principles for distinguishing genera in time to influence the final orchid classification in *Genera Plantarum*—which appeared in Volume III in 1883 without feedback from Reichenbach. Bentham included the genera *Apostasia* and *Neuwiedia* in the tribe Cypripedieae, but he did not recognize a funda-

Table I.1. A tabular overview of major orchid classifications. The sequence of the groups in some systems has been changed to facilitate comparison. Groups may differ considerably in content, making a linear comparison impossible (continued).

DRESSLER 1993	SZLACHETKO 1995
APOSTASIOIDEAE subf.	**Apostasiaceae** fam.
CYPRIPEDIOIDEAE subf.	**Cypripediaceae** fam.
ORCHIDOIDEAE subf. [3 TRIBES]	**Orchidaceae** fam. ORCHIDOIDEAE subf. [5 TRIBES]
	THELYMITHROIDEAE subf. [7 TRIBES]
SPIRANTHOIDEAE subf. [3 TRIBES]	SPIRANTHOIDEAE subf. [3 TRIBES]
	NEOTTIOIDEAE subf. [2 TRIBES]
EPIDENDROIDEAE subf. (No phylad) [6 TRIBES INCL. NEOTTIEAE]	TROPIDIOIDEAE subf. VANILLOIDEAE subf. [6 TRIBES]
	EPIDENDROIDEAE [9 TRIBES]
Cymbidioid phylad [4 TRIBES]	
Epidendroid phylad (no subclade) [4 TRIBES]	VANDOIDEAE subf. [12 TRIBES]
Dendrobioid subclade [3 TRIBES]	

mental distinction between monandrous and diandrous orchids. The quest for a 'natural system' and Bentham's principles of classification were reviewed by Stevens (1997).

Ernst Pfitzer, director of the botanical garden in Heidelberg, had already published outstanding works on vegetative and floral morphology of orchids when he agreed to describe this family for Engler and Prantl's ambitious project *Die natürlichen Pflanzenfamilien*. In 1887 he published his own system of classification, *Entwurf einer natürlichen Anordnung der Orchideen*, which in several aspects was a radical deviation from the tradition established by Swartz, Lindley, and Bentham. The basal division Diandrae–Monandrae was maintained, but the monandrous orchids were in turn dichotomously subdivided into Basitonae with basal pollinium stalks (orchidoids in modern jargon) and Acrotonae with apical pollinium stalks. From this level, the distinguishing characters are vegetative: the Acrotonae branches into Acranthae

with terminal inflorescences and Pleuranthae with lateral inflorescenses. At the next subordinate level the vernation (arrangement of the leaves in bud) are used for forming the groups Convolutae and Duplicatae. The Monandrae–Pleuranthae–Convolutae are further classified into Homoblastae and Heteroblastae according to the number of internodes in the pseudobulbs, the Pleuranthae–Duplicatae into Sympodiales and Monopodiales referring to vegetative architecture of the plants (Table I.1). Below these categories are found two levels with names derived from names of genera with the suffix *-inae* and *-eae* (the latter ranking lower, contrary to modern usage).

Pfitzer's system was used for Orchidaceae in *Die natürlichen Pflanzenfamilien* (1889) and for his contribution to *Das Pflanzenreich*, in which only the Orchidaceae-Pleonandrae appeared with Pfitzer as sole author (1903). 'Pleonandrae' was a substitute name for 'Diandrae' because Pfitzer included the apostasioids of which *Neuwiedia* has more than two stamens. However, Pfitzer's system was difficult to use for a number of reasons. There are several exceptions in the 'vegetative groups' as was quickly pointed out by Reichenbach (1885). The hierarchy was complicated and not always used consistently, and the many categories were difficult to compare with suprageneric categories used in other classifications. Pfitzer's lowest level (e.g. Aerideae) is now interpreted as subtribus, the next higher level (e.g. Sarcanthinae) as tribus. The nomenclatural standing of these names and how their suffixes must be changed to conform to the *International Code of Botanical Nomenclature* was analysed by Butzin (1971).

Contrary to Bentham, Pfitzer addressed phylogenetic questions directly and discussed the evolution of characters, and he attempted to place 'primitive' groups before more 'advanced' taxa in a kind of evolutionary sequence. It is thus significant that he made the Diandrae (later Pleonandrae) the first subdivision, and his classification may be interpreted as reflecting a phylogenetic view. The hierarchical arrangement, which later authors have criticized for being merely an artificial key, is in fact justified by hypotheses of character evolution discussed in the *Entwurf* (1887). However, despite the new evolutionary awareness after Darwin, the principles employed by taxonomists did not change radically.

Heinrich G. Reichenbach (also known as Reichenbach *fil.* because his father, H. G. L. Reichenbach, was also a well-known botanist) produced a number of articles and monographs on orchids. The most impressive single work is his large-format *Xenia Orchidacea*, published in three volumes (1854–1900) and completed posthumously by F. Kränzlin. Reichenbach did not set forth his own classification of orchids, but in 1885 he published an essay on orchid systematics in which he respectfully criticized Pfitzer for placing too much emphasis on vegetative structures. Bentham was (less respectfully) criticized for being superficial and for paying too little attention to the details of the column, which to Reichenbach was the taxonomically most important structure in orchids. His approach to systematics was generally conservative, and he warned against excessive splitting of genera.

Reichenbach's very large collection of orchid material was left to the National History Museum in Vienna on the condition that it should not be accessible for 25 years after his death. This strange request is usually explained as an eccentric expression of vanity, but later users of the Reichenbach Herbarium have observed that it contains a remarkable number of flowers and bits pilfered from other collections, often with just a few letters or numbers that will reveal their origin only to experts. Perhaps Reichenbach simply did not want to upset his contemporary colleagues by untimely disclosure of this practice (Seidenfaden, personal communication.)

Fritz Kränzlin was a contemporary of Bentham, Reichenbach, Pfitzer, and Schlechter. He wrote a number of orchid monographs; some of the most important appeared in the work *Orchidacearum Genera et Species I* and *II* (1897–1904). He concluded the third volume of *Xenia Orchidacea* that was begun by Reichenbach (1900) and continued Pfitzer's writings for Engler's *Das Pflanzenreich*. He did not attempt his own systematic arrangement of the family. Robert A. Rolfe continued the orchid tradition in Kew. Besides contributions to large flora projects he wrote numerous short papers on genera and species and a series on the evolution of orchids (1909–1912).

The most prolific of all orchid systematists was probably Rudolf Schlechter, who published over 300 papers and books. He spent the early years of his career travelling and collecting in Africa and the Far East. The most impressive of his voluminous floristic works may be *Die Orchidaceen von Deutsch Neu Guinea* (1911–1914, figures published 1923–1928), which deals with about 1400 species, 1102 of which were new. Schlechter also published monographs on larger groups as the Polychondreae (neottioids) (1911), the Spiranthinae (1920), and the 'angraekoiden Orchidaceen' (1918). The first edition of his famous gardening manual of orchids, *Die Orchideen*, appeared in 1914–1915, the second edition in 1927. (The third edition of Schlechter's *Die Orchideen* initiated in 1970 is in reality a completely new work, to which Schlechter did not contribute.) Schlechter's main contribution to the general classification of orchids is, however, the paper *Das System der Orchidaceen*, published posthumously in 1926. In this work he incorporated his comprehensive knowledge of orchids from all continents into a classification and key to all accepted genera.

Schlechter did not state any principles or philosophy of classification, but his system was an amalgamation of the views of Bentham and Pfitzer. The Diandrae (without the apostasioids) and the Monandrae were ranked as subfamilies, the Basitonae and

Acrotonae as 'divisions'. The Division Acrotonae were divided into two tribes: the Polychondreae with soft pollinia, corresponding to neottioid orchids in the broadest sense, and the Kerosphaereae with harder pollinia, comprising Bentham's Epidendreae and Vandeae. The Kerosphaereae were in turn subdivided into two series: Acranthae and Pleuranthae, the Pleuranthae in the subseries Sympodiales and Monopodiales (Table I.1). The consistent bifurcation of one branch at each level gives this system a peculiar asymmetry, as pointed out by Dressler (1981). Each tribe or series was divided into subtribes with names ending in –eae, of which Schlechter used 81. Subtribes with many genera, such as his Sarcantheae, were sometimes subdivided into two informal levels. The classification was accompanied by a key in Latin down to subtribe. A slightly modified English translation of the key was published by Schweinfurth (1959). Schlechter's (1926) classification of orchids was the last complete system with all accepted genera enumerated until Dressler and Dodson (1960), and it became a de facto standard in orchid literature for the next two decades.

Schlechter, who has been characterized as a splitter, recognized 610 genera. His system and the numbers he assigned to the genera are still used for intercalation and filing in some of the world's large orchid herbaria. Schlechter's own herbarium with thousands of unique holotypes was kept in Berlin, until it was destroyed by wartime bombing in March 1943.

Rudolf Mansfeld (1937a, b) suggested a revision of Schlechter's system. As an introduction to these papers, he complained that all the known characters in Orchidaceae are of different value in different groups, making delineation of natural groups difficult. (Other systematists might have regarded this as an inherent difference between artificial and natural systems.) Under these conditions, he argued, all systems must depend on the author's personal experience. The most remarkable innovation in Mansfeld's system was the 'Tribengruppe' Thraurosphaereae (Table I.1), comprising the tribes Neottieae and Ophrydeae and contrasting with the Kerosphaereae. This division is comparable to the groups A and B in Lindley's key to monandrous orchids (1830–1840). He moved some genera between subtribes, reduced the number of monandrous subtribes to 63, and tried to arrange them into informal 'Verwandschaftskreise' (circles of relationship). He published a survey of characters used

for his arrangement in 1955. Mansfeld's amendments were, however, largely ignored by the users of Schlechter's system.

Hatch (1954) pointed out that the names used in Schlechter's classification were not in accordance with the *International Code of Botanical Nomenclature* (*ICBN*). DeWolf (1956) reviewed the nomenclatural problems in the orchid categories between family and genus and proposed formal conservation of some of the names. A thorough assessment of the many categories and names used in orchid systematics were published by Butzin (1971).

Recent orchid systematics—Dressler and cladistics

The pre-eminence of Schlechter's classification was not challenged until 1960, when two papers revitalized the interest in orchid evolution and systematics. Robert Dressler and Calaway Dodson (1960) presented a thorough review of Schlechter's system in the view of accumulated new information of characters. Although emphasizing the preliminary nature of their work, they did in effect propose a complete new classification (Table I.1) with a key to the tribes and subtribes. This orchid classification is the first to follow the *ICBN* consistently. The diandrous or pleonandrous orchids thus became the subfamily Cypripedioideae (with the tribes Apostasieae and Cypripedieae) and the monandrous orchids the subfamily Orchidoideae. The latter were divided into three tribes: Neottieae (with nine subtribes), Orchideae (four subtribes), and Epidendreae (27 subtribes) plus two tribes of uncertain affinity. Some of the subtribes were subdivided in informal 'alliances'. Dressler and Dodson did not accept Mansfeld's implication of a closer relationship between the orchidoid and the neottioid orchids, but the subtribe Epipogiinae was transferred to the tribe Orchideae.

In a simultaneous essay 'On the origin of Orchidaceae' Leslie Garay (1960) suggested dividing the orchids into five subfamilies (Table I.1), abandoning the traditional hierarchy. Garay's subfamilies seem to match Dressler and Dodson's tribes, but at the subfamily level Orchidoideae *sensu* Dressler and Dodson is, of course, a broader concept than Orchidoideae *sensu* Garay. Despite the erratic character of Garay's paper (see Dressler 1960), the idea of using 5–6 subfamilies as practical entities caught on and has been widely adopted in the orchid literature.

Pieter Vermeulen (1966) proposed a different use of higher taxonomic categories of orchids. He distinguished three families in the order Orchidales: Apostasiaceae, Cypripediaceae, and Orchidaceae (Table I.1) as did Lindley (1833). The Orchidaceae *sensu stricto* were classified in two subfamilies, Orchidoideae (orchidoid orchids, corresponding to Pfitzer's Basitonae or tribe Orchideae by other authors) and Epidendroideae (the remainder of the monandrous orchids, corresponding to Acrotonae as used by Pfitzer and Schlechter). The Epidendroideae were in turn divided into the contribes (a new category, inspired by Mansfeld's informal '*tribengruppe*') Neottianthae and Epidendranthae, equalling Schlechter's tribes Polychondreae and Kerosphaereae. Contribe Epidendranthae contained two tribes, Epidendreae and Vandeae, distinguished by the presence or absence of a stipe. Levels below tribe were not discussed.

Vermeulen's system differs only from Schlechter's by upward shifting of the higher ranks and by inclusion of the apostasioids but not associating them with the slipper orchids in a special pleonandrous group. The three-family scheme became adopted by the Dahlgren school of monocot systematics which generally applied a narrower family concept (Dahlgren 1977, 1980; Rasmussen 1985). Vermeulen declared his preference for a hierarchical system 'because it emphasizes the assumed greater or smaller affinity so much better' (p. 237). It may be added that hierarchy is indeed necessary if a classification is meant to express phyletic branching. Metaphors implying reticulate relations between peer groups are unsuited for this purpose (Stevens 1984, 1987).

The differences of opinion were accentuated by Garay, who in 1972 defended the non-hierarchical view on orchid systematics. Garay pointed out that he considered his five subfamilies as natural entities in some evolutionary sense, although he regarded the subfamily Epidendroideae to be derived from the Neottioideae. Brieger (1971, 1976) also adopted a non-hierarchical system with five subfamilies (Apostasioideae excluded, but Vandeae elevated to Vandoideae). As he pointed out (1976), this debate could be viewed as a fresh outbreak of the rivalry between Bentham's system with five entities treated as peers versus Pfitzer's bifurcating hierarchy. However, Pfitzer's subdivisions, based on vegetative characters, had mostly been replaced by a hierarchy based on other criteria.

In a discourse on orchid systematics in 1974 Dressler criticized Garay and Brieger as well as Vermeulen. He preferred to keep the orchids as one 'very natural' family divided into only three subfamilies (but with 'reticulate relationships'). The monandrous orchids were then subdivided in three informal groups referred to as neottioid tribes, epidendroid tribes, and vandoid tribes. The Orchideae (later called orchidoids) were included in the neottioid group of tribes, as treated by Mansfeld (1937*a,b*). Dressler stressed that these groups represented levels of evolutionary specialization rather than 'natural phyletic groups' (p. 261). In 1979 Dressler again criticized earlier systems for being inconsistent and containing groups defined solely on one or two shared primitive features. However, he gave up his resistance to proliferation of subfamilies and presented a new classification with six such entities, accepting Brieger's Vandoideae and adding the new Spiranthoideae to accommodate the more distinct tribes of the Neottioideae *sensu* Garay. The remainder of the neottioids (the mainly Australasian terrestrial tribe Diurideae and a few widespread genera with soft pollinia including *Neottia*) went to the Orchidoideae. The subfamily Spiranthoideae were soon accepted as a distinct and most probably monophyletic entity.

There has never been a generally accepted definition of 'naturalness' in botanical classification (Stevens 1984, 1997). The term 'natural' has been used by virtually all orchid systematists since Lindley. It seems, however, that natural classification to evolutionists such as Pfitzer and Schlechter implied a more hierarchical prospect than to Lindley and Bentham. In this century, 'natural' has often been used for groups that are maximally homogeneous in characters. 'Overall similarity' has been invoked as a measure of naturalness or at least a relevant parameter for taxonomic usefulness (Stevens 1986). When computers became available tools in the 1960s, methods for numerical estimates of similarity were applied, giving rise to a school of systematics known as phenetics. Phenetics has not played a significant role in high-level orchid systematics, but Lavarack (1971) presented a phenetic study of 83 exemplar orchid genera.

The breakthrough of phylogenetic systematics that followed the publication of Hennig (1966) and became known as 'cladistics' had an immense impact on the methods for estimating phylogenies as well as for classification. It became clear that measures of similarity and patterns of phylogenetic branching cannot be conveyed unequivocally in a single class-ification. It is still debated to what extent class-ification should reflect phylogeny in practice, but most recent taxonomists will aim at arrangements such that as many taxa as possible are monophyletic and such that paraphyletic taxa are acceptable only as convenience groups, awaiting character inform-ation that may resolve them. Classification based on recognition of taxa with minimal internal variation and reticulate affinities has disappeared from main-stream systematics, and a 'natural' taxon to modern systematists usually means a clade.

Dressler's book *The Orchids: Natural History and Classification* in 1981 was a milestone in orchid sys-tematics. Its impact can be compared to Lindley's *Genera and Species* and Schlechter's *Die Orchideen*, although Dressler did not take the discussion to the level of genus. Like Lindley's and Schlechter's works it was a result of many years of experience with orchids. It widely influenced other works on orchids, although it was amended later by Dressler himself.

The subfamilies used in *The Orchids: Natural History and Classification* conformed to Dressler (1979). The subfamilies were shown as branches (p. 155). The basal relationships between them were not indicated, but it is possible to extract hypotheses about relationships between the tribes. The para-phyletic status of Neottioideae *sensu* Garay was illus-trated (p. 157). Dressler discarded the Vandoideae as being probably polyphyletic already in (1983), when he published the first corrections to his system. Dressler's system (Table I.1) was adopted with slight modifications by Rasmussen (1985), who illustrated the assumed phylogenetic status of the major taxonomic groups by depicting them as nested sacks.

Pamela Burns-Balogh and Vicki Funk (1986) were the first to attempt a large-scale cladistic assessment of the Orchidaceae, evaluating the distribution of 68 characters of 37 supposedly monophyletic terminal groups. The cladogram presented was not a result of an algorithmic treatment of the data matrix but a manual optimization of characters and a phylo-genetic hypothesis. The classification based on the cladogram used seven subfamilies (Table I.1) arranged in a sequence so that each subfamily is the phylogenetic sister group to the sum of the follow-ing groups. (This kind of sequencing was suggested by Wiley (1979, 1981) to save levels in phylogenetic classifications.) The most notable innovations were the creation of a new subfamily (elevated to family by Reveal and Hoogland (1991)) comprising the genus *Neuwiedia*, construed as sister group to all other orchids, and the acceptance of Spiranthoideae as well as Neottioideae, the former being sister to the rest of monandrous orchids. The disputed Australasian terrestrials in the tribes Diurideae, Prasophylleae, Thelymitreae, and Geoblastae were distributed among these subfamilies and not associ-ated with the Orchidoideae. The tribe Vanilleae were placed some way up in the epidendroid sub-family. This feature in particular, and the whole approach of Burns-Balogh and Funk's paper, was criticized in most acerbic terms by Garay in a paper advocating a splitting of *Vanilla* and relatives into a number of segregate genera (1986).

Dressler's (1993) second book on orchid class-ification, *Phylogeny and Classification of the Orchid Family*, contained a substantially revised system based on new character information and significantly influenced by cladistic reasoning. The most remark-able change was that the tribe Neottieae in the strict sense (*Aphyllorchis*, *Cephalanthera*, *Epipactis*, *Limo-dorum*, *Listera*, and *Neottia*) were included in sub-family Epidendroideae. Discarding the Vandoideae, the system contains five subfamilies (Table I.1). The arrangement of tribes, however, was generally more informal than earlier Dressler systems. Two tentative 'phylads', distinguished by stem characters, are pro-posed in the subfamily Epidendroideae, each contain-ing some of the earlier vandoid groups. There is no phylad for the remaining six basal tribes (including Neottieae and Vanilleae)—they may be regarded as constituting a paraphyletic 'null-phylad'. Within the epidendroid or reed-stem phylad we find a 'dendro-bioid subclade' containing the tribes Podochileae, Dendrobieae, and Vandeae. The last group, aptly named 'misfits and leftovers', contains three subtribes and four orphan genera not referable to any tribe with present information. Dressler's (1993) system is the most comprehensive presentation of hypotheses of orchid phylogeny at different levels presented so far. Because of the relaxed organization of the treat-ment and many loose ends, it may not gain status as reference classification like his earlier systems, but it is a pivotal work for discussions of orchid phylogeny.

The most recent complete classification is the *Systema Orchidalium* by Szlachetko (1995). Szlachetko accepted the three-family scheme of Vermeulen (1966) and divided the Orchidaceae *sensu stricto* (the monandrous orchids) into eight subfamilies, 47 tribes, and 140 subtribes (Table I.1). New in the subfamily category were Thelymitroideae, (published already in Szlachetko (1991)), Tropidioideae, and Vanillioideae. Szlachetko's groupings are stated to be polythetic, i.e. not defined by any single shared character. He criticized classifications based on synapomorphies and found that genera as well as suprageneric taxa of orchids in other systems often were heterogeneous, especially in reproductive traits. Although the objective of the system was to form homogeneous groups, cladogram-like figures were used to illustrate relations between subtribes and questions of phylogeny. This system is incompatible with the recent explicitly phylogenetic schemes and may not become widely accepted, but the numerous new taxa and combinations at all levels makes it an indispensable nomenclatural reference.

At the same time as the neo-phylogenetic revolution in taxonomy, another revolution was happening in molecular biology. New methods for investigating DNA sequences have created a source of characters at a different scale. Molecular phylogenetics is expected to answer many of the persistent questions in systematics and has already had an immense impact on the understanding of plant phylogeny. In orchids, molecular phylogenies at family or subfamily level have just begun to appear (Chase *et al.* 1994; Dressler and Chase 1995; Neyland and Urbatsch 1996; Kores *et al.* 1997; Cameron *et al.* 1998). The morphological and anatomical characters, which have been the basis for all orchid systems up to now, have only recently been used in a cladistic analysis at family scale (Freudenstein and Rasmussen 1999).

The combination of cladistic methods and molecular characters will no doubt provide new insight into phylogeny. However, classifications are man-made arrangements and will, like units for measuring, always depend on principles and conventions. The usefulness of a classification depends on many factors. A good classification should at least conform to its own premises, which should be unambiguously declared. To obtain wide usage, a classification must produce groups relevant to many users so that a consensus about its application can be reached. It seems that reference to phylogeny and monophyly of taxa will be standard requirements in future classification, but this will not settle disputes about ranks and inclusiveness of taxa. Classification of large and diverse groups of organisms like the orchids will no doubt continue to engage scientists in exchanges of views.

FINN N. RASMUSSEN

II

MORPHOLOGY

Orchid morphology has formed the basis or a substantial element of all orchid classifications since those of Swartz (1800*b*), Richard (1818), and Lindley (1826, 1830–1840). These early classifications concentrated on floral structure to delimit genera and higher taxa. Not until the work of Pfitzer (1889) were vegetative features considered significant in the construction of classifications. Dressler (1981, 1993) assessed information from both macro- and micromorphology, together with increasing amounts of other cytological, embryological, chemical, and other data in his successive classifications.

Comprehensive surveys of orchid morphology have been published by Pfitzer (1889), Adams (1959), and by Dressler (1981, 1993). Substantial variation in both vegetative and floral morphology provides an immense store of information for the botanist attempting to construct classifications of this enormous family.

Plant habit and size

The major and remarkable adaptations seen in orchid vegetative morphology have evolved in response to adverse environmental conditions, in particular the problems of water conservation on a daily and seasonal basis. Orchids can be found in almost every situation: on the permanently moist floor of the lowland tropical rain forest, in the uppermost branches of tall forest trees where heavy rainfall is followed by scorching sun for hours on end, on rocks near the summit of tropical mountains, in grassy and marshy areas, on landslips and roadsides, and even on the margins of deserts.

Orchids can be terrestrial, epiphytic or lithophytic. In the tropics the majority of orchids are epiphytic on forest or woodland trees, or lithophytic on rocks and cliffs, but a significant minority are terrestrial, growing on the forest floor in seasonal woodland and in natural and secondary grasslands. The epiphytes can be found on the trunks, branches or twigs of trees (Plate 1A). Large plants are usually trunk or branch epiphytes (Plate 1B), whereas epiphytes growing on twigs are small, often minute (e.g. *Psygmorchis*). Epiphytes can be lithophytic where the rainfall is sufficient. It is not uncommon to find the same species growing on trees and rocks in the same locality. Freshly made roadside banks are also a rich habitat for orchids which can form the pioneering vegetation in such competition-free places; in such cases plants that are usually epiphytic can grow terrestrially. Some orchids such as the Australian *Dendrobium speciosum* and *D. kingianum* are usually lithophytic (Plate 1C), rarely epiphytic.

In temperate regions the terrestrial habit predominates. Indeed in Europe and Chile all species are terrestrial. In Australia, New Zealand, east and south-central tropical Africa, South Africa, Argentina, and in North America (outside of Florida) the majority are terrestrial (Plates 1D, 2A). Grassland and woodland are favoured habitats for temperate terrestrials, but some occupy wet meadows, fens, and bogs. In Europe, *Hammarbya paludosa* grows on sphagnum cushions in bogs. *Liparis loeselii* can grow in floating vegetation in calcareous fens or dune slacks. The North American *Spiranthes cernua* also grows in floating vegetation mats near the edges of ponds and streams.

Plant habit can be a useful taxonomic guide. All Apostasioideae are terrestrial. The Cypripedioideae, Orchidoideae, and Spiranthoideae (*sensu* Dressler 1993) are also predominantly terrestrial. In contrast the Epidendroideae are predominantly epiphytic. There are, however, many exceptions. In the Cypripedioideae three species of *Paphiopedilum* and one *Phragmipedium* are usually epiphytes. A few species of Orchidoideae such as *Stenoglottis fimbriata* and a number of Spiranthoideae (*sensu* Dressler 1993) such as *Goodyera viridiflora* are commonly epiphytic but can also grow terrestrially. Many Epidendroideae are terrestrial, notably *Sobralia*, most *Elleanthus*, and many *Epidendrum* species. Many vanilloid orchids grow as lianas, some of considerable size (Plate 2B). These scramble among trees and shrubs or climb into the canopy of trees. The achlorophyllous *Galeola* can scramble for many metres across the ground and on surrounding shrubs.

A number of terrestrial orchids are achlorophyllous, bearing scale leaves or lacking leaves altogether and reliant on their mycorrhizal fungi for nutrients throughout their life. They are usually referred to as saprophytes but are more probably parasitic, at least on their mycorrhizal fungi. Most appear above ground only at flowering time for pollination and seed dispersal. Two species of the temperate Australian genus *Rhizanthella* never appear above the surface of the ground; their flowering heads lie just beneath the surface and are exposed to air only when the subtending bracts push aside the surface leaf litter (Plate 2C). The 'saprophytic' orchids are found in a wide range of genera, throughout both temperate and tropical zones. Some genera have been defined using this character, for example *Aphyllorchis*, *Corallorhiza*, *Chamaegastrodia*, *Epipogium*, *Limodorum*, *Neottia*, *Galeola*, and *Gastrodia*. Others have been assigned to genera that are predominantly autotrophic, e.g. *Cymbidium*, *Eulophia*, *Habenaria*, *Cystorchis*, *Tropidia*, and *Cephalanthera*. The relationships of some of the 'saprophytic' genera to allied autotrophic genera need further investigation, e.g. that of *Chamaegastrodia* to *Zeuxine* and *Hetaeria*, and of *Neottia* to *Listera*.

The diversity of size found in the orchid family rivals that of most other plant families. The smallest orchids are possibly some of the tiny *Bulbophyllum* species such as the Australian *B. minutissimum*, the pseudobulb, leaf, and inflorescence of which are no more than 3–4 mm tall. At the other end of the spectrum the liana-like *Vanilla* species that climb up into the tallest forest trees may reach 30 m or more long. Vanillas are by no means the bulkiest of orchids because some of the giant specimens of the *Grammatophyllum speciosum* from the Malay archipelago may weigh many hundreds of kilograms.

Roots

Orchid roots may have one or more functions: attachment to the substrate, anchorage in a substrate, absorption of water and nutrients, photosynthesis, litter-gathering, and as a link between orchid and mycorrhizal fungus (Plate 2D). Not surprisingly

they can be diverse in their morphology depending on which functions they perform. The entire root system is made up of secondary or adventitious roots. These can vary greatly in thickness according to species. Mycorrhizal fungi enter the roots through the root hairs, and other non-mycorrhizal fungi can be present in mature roots.

Although most terrestrial orchids have a simple rhizodermis, some terrestrials and virtually all epiphytes have a **velamen**, epidermal tissue comprising one to eighteen layers of cells which die as the root matures. It forms a spongy, whitish sheath around the root, separated from the cortex by an exodermis of long empty cells and shorter, living passage cells. The structure and physiology of the velamen and underlying **exodermis** are adapted for at least two functions: attachment to the substrate and water and nutrient uptake in a periodically dry environment (Pridgeon 1987).

Capesius and Barthlott (1975) and Barthlott (1976*b*) have shown that the velamen absorbs moisture and nutrients (except in *Vanilla*). Blue-green algae or cyanobacteria have been found within the velamen, and it has been postulated that these can fix nitrogen for the orchid. A velamen is not unique to the orchids but is also found in some Liliaceae and Araceae (Pridgeon 1987).

Epiphytic orchids can have several types of roots. Aerial roots are generally cylindrical, whereas those on the substrate are dorsiventrally flattened to a greater or lesser degree. However, in some such as *Phalaenopsis* and *Dendrophylax* all of the roots are somewhat flattened. The side of the flattened roots in contact with the substrate often develops short outgrowths or becomes intimately affixed to the surface of the substrate, which allows maximum surface area for absorption.

Erect basket roots that have a litter- and detritus-gathering function occur in some orchids, notably *Ansellia*, *Grammatophyllum*, *Graphorkis*, *Cyrtopodium*, and some *Cymbidium* species. Some epiphytes and lithophytes have fleshy storage roots, e.g. *Sobralia*, *Ponera*, *Isochilus*, and some *Epidendrum* species, but the species with such roots usually have thin stems.

In many terrestrial orchids the roots are storage organs and consequently are swollen and often tuberous. The Greek word *orchis* refers to the spherical **tubers** of *Orchis* and some other European orchids having a fancied resemblance to human testicles. In Spiranthoideae (*sensu* Dressler 1993) the root can be fleshy. In *Cleistes*, *Apostasia*, and some *Tropidia* part of the root can be swollen (nodular tubers). Orchidoideae, Diurideae, and Diseae have tubers which are formed largely from root tissue but have a core of stem structure including a dormant bud. These are sometimes produced at the end of stolons, e.g. in *Platanthera*.

The roots of some saprophytic orchids can be abbreviated and even vestigial. *Neottia* and *Corallorhiza* have coralloid roots, in which the the main absorptive function is assumed by the mycorrhizal fungi.

Roots can produce adventitious buds which form new shoots, e.g. *Listera*, *Pogonia*, *Psilochilus*, or *Phalaenopsis* (Stoutamire 1974).

Many orchids have photosynthetic roots, the growing tip commonly green. In some vandaceous orchids the roots are entirely photosynthetic and, in suitable conditions, the chlorophyll can show through the velamen, giving the root an overall greenish appearance. In the leafless orchids of such genera as *Microcoelia*, *Taeniophyllum*, *Campylocentrum*, and *Chiloschista*, the roots have entirely taken over the photosynthetic function (Plate 3A).

Stems

Orchids can have one of two growth patterns, either **sympodial** or **monopodial**. This distinction was first used systematically by Pfitzer (1889). In sympodial growth the apex of each shoot terminates in a flowering axis, the subsequent vegetative shoot emerging from a node below the apex of the new shoot. Usually the new growth is produced annually or seasonally (Plate 3B). Monopodial growth, the derived condition, is the result when the stem apex is of indeterminate growth and the inflorescences are borne laterally, as in most vandaceous orchids. Monopodial growth is found in vanilloid and vandaceous orchids (Plate 3C).

Rhizomes are horizontal or ascending stems of sympodial orchids, either below or on the substrate surface, and are found in a wide variety of orchids (Plate 3B). These are compound organs, consisting of the first few nodes of many stem bases, either fleshy and succulent or tough and woody. Along the rhizomes are distichously arranged scale leaves; among these are the renewal bud, which ensures the continuance of the sympodium, and the reserve bud, which represents the potential branching point

of the rhizome (Rasmussen 1986). Aerial or erect shoots arise above the renewal bud and may become pseudobulbous (Plate 3B). *Cheirostylis* and other Goodyerinae some of the species have swollen rhizomes that lack roots but have root ridges which bear root hairs.

Species of *Bletia*, *Eulophia*, and *Spathoglottis* have **corms,** underground swollen storage organs the leaves of which bear little reserve material.

Pseudobulbs are swollen or thickened stems, essentially water-storage organs of sympodial orchids (Plate 3B). These often resemble swollen eggs or clubs. They are common in tropical epiphytic orchids but are also found in many tropical terrestrial species, growing either above the ground as in *Calanthe* or underground as in many *Eulophia* species. Two sorts of pseudobulbs are recognized: heteroblastic ones, which comprise a single node, and homoblastic ones, comprising two or more nodes. This distinction has been used to distinguish genera, such as *Oeceoclades* from *Eulophia*. Homoblastic pseudobulbs may bear leaves at each node, toward the apex or only at the apex.

That tropical orchids might suffer from periodic water deficits is not immediately obvious. Rainfall is not continuous even in the wettest habitats and, in many places, including the tropics, the rainfall patterns are markedly seasonal. Furthermore, most tropical orchids are epiphytic or lithophytic, growing on the trunks, branches, and twigs of trees or on rocks. In these situations water run-off is rapid, and the orchids will dry quickly in the sunshine that follows the rain. Many orchids have therefore developed marked adaptations of one or more organs to allow them to survive these periodic droughts. Some of these adaptations are as dramatic as those encountered in the Cactaceae.

In a few terrestrial orchids a different approach to surviving adverse conditions has been adopted. *Orchis*, *Disa*, *Diuris*, *Habenaria*, and *Platanthera*, for example, lack pseudobulbs but have underground tubers as perennating organs to survive drought or cold weather. The new growth grows from one end of the tuber in suitable conditions. In others such as *Cypripedium*, *Zeuxine*, and *Goodyera* the rhizomes are succulent or somewhat swollen but not tuber-like.

In many orchids the pseudobulbs or stems are green and photosynthetic. In *Bulbophyllum minutissimum* the leaf is so small that the pseudobulb functions as the main photosynthetic organ.

Leaves

The leaves of orchids perform the main photosynthetic function in most species. In most orchids the leaves are arranged distichously. However, in some they can be spirally arranged on the stem or reduced to a single terminal leaf. Distichous leaves are considered to be derived from spirally arranged ones (Dressler 1993). In a few orchids such as *Codonorchis* and *Isotria* the leaves are borne opposite one another.

The leaf is another organ that has undergone many modifications in the orchids. Orchid leaves are typical monocotyledon leaves with parallel longitudinal venation, usually linked by less obvious commissural veins. Leaf venation can be systematically informative. In *Cypripedium japonicum* and *C. formosanum* and in some other terrestrials such as *Nervilia*, the venation appears fan-shaped because of the leaf-shape, but this is merely a variation on the typical venation pattern adapted to the leaf shape. Reticulate venation is rare in orchids, but in South American *Epistephium* and the New Caledonian *Clematepistephium* the ovate to elliptic leaves have prominent reticulate venation (Cameron 1996).

Leaf **vernation** has been used as a systematically useful character. Orchids may have leaves with either (1) plicate (Plate 3D) or soft herbaceous or (2) conduplicate vernation (Plate 4A). The former are found in most terrestrial species, whereas both types are found among epiphytes. Both can be found in a single genus such as *Eulophia*, but it is more usual to find a genus having only one or the other type.

Dorsiventrally flattened leaves are common. The leaves of species found in the drier places are often more or less succulent or leathery. For example, those of *Cymbidium finlaysonianum*, *Cattleya labiata*, *Oncidium lanceanum*, and most *Phalaenopsis* species are thick and leathery (Plate 4A). Genera such as *Luisia*, *Paraphalaenopsis*, and *Papilionanthe* are characterized by terete leaves, but terete-leaved species are also found in genera with dorsiventral leaves (*Oncidium*, *Bulbophyllum*, *Polystachya*). Laterally flattened leaves are characteristic of some genera such as *Podangis* and *Bolusiella*, but dorsiventral or laterally flattened leaves can both be found among the species of *Angraecum* and *Maxillaria*. *Campylocentrum*, *Taeniophyllum*, and *Chiloschista* are among the few epiphytic genera in which the leaves have

been reduced to scales with the photosynthetic function assumed by dorsiventral green roots.

A few 'saprophytic' terrestrial orchids are leafless and lack chlorophyll, or else the leaves are reduced to colourless or buff scales. The most impressive of the saprophytes is undoubtedly *Galeola*, which can form liana-like plants many metres in length. Lacking chlorophyll the plants cannot photosynthesize and must obtain all of their nutrition from the mycorrhizal fungus with which they are associated. In the case of *Galeola* the fungal associate is said to be the wood-rotting fungus *Armillaria*. The Australian genus *Rhizanthella* comprises two species usually referred to as the 'underground' orchids. These have subterranean rhizomes with a short, branching horizontal part, erect shoots bearing sterile bracts, and a capitulate inflorescence (Plate 2C).

Leaf shape in orchids is relatively uniform for many genera. Most orchids have simple, linear, lanceolate, oblanceolate, ovate, or elliptic leaves. However, fan-shaped leaves are found in *Cypripedium formosanum* and *C. japonicum*, palmate leaves in a few New Caledonian *Acianthus*, sagittate leaves in *Pachyplectron*, and spirally twisted, linear leaves in *Thelymitra spiralis*.

Leaf size is variable, ranging from 1 mm long (*Bulbophyllum minutissimum*) to more than a metre long (*Phaius tankervilleae* and *Bulbophyllum fletcherianum*).

Leaf indumentum is often a taxonomically useful character. Glabrous leaves are common, especially in epiphytes, the Goodyerinae and some other Orchidoideae, but hairy leaves occur in genera as diverse as *Cypripedium*, *Calanthe*, *Nervilia*, and *Pilophyllum*. Margins of leaves can be ciliate or not. Serrations or other marginal dissection are rare, but the terminal lobes of some *Ancistrorhynchus* species are serrate.

Most orchids have unmarked green leaves, often a darker and glossier shade above than below. However, some orchids have varying degrees of markings on their upper (adaxial) or lower (abaxial) surfaces. Tessellations, differentially pigmented areas of mesophyll and veins, are found in some orchids, for example in *Paphiopedilum* subgenera *Parvisepalum* and *Brachypetalum* and in section *Barbata* of subgenus *Paphiopedilum*. The Ecuadorian *Lepanthes calodictyon* has a tessellated leaf of dark and grass-green with strongly undulate leaf margins. Dark purplish black spotting is found on the adaxial surface of leaves of species in *Cypripedium* sect.

Trigonopedium. Dull purple blotches mark the heart-shaped leaf of *Nervilia aragoana.* Yellow spots sometimes mark the adaxial surface of the leaves of *Phaius flavus.* Purple spotting is more common on the abaxial surface, especially in genera such as *Paphiopedilum, Nervilia,* and *Psychopsis.*

Some of the terrestrial species of the forest floor, especially in the genera *Goodyera, Anoectochilus, Macodes, Ludisia* and their relatives, have entirely purple or green or purple leaves that have beautifully marked veins (Plate 4B). The purple leaves of some are reticulately veined with silver, gold or red. They are popularly known as the 'jewel orchids'.

In many orchids the leaves have a sheathing base which closely clasps the stem. Leaf sheaths can be variously marked with purple or brown and can be pubescent or less commonly glandular. Leaf articulation is found only in the Epidendroideae of the five subfamilies of orchids. The abscission layer is usually between the leaf blade and leaf sheath. Leaf articulation or the lack of it has also been used systematically in genera such as *Dendrobium* such that presence or absence defines the subgenera.

Inflorescences

Orchids carry their flowers in a variety of ways. Even within the same genus different species have different ways of presenting the flowers. Inflorescences may be produced from any part of the stem. The ancestral state is terminal, the termination of a shoot axis. In other cases the inflorescence is lateral, produced from the side or base of a shoot or from a rhizome. The most common inflorescence type is the **raceme** (Plate 4C), bearing one to many flowers, usually axillary on a more or less elongate floral axis comprising a stalk called the **peduncle** and a portion bearing the flowers, the **rachis**. The peduncle and rachis can be variously shaped and ornamented, and this can be taxonomically significant. They are most frequently terete in cross-section but may be angular or even winged. They can be glabrous, hairy, or glandular. In all cases the flowers are subtended by bracts which may or may not be conspicuous and coloured. Sterile bracts can be present or absent; if present they can cover the axis or not. For example, the common African species *Eulophia odontoglossa* is easy to identify because its peduncle is covered by papery bracts. In

Cymbidium lowianum the flowers are borne on an elongate pendent raceme which is unbranched with the flowers arranged in a lax spiral around the rachis. In some other *Cymbidium* species the raceme is still many-flowered but erect, and in *C. lancifolium* the inflorescence is reduced to two or three flowers on an erect raceme. On a raceme the individual flowers are attached to the floral axis by a stalk called the **pedicel**. In some genera such as *Peristylus* and *Neuwiedia* pedicels are absent, and the flowers are sessile on the axis; such inflorescences are termed **spikes**.

Compound inflorescences with many flowers are also not uncommon in orchids, particularly among the epidendroid and vandoid genera. In genera such as *Trizeuxis, Oncidium, Renanthera,* and *Pomatocalpa* we find branching inflorescences which are termed **panicles** (Plate 4D). Panicles occur in many epiphytic genera but are rare among terrestrials. The strange determinate inflorescence of *Lockhartia* is a **cyme** (Plate 5A), an unusual type in monocots although frequent in dicots. The terminal flower opens first, followed by one or two subtending buds which also produce branches ending in flowers, which in turn produce branches ending in flowers and so on.

Some inflorescences are more or less globose, with flowers densely and spirally arranged at the apex of the rachis, as in *Glomera* and its relatives (Plate 5B).

In the genus *Bulbophyllum* we find some interesting variations on the multiflowered inflorescence. In several species the flowers are borne all facing to the same side of the rachis; such an inflorescence is termed secund. However, the most spectacular group is that in which the rachis is so contracted that the flowers all appear to come from the top of the flower stalk in a false umbel, with the inflorescence resembling the head of a daisy. For this reason these bulbophyllums, such as *Bulbophyllum lepidum,* have been considered by some to belong in the separate genus *Cirrhopetalum* (Plate 5C).

Solitary flowers borne one at a time, either sessile or on a shorter or longer peduncle, are found in many genera such as *Bulbophyllum* (*B. lobbii*), *Corybas,* and some paphiopedilums such as *P. hookerae* and *P. lawrenceanum.* An aborted flower bud can often be found in the bract of tropical slipper orchids and will occasionally develop to produce a second flower.

Flowers

An interesting feature of the development of most orchid flowers is the phenomenon of **resupination**. In bud, the lip normally lies uppermost in the flower and the column lowermost. In most species, however, the lip is lowermost in the flower. This is achieved by a twisting of the flower stalk or ovary through 180° as the bud develops or by a bending of the pedicel in some solitary-flowered genera. In resupinate species with a pendent inflorescence, such as *Cymbidium lowianum*, the lip will still be lowermost in the flower when it opens. In the terrestrial genus *Geodorum* the inflorescence bends just below the first flower, with the effect that the rachis is pendent. Nonresupinate flowers are those that either do not twist during development (e.g. *Satyrium*) or else twist through 360° (e.g. *Angraecum*), so that the lip is uppermost. There are many examples of nonresupinate flowers in Orchidoideae and Epidendroideae.

Orchid flowers are simple in structure and yet highly modified from the more typical monocotyledon flower as exemplified by *Lilium*, to which orchids are distantly allied. These characteristically have their floral parts arranged in threes or multiples of three. Orchids are no exception, and this can most easily be seen in the two outer whorls of the flower, such as in the common and widely grown Asiatic orchid *Cymbidium lowianum* and the South American *Cattleya amethystoglossa*, which more or less represent the floral structure of the majority of orchids. Their floral parts are situated at the apex of the **ovary**, which is inferior and usually borne on a pedicel. In a few orchids, for example in the Pleurothallidinae, there is an abscission layer between the pedicel and ovary. Placentation in most orchids is parietal (Plate 5D), but in a few orchids, such as *Selenipedium* and *Phragmipedium*, it is axial. In *Epistephium* and *Clematepistephium* a bract-like whorl is found at the apex of the ovary but below the calyx.

The outermost whorl of the flower is the **calyx** which, in *Cattleya amethystoglossa* (Plate 6A) and *Cymbidium lowianum* (Plate 6B), consists of three sepals. The two lateral sepals differ slightly from the third, called the **dorsal** or **median sepal**. In some orchids such as *Dendrobium*, *Bulbophyllum*, *Polystachya*, and *Eria* the **lateral sepals** form at the base a 'chin' called a **mentum**. This is usually conical, but in some tropical American genera such as *Trichocentrum* the mentum is slender, elongated, and spur-like. In some orchids such as the Cypri-

pedioideae the lateral sepals are united to the apex, forming a synsepal. In others the degree of fusion of the lateral sepals varies from very little to almost complete, and the dorsal sepal may also be variously united to the lateral sepals to form a short to long tube. In *Zootrophion*, the lateral sepals are fused longitudinally with each other and also partially with the dorsal sepal at the base and at the apex, leaving two lateral windows providing access for pollinators to the inner parts of the flower. The sepals protect the delicate inner organs and can be adorned with hairs (e.g. *Dendrobium* section *Latourea*) or glands (e.g. *Scaphosepalum*, *Caladenia*). Sepals are usually relatively drably coloured compared with the petals, although they are often more brightly coloured within than without. They can assume the function of petals as in *Bulbophyllum*, *Masdevallia*, and *Dracula*.

The **corolla** of *Cattleya amethystoglossa* (Plate 6A) and *Cymbidium lowianum* (Plate 6B) comprises three **petals** which are brightly coloured. The two lateral petals are uppermost in the flower and differ markedly from the third petal which lies at the bottom of the flower (in these and other resupinate species). The lateral petals are often large and showy and variously coloured. In some orchids they can bear hairs or glands on their surface and ciliae on their margins. Rarely the petals are much smaller than the sepals, and in some *Bulbophyllum* species can be much reduced or even vestigial. The third petal, called the **lip** or **labellum**, is highly modified, three-lobed, and bears a callus on its upper surface. In other orchids the upper surface of the lip may be adorned with a callus of raised ridges, lamellae or tufts or areas of hairs or glands. The lip is an important adaptation of the orchid to facilitate cross-pollination. It can be imagined as a brightly coloured or otherwise attractive flag to tempt potential and specific pollinators which are then guided toward the pollen and stigmatic surface by the form of the callus. The lip, including visual cues and osmophores (fragrance glands), acts as an attractant and/or landing platform. Lip shape and accessory features such as calluses can be diagnostic in orchid classification, particularly in species delimitation.

In many species, but not in *Cattleya amethystoglossa* and *Cymbidium lowianum*, the lip is extended at the base to form a saccate to filiform **spur** (Plate 6C) which may or may not contain nectar. This is also important in attracting a potential pollinator, either by reward or deceit. The spur can exceptionally reach 30 cm long, as in the Madagascan *Angraecum sesquipedale*, the hawkmoth pollinator of which was predicted by Darwin (1862). In some of the vandoid orchids the spur itself may contain fleshy callosities within, which are important for identification purposes. A spur can also be formed by the mentum, from the extension of the base of the column and the fused lateral sepals.

The central part of the orchid flower shows the greatest modifications to the basic monocotyledon pattern. Reduction in the number of parts and fusion of the male and female organs into a single structure have been the major evolutionary trends. The fused organ in the centre of the flower is called the **column** (Plate 6D, 7A–D). The column is immensely variable in orchids, a simple stalk-like organ in the many of orchids but variously more complex in many. It is very difficult to interpret homologies of column structure without careful developmental studies such as those by Kurzweil (1987*a,b*, 1988, 1993) and Rasmussen (1982).

In Apostasioideae *Neuwiedia* has three functional **anthers**, but the median stamen in *Apostasia* is sterile or absent. Similarly, only the two lateral anthers in Cypripedioideae are fertile (Plate 6D). In *Cattleya* and *Cymbidium* (and most other orchids), a single median anther lies at the apex of the column with the other two either sterile or absent. The anther of most orchids does not contain powdery pollen as in other angiosperms but discrete waxy masses called **pollinia** (Plate 7B–D, 8A). The pollinia are attached by a stalk to a well-defined sticky organ called the **viscidium**. The pollinia, stalk, and viscidium together comprise a pollination unit termed the **pollinarium** (Plate 7B–D, 8A).

The anther, its development, and its contents are variable in the orchids and of considerable significance in their classification at all levels. According to Dressler (1993) the ancestral anther, as in the Spiranthoideae (*sensu* Dressler 1993) and many Orchidoideae, is erect, parallel to the axis of the column, and firmly fixed to it. In the Diseae it may bend backwards to such an extent that the apex of the anther is basal. In the less specialized Epidendroideae the anther is erect in bud but bends forward and downward through 90° or 120°, becoming incumbent, so that the anther rests like a cap on the apex of the column. In epidendroid orchids the versatile anther is hinged to the column apex and can swivel. In the advanced vandoid orchids the anther sits at the column-apex like a cap as in other

Epidendroideae, but its development is quite different because the anther does not bend. This has been used in the past to separate the vandoid orchids at subfamilial level. The basal attachment of the anther in the Orchideae and Diseae has been termed basitony and used to distinguish them from the Epidendroideae which have the anther fixed at the apex (acrotony). However, in the Australasian Diurideae we find all degrees of fusion from basitony to acrotony, whereas it is apparent that the vandoid orchids, now considered part of the epidendroid clade, are acrotonic (Dressler 1993; Freudenstein and Rasmussen 1999).

Pollen masses or pollinia are often cited as characteristic of orchids, but they are also found in the dicot family of milkweeds or Apocynaceae. In all orchids the pollen grains are aggregated in some way into discrete pollen masses or pollinia, but the aggregations in Apostasioideae and Cypripedioideae are less well defined. Most other orchids shed pollen in tetrads or in larger masses.

Pollen is held together in those orchids with softer pollen by an elastic substance derived from the anther tapetum or pollen exine. Dressler (1981) termed this elastoviscin but admits that its origins and nature need further investigation. Soft pollen may be packed relatively uniformly into masses, as in Diurideae, Cranichideae, and Vanilleae, or aggregated into packets or **massulae**. The latter are found in the Orchideae, Diseae, Erythrodeae, Gastrodieae, and some Diurideae. Pollinium texture may be either solid (if exine is deposited on internal grains of a pollinium) or granular (Wolter and Schill 1986; Freudenstein and Rasmussen 1999). Further, the orientation of pollinia may be juxtaposed or superposed in the more derived taxa, although there is more than one developmental pathway for superposed types (Freudenstein and Rasmussen 1999).

Pollinium number is moulded by the form of the anther and the number of loculi. In many orchids there are four pollinia, representing four anther cells. In the Epidendreae and Arethuseae, the ancestral number is eight, and these may be clavate or bilaterally flattened. A reduction series from eight to six to four to two can be traced independently in the Laeliinae and Pleurothallidinae of the Epidendreae, although the series is probably not linear. Freudenstein and Rasmussen (1996) showed developmentally for Orchidoideae, Spiranthoideae (*sensu* Dressler 1993), and Epidendroideae that the orchid anther has a single meristematic region, septation of which results in four or eight pollinia per anther, the latter from partitioning either by two longitudinal septa or one longitudinal and one transverse septum. Lack of septation in Epidendroideae yielded only two pollinia.

Pollen stalks that attach the pollen masses to the viscidium can have various origins. Those that originate in the anther are called **caudicles** and consist of material from aborted pollen mother-cells as well as perhaps the tapetum and pollen tetrads (F. N. Rasmussen 1986). These are found in the Orchidoideae, some Arethuseae, and some Epidendreae such as *Coelogyne*, in which they are distinct, and *Epidendrum*, in which they are not well developed. Most vandoid and a few other orchids have a strap of non-sticky tissue called the 'stipe' (L. *stipes, stipites*), derived from rostellar tissue (see below) instead of anther tissue. The most common kind of stipe is the **tegula**, a plate formed by the periclinal walls of the epidermis of the rostellum, and characterizing the Vandeae (F. N. Rasmussen 1986). The presence of a stipe attaching the pollinia to a single viscidium, two stipites attached to a single viscidium or two stipites each attached to its own viscidium can all be diagnostically useful at generic or infrageneric rank. A second kind of stipe, the **hamulus**, is the recurved apex of the rostellum (F. N. Rasmussen 1982), found in Prasophyllinae, some Neottieae, Tropidieae, and *Bulbophyllum ecornutum* (F. N. Rasmussen 1985, 1986). A third kind of stipe with a 'hammer-like' morphology occurs in *Sunipia* and *Genyorchis* (F. N. Rasmussen 1986).

The **pollinarium** (Plate 7B–D, 8A) is the unit of pollinia, stalk, and viscidium carried by the pollinator. In some Vandeae and many Orchideae two pollinaria are produced by each flower. The features of the pollinarium have long been considered of taxonomic significance, useful at specific level and above as in the Catasetinae, Stanhopeinae, and Vandeae. It is often possible to identify a species from the morphology of its pollinarium, using characters related to size, shape, and texture of the constitutents (Dressler 1976, 1993). The pollinarium can often change shape following its release from the flower and its attachment to the pollinator. For example in the well-studied genera *Ophrys* and *Orchis*, differential drying causes the caudicles to bend and align the pollinia to prevent self-pollination and promote outcrossing (Dressler 1981).

The **stigma** in *Cymbidium lowianum* and *Cattleya amethystoglossa* is also positioned on the ventral surface of the column in the centre of the flower. The stigma is a sticky lobed depression situated below and behind the anther, but in some terrestrial genera such as *Habenaria* the stigma is bilobed with the receptive surfaces at the apex of each lobe. In many species the pollen masses are transferred to the stigmatic surface by a modified lobe of the stigma called the **rostellum**. This is developed in our examples of *Cymbidium lowianum* and *Cattleya amethystoglossa* as a projecting flap that catches the pollen masses as the pollinator passes beneath on its way out of the flower, directing pollinia into the stigmatic cavity.

Dimorphic flowers

A few genera produce dimorphic or polymorphic flowers. In the Bornean *Dimorphorchis* species and in *Grammatophyllum speciosum* two different flower types are found on a single inflorescence (Plate 8B). The basal two or three flowers in the former are distinctively coloured, creamy rather than yellow, strongly scented but with a sterile column; in the latter the basal flower is sterile and may have only four or five segments, rather than six in the upper flowers. The tropical American *Oncidium heteranthum* and related species also have inflorescences of dimorphic flowers with the lower flowers of each branch of the inflorescence sterile and much reduced to vestigial tepals, and only the apical flowers fully formed and fertile. Dimorphism is most highly developed in the tropical American members of the Catasetinae which are pollinated by euglossine bees. Inflorescences produce either male or female flowers but also more rarely hermaphroditic ones. Male flowers have functional pollen and often a trigger mechanism on the column; female flowers have a functional stigma but no pollen or aborted pollen; hermaphroditic flowers have both functional pollen and stigma. Within a single species flowers of different gender look quite different, so much so that Lindley described the male, female, and hermaphroditic flowers of *Catasetum* in different genera: the male flowers as *Myanthus*, the female as *Monachanthus*, and the hermaphrodites as *Catasetum*. Darwin (1862) first realized the congeneric nature of the widely differing morphs. Another genus with dimorphic flowers is *Cycnoches*, in which the female flowers are non-resupinate in an erect inflorescence and have an entire, convex lip; the male flowers are resupinate in a pendent inflorescence and have a dissected lip.

Although most orchids are outcrossing, self-pollination (autogamy) is not uncommon. One of Darwin's earliest observations was on *Ophrys apifera*, which is self-pollinating in Britain. Although the lip closely resembles a bee in shape and texture, it is not usually bee-pollinated. Instead, the anthers drop out of the anther loculi into the pit-like stigma, and self-pollination ensues. Morphological changes often accompany autogamy (Catling 1990). For example in the autogamous species of *Epipactis* such as *E. phyllanthes* and *E. muelleri*, the rostellum is vestigial compared with that of *E. helleborine* and *E. leptochila*. The lip can also be morphologically different, often reduced in complexity of form and with a vestigial callus. Usually the autogamous forms are treated as varieties or forms of the autogamous species, but some allogamous orchids have been recognized as distinct species and even as distinct genera. Lindley described *Paxtonia* as a new genus but later realized that it was a form of *Bletia* in which the lip was petaloid.

Teratological flowers occur with surprising frequency in orchids. This may take the form of doubling of floral parts. For example, two lips are occasionally seen in some slipper orchids as are double dorsal sepals and double columns. More difficult to interpret are the teratologies that result in more or less regular flowers, such that the lip can become petaloid or the petals resemble a lip. These are often called peloric flowers. Although the sepals and petals may be regular, the column often is not. Taxonomically these have been treated in a variety of ways: as monstrosities not warranting taxonomic recognition, as varieties or forms or even as distinct genera. One of the main causes for dissent is that such teratologies can occur throughout a population, giving the population or populations a uniformity which suggests that the morphotype is fixed and reproducing true to type. This has led Chen (1965, 1978, 1979) to describe three genera *Tangtsinia*, *Sinorchis*, and *Diplandrorchis*, all with regular flowers, as distinct and to discuss their 'primitive' nature. *Tangtsinia* is, in our opinion, a teratological *Cephalanthera*. Both of the other genera are saprophytes, and it seems clear that *Sinorchis* is an abnormal *Aphyllorchis* with a petaloid lip and that *Diplandrorchis* is an abnormal *Neottia* with a petaloid lip and two anthers. The two anthers of *Diplandrorchis* lie opposite each other and possibly derive from each of the staminal whorls.

Fruit

Orchid fruits termed **capsules** (Plate 8C) are seldom described by authors of orchid monographs, floras, or general accounts. This is because the capsules may be unknown, poorly preserved or because the author considers that the fruits have little diagnostic value. Notable exceptions are Beer (1863) and Hallé (1977), who illustrated the fruits of New Caledonian orchids in his account of that strange orchid flora.

Dressler (1981) provided a succinct account of orchid capsule morphology. In most orchids the flower changes colour after fertilization and drops from the tip of the ovary. However, in a significant number of orchids from *Paphiopedilum* to *Phalaenopsis*, *Promenaea*, and *Zygopetalum* the flower is persistent, changes colour to green and photosynthesizes, thereby helping the capsule to develop.

The capsule comprises three carpels which can be seen most easily in cross-section. It dehisces by splitting along the mid-line of each carpel rather than between them, and the midvein with some adjacent tissue also separates, leaving three broad and three narrow sutures separating the six half carpels. In some species the valves are linked by transverse connecting fibres through which the seeds must pass before they can be released to be dispersed by air. Occasionally the six half-carpels separate completely at the apex as in some *Maxillaria* and *Lockhartia* species, but in most species they remain joined at the apex. Capsules of some orchids such as *Angraecum* split along a single line, whereas those of *Dichaea*, *Pleurothallis*, and other genera split into two unequal valves.

Capsules can be glabrous or pubescent, although hairy ovaries can develop into glabrous fruits as the indumentum falls off or is rubbed off. They can also be variously ornamented with papillae, setae, spines, ridges, or beaks. The beak usually comprises a floral tube or cuniculus.

Young and developing fruits are usually green and photosynthetic (Plate 8C). Mature capsules of terrestrial orchids are generally erect, thin-walled, rather dry and papery, whereas those of epiphytes are usually pendent, thicker-walled, and often somewhat fleshy. Posture of the capsule ensures that the seeds are released into moving air-streams for efficient dispersal.

A number of orchids have fruits borne on a peduncle or pedicel that elongates after fertilization.

In *Cypripedium margaritaceum* and its allies the pedicel elongates 1–3 (rarely up to 15) cm, and in *Nervilia* and *Corybas* the peduncle elongates greatly as the capsule nears maturity.

Capsules of epiphytic orchids often have elaters or spring hairs in the capsule among the seeds. These are elongated and, at least in some orchids such as the Vandeae, hygroscopic. The theory is that by absorbing water the movements of the hairs assist the release of the seeds. Hairs appear to be absent in terrestrial species.

Dry fruits shedding dry seeds are the norm in most orchids. Dressler (1981) suggested that two features, fleshy fruits with a moist pulp and long, slender fruits, are characteristic of relatively primitive genera. Fleshy fruits are found in *Neuwiedia*, *Selenipedium*, *Galeola* sect. *Cyrtosia*, and *Vanilla* and long, slender fruits in *Apostasia*, *Selenipedium*, and remaining Vanilloideae.

Seeds

Orchid seeds have been widely study by taxonomists, and Beer (1863) gave an early indication of their diversity. More recently, the works of Clifford and Smith (1969) and Barthlott (1976a) suggested that seed morphology is useful particularly at tribal and subtribal levels.

The majority of orchids have minute seeds with a loose, thin-textured seed coat around the embryo, usually ranging from 8–10 μm in length (Stoutamire 1974; Arditti 1992), but up to 5 mm (Dressler 1981), and containing from 10–120 cells (Johansen 1950; Nishimura 1991). Until recently the embryo was considered an undifferentiated mass of cells, but in some species Clements (1988, 1995) distinguished a suspensor, epidermis, cortex, and meristem.

The shape, size, and ornamentation of the seed coat are variable and can be useful taxonomically. Ornamentation is associated with the structure of the cell walls of the testa. The outer wall may have longitudinal, transverse or reticulate patterning, or may bear wax deposits, for example in the Epidendreae. Barthlott and Ziegler (1981) described the extraordinary seed morphology of the leafless epiphyte *Chiloschista lunifera* in which the seed coat at one end has spirally thickened cell walls. When the seed is moistened the spiral thickenings unravel into threads up to 4 mm long. They postulated that these threads anchor the seed in suitable places for germination.

Species of *Apostasia*, *Neuwiedia*, *Selenipedium*, and *Vanilla* have seeds with a hard, crustose, dark brown or black seed coat. Crustose dark, glossy seeds are also found in the underground orchids of the genus *Rhizanthella*. Seeds of *Apostasia* are pitted and sticky, those of *Selenipedium* angular and glossy, and those of *Vanilla* lenticular and glossy. Other Vanilloideae such as *Galeola* and *Epistephium* have a hard seed coat with a more or less developed wing around the seed (Cameron and Chase 1998). Neuwiedias have small seeds with the seed coat inflated at each end (Dressler 1993).

Protocorms and seedlings

Until recently the lack of access to **protocorms** of named species other than those commonly cultivated has meant that protocorm morphology has scarcely been mentioned in accounts of the family, nor has it been used in the construction of classifications. Descriptions of protocorms of a limited number of species have been given by Burgeff (1932), Stoutamire (1963, 1974), and Veyret (1965). The recent review by Clements (1995) was much more comprehensive and systematically significant.

The term 'protocorm' was coined by the French botanist Bernard (1899) to designate a stage of development of the orchid embryo. Clements (1995) compared it to the hypocotyl, but it differs in possessing rhizoids/trichomes and a mycorrhizal fungus. The protocorm (Plate 8D) is an ephemeral tubercle that develops after germination of the orchid seed and associates with the appropriate mycorrhizal fungus before the production of the first seedling leaf and root. Protocorms vary in size, from 2–24 mm long and from 2–10 mm wide. Exceptionally they can be much larger, up to 10 cm long and 4 cm wide in *Cymbidium* and *Geodorum* (Clements 1995). A typical protocorm comprises a meristematic region at the chalazal end, a central region, a basal or suspensor region at the micropylar end, and an epidermis with rhizoids, often of several types. The rhizoids may be elongated single or multicellular trichomes with a swollen base, although structure may also be more complex, with a rhizoid arising from a basal cushion along with numerous echinoid processes. Protocorms may be negatively geotropic, positively geotropic or random in this respect.

Veyret (1974) attempted to relate protocorm morphology to orchid classification. Following the classification of Schlechter (1926), she recognized and illustrated a number of different protocorm types:

(1) an elongated, relatively slender protocorm in the Polychondreae (Neottieae, Cranichideae, Epipogieae);

(2) a dorsiventral type, sometimes with a distinct dorsal crest, in the Kerosphaereae-Pleuranthae-Monopodiales (Vandeae);

(3) a top-shaped, discoid, or rarely a narrow protocorm in the Kerosphaereae-Acrantheae (Epidendreae) and Kerosphaereae-Pleurantheae-Sympodiales (Epidendreae); and

(4) a top-shaped type in the Ophrydeae (Orchideae) and Cypripedioideae.

Veyret's analysis and Dressler's (1981) brief description and treatment have been much amplified by Clements' (1995) work. Clements surveyed the protocorm morphology and protocorm/seedling ontogeny of 252 species in 95 genera from all subfamilies except Apostasioideae, 16 tribes, and 29 subtribes of the family. He recognized and illustrated 11 major types of protocorms based on gross morphology and developmental pattern:

(1) obovoid (Habenariinae, Diseae, some *Pterostylis*, Chloraeinae, Spiranthinae);

(2) obovoid-deorsum (Orchidiinae, most Diurideae);

(3) elongate (some Bletiinae, *Nervilia*, *Galeottia*, *Polystachya*, some Epidendrinae, some Dendrobiinae, Coelogyninae, some Stanhopeinae);

(4) reptant (Goodyerinae);

(5) discoid (some Dendrobiinae and Epidendrinae);

(6) bracteate (Malaxideae, *Eria*);

(7) rhizomatous (Cypripedioideae, Neottieae);

(8) callus (Vandeae);

(9) isobilateral (*Pseudovanilla*, *Corallorhiza*, some Bletiinae, Lycastinae, some Maxillariinae, some Stanhopeinae, *Acriopsis*, Catasetinae, Cyrtopodiinae, *Sigmatostalix*);

(10) echinate (*Vanilla*); and

(11) globular (most Oncidiinae).

Clements argued that the occurrence of distinct protocorm types within a genus, e.g. *Dendrobium* and *Epidendrum*, and within currently recognized higher taxa has phylogenetic implications. The common occurrence of type 1 protocorms in the Spiranthinae and some Orchidoideae is also considered significant.

This is only a small sample of the morphological variation present in Orchidaceae, which will be expanded in future volumes genus by genus and used systematically in association with molecular data at different hierarchical levels of this phylogenetic classification.

PHILLIP J. CRIBB

III

ANATOMY

Historical account

Following the first description of the cell nucleus in *Cymbidium* flowers by Robert Brown (1831), certainly one of the most significant discoveries in the history of science, anatomical study of Orchidaceae was dominated by German botanists and remarkable illustrators. Heinrich Friedrich Link (1824) had already reported the unusual outer covering of roots of epiphytes (named *velamen radicum* by Schleiden (1849)) and later provided one of the first overviews of organography of both terrestrial and epiphytic orchids (Link 1851). It was in this seminal work, read to the Königlichen Akademie der Wissenschaften in 1849 and 1850 but not published until 1851, that stegmata in orchids were reported and illustrated for the first time. There also Link made the first reference to the occurrence of multiple discrete steles in tubers of *Orchis*, each bound by an endodermis in a common cortex. These were later to be misinterpreted as 'polysteles' (White 1907; Arber 1925; Sasikumar 1975) rather than as simply meristeles, or individual bundles of a dictyostele. Link (1851) even remarked on the secondary thickenings of velamen cells ('*spiralfasern*') of aerial roots of *Stanhopea eburnea* (although Meyen (1837) had already observed them) and described the vascularization of buds of *Orchis* and columns of *Orchis* (=*Dactylorhiza*) *sambucina* and *Maxillaria* (=*Lycaste*) *cruenta*.

Until the end of the nineteenth century the philosophical focus of anatomical studies was interpretation of homologies of tissues and underground organs as well as speculation on functions of velamen. The debate over the nature of the tuber characterizing subfamily Orchidoideae is attributable to its mode of development, originating as an axillary shoot that produces a basal extension resembling an adventitious root (Bell 1991; Rasmussen 1995). Irmisch (1850) conceived of it as a root tuber, the product of a close connection between a shoot and the coalescence of several roots. Schleiden (1845–1846) and Fabre (1856), however, viewed it solely as a swollen shoot formed laterally from an axillary bud. The issue was not settled among morphologists until the following century and Moreau's (1913) study of African Orchideae and Diseae and Arber's (1925) classic work on monocot structure, both of which treated them as root tubers. This interpretation was supported recently by Kurzweil *et al.* (1995) for tribe Diseae, Pridgeon and Chase (1995; Fig. III.1) for tribe Diurideae, and Stern (1997*a*, *b*) for tribe Orchideae of Orchidoideae on the basis that the mature tubers have a velamen and exodermis, an endodermis with casparian strips, pericycle, and a characteristic stelar configuration of alternating xylem and phloem.

Following its discovery by Link (1824), whether the velamen was epidermal or extraepidermal was argued by the some of the leading anatomists of the day. Meyen (1837), Schleiden (1849), Chatin (1856), and Fockens (1857) conceived of it as a root sheath covering the epidermis, which had thick-walled long cells and thin-walled short cells. In contrast, Schacht (1856) believed that the outermost layer of the velamen is the epidermis because root hairs arise from it and that what the others were calling an epidermis was a cortical layer that Oudemans (1861) unfortunately termed an endodermis. The issue was

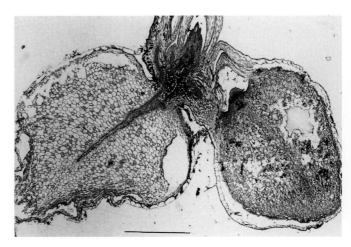

Fig. III.1. *Prasophyllum parvifolium.* Old (L) and new (R) tubers (LS). Light micrograph. Scale bar = 500 μm. With permission of A. M. Pridgeon.

finally resolved when Hubert Leitgeb (1865) undertook ontogenetic studies and determined that the velamen is actually a multiple epidermis. The term 'endodermis' for the outermost layer of the cortex consisting of long cells and short cells persisted until after the turn of the century when it was replaced by the more appropriate 'exodermis'. Leitgeb's (1865) other main contribution was a benchmark survey of root anatomy of 69 species, comparing the number of layers of velamen, secondary wall thickenings ('spiral-fasern'), pores, and wall thickness. This impressive survey set a standard, both in quantity and quality, that influenced work for the next half-century.

Between 1880 and the outbreak of World War I the focus of German research shifted to anatomical surveys and monographs of different taxonomic groups, particularly at the University of Heidelberg under the direction of the great orchid taxonomist, Professor Ernst Hugo Heinrich Pfitzer (1846–1906). Just as Leitgeb's contribution was the most influential monograph of orchid roots up to that time, so was the anatomical survey of orchid leaves by Martin Möbius (1887), who dedicated his work to Pfitzer. Möbius not only surveyed 193 species in 95 genera but also interpreted results in the light of Pfitzer's system of classification that gave more emphasis to vegetative morphology than prior systems. He was one of the first to try to correlate anatomical organization and presence of tissue types with ecological and climatic factors, work which was later continued by Tominski (1905), one of his students. Meinecke (1894) continued the research on root anatomy begun by Leitgeb (1865), examining 70 species and listing those features shared by members of Cypripedioideae and within six individual subtribes. In that same work he provided a brilliant study of the ontogeny, morphology, and systematic occurrence of Stabkörper, renamed tilosomes by Pridgeon et al. (1983). Completing the major surveys overseen by Pfitzer, Weltz (1897) examined the stems, rhizomes, and/or Luftknollen (pseudobulbs) of 130 species of monandrous sympodial orchids, but he found few groups with consistent patterns of organization. This observation continues today, and compared to the leaf and root the axis is the least useful organ in systematic studies. Hering's (1900) anatomical survey of stems or inflorescence axes of monopodial orchids complemented Weltz's, although the coverage was much smaller, only 50 species in 17 genera.

In the next decade the emphasis in Pfitzer's laboratory at the University of Heidelberg was on anatomical monographs of different subtribes: Siebe (1903) on Apostasia and Neuwiedia, Faber (1904) on Cypripedium and Paphiopedilum, Hünecke (1904) on Pleurothallidinae, Sprenger (1904) on Bulbophyllinae, and Zörnig (1903) on Coelogyninae. Of these only Hünecke and Zörnig put the data to systematic use and constructed artificial keys to the species studied.

The impetus for anatomical studies seemed to have collapsed, initially due to the death of Pfitzer in 1906, but compounded by the onset of World War I and its harsh aftermath. Even after the war few new works on orchid anatomy appeared. Staudermann (1924) surveyed the hairs of monocots, among them the trichomes of several genera including Cypripedium, Epipactis, Glossodia, and Renanthera. Fuchs and Ziegenspeck (1925, 1926a,b, 1927a,b,c) published comprehensive, exceedingly detailed comparative and developmental studies of European orchids. Solereder and Meyer (1930) synthesized most of the comparative studies to that time and added many original observations for almost 150 genera. Despite the inevitable taxonomic changes since 1930, their work is still authoritative and extremely useful to anyone studying aspects of vegetative anatomy.

Floral anatomy

The few early anatomical studies of floral axes of orchids are referenced in Solereder and Meyer (1930). In addition, Ziegenspeck (1936) examined the rachises of such genera as Cephalanthera, Cypripedium, and Cattleya. Such studies of flowers themselves are limited as well. Some of the earliest are by von Guttenberg (1926), Pohl (1931), and Daumann (1933), who described the anatomy of the perianth of a few species.

More recently Dannenbaum et al. (1989) prepared an elegant survey of stigma types in Orchidaceae, looking at 180 species in 95 genera. There are three basic types. The first, characterizing Paphiopedilum of Cypripedioideae, is convex to dome-shaped with three distinct lobes. Surface cells are smooth and are not developed into papillae. The second type, occurring in Spiranthoideae and Orchidoideae, has two subtypes. The convex stigmatic surfaces of most taxa in these two subfamilies are either indistinguishable or else the lateral lobes and only the basal part of the median lobe comprise the receptive surface. Finger-shaped papillae protrude from the surface. In the second subtype, a pellicle covers the stigma surface in, for example, *Ophrys* and *Barlia*. The third stigma type, observed in Epidendroideae (including Vandoideae), is concave with barely discernible stigma lobes. Receptive cells of the surface can be toothed or have sculptured walls. A subtype of this kind of stigma is similar to the above except for a pellicle covering the stigmatic cavity.

The study of floral glands, discussed here only briefly, was popularized by Stefan Vogel, whose classic works on osmophores (fragrance glands) and elaiophores (oil glands) opened new vistas in developmental and systematic research. His morphological and anatomical studies on osmophores (1962) covered subtribes Chloraeinae, Orchidinae, Pleurothallidinae, Dendrobiinae, Bulbophyllinae, Catasetinae, Stanhopeinae, and the genus *Phragmipedium*. Pridgeon and Stern (1983, 1985) and Stern *et al.* (1987) supplemented Vogel's work with ultrastructural studies of development of osmophores in *Restrepia*, *Scaphosepalum*, and *Stanhopea*, respectively. Curry *et al.* (1991) used characters related to osmophores of *Stanhopea* along with lip characters in a systematic study of the genus: texture of the osmophore surface, number of cell layers, nature of lipid inclusions, and presence of plastoglobuli in osmophore amyloplasts. They found that osmophore characters generally support clades based on floral morphology, although they noted the problems associated with using ultrastructural characters.

Vogel (1974) also described and illustrated the elaiophores of *Zygostates*, *Ornithocephalus*, *Oncidium*, and *Sigmatostalix*, recently updated by Toscano de Brito (1994) in his systematic study of Ornithocephalinae. Steiner's exposition of the elaiophores of *Disperis* and oil-gathering bees (1989) led to an interesting phylogenetic study of that genus by Manning and Linder (1992), who used some characters related to type and position of elaiophores along with other (mainly floral) characters to show that interspecific reproductive isolation may be explained by differential deposition of pollinia on pollinators.

Swamy's well-known analysis (1948*c*) of floral vasculature of 24 genera, overshadowing earlier, more detailed works by Oesterberg (1883) and Raunkiaer (1895–1899), is still regarded as the starting place for all future research. He recognized three types of vascular supply to the staminal and stigmatic whorls from the ovary, one characterizing Cypripedioideae, one the Ophrydinae (=Orchidinae), and the last and most complex all other monandrous orchids. In this latter type there are anastomoses, cohesion, and

adnation paralleling the evolutionary trend toward increased complexity of the orchid flower. The two most important conclusions from the work were that (1) the labellum receives the same vascular supply as other perianth parts and so is not a compound structure as argued by Robert Brown (1831) and Charles Darwin (1877), and (2) the ancestral orchid flower had six functional stamens but that in time the posterior three ceased to function. Garay (1960) reported that, as in Cypripedioideae, the six traces from the ovary of Apostasioideae do not differentiate, which he said supported the 'relative primitiveness of these two groups'.

Secondary thickenings of the endothecium, the subepidermal layer of the anther, have been used in systematic studies of several plant families, including orchids. Freudenstein (1991) surveyed 210 orchid genera and demonstrated that there are four basic thickening types:

(1) tightly packed channels of loops or helices, found in the 'neottioid genera', Apostasioideae, and presumably early-diverging genera in remaining subfamilies (Fig. III.2);

(2) scattered loops, occurring in all genera of Orchidoideae sampled except *Disperis* (Fig. III.3);

(3) circular thickenings (three subtypes), as in Cypripedioideae and some taxa in Spiranthoideae and Epidendroideae (Fig. III.4); and

(4) scattered bars (also three subtypes), as in most Epidendroideae and some Spiranthoideae (Fig. III.5). Generally, terrestrial genera have regularly arranged, well-developed thickenings, whereas epiphytic genera have congested, irregular, and less-developed thickenings.

Vegetative anatomy

In addition to anatomical studies as part of generic monographs, which will be discussed throughout this series in the context of the genus in question, modern researchers have surveyed higher taxa—subfamilies, tribes, subtribes, or even the entire family—searching for characters that might be useful systematically at those levels. We will focus on those works briefly, mentioning useful systematic characters and their occurrence throughout Orchidaceae.

Stern *et al.* (1993*a*) examined leaves, stems, and roots of *Apostasia* and *Neuwiedia* (Apostasioideae) to determine if there is any anatomical justification for

Fig. III.2. *Chloraea alpina.* Type I endothecial thickening. Scale bar = 0.1 mm. (Photo: J. V. Freudenstein. Reproduced, with permission, from Freudenstein, J. V. (1991). *American Journal of Botany* **78**, 766–82.)

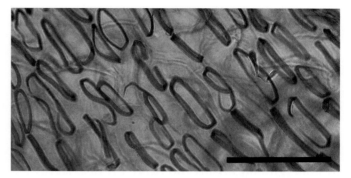

Fig. III.3. *Disa filicornis.* Type II endothecial thickening. Scale bar = 0.1 mm. (Reproduced, with permission, from Freudenstein, J. V. (1991). *American Journal of Botany* **78**, 766–82.)

Fig. III.4. *Corallorhiza striata.* Type III endothecial thickening. Scale bar = 0.1 mm. (Reproduced, with permission, from Freudenstein, J. V. (1991). *American Journal of Botany* **78**, 766–82.)

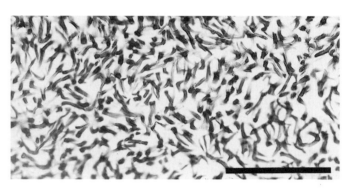

Fig. III.5. *Gomphicis valida.* Type IV endothecial thickening. Scale bar = 0.1 mm. (Reproduced, with permission, from Freudenstein, J. V. (1991). *American Journal of Botany* **78**, 766–82.)

thinking that diandrous and monandrous orchids arose from them or something like them, as some have claimed on the basis of the putatively plesiomorphic number of stamens (two or three). Although there are many characters that unite the two genera, such as tetracytic stomata, uniseriate velamen, cauline stegmata, and root tubercles, there is none that indicates that they are close to the ancestral lineage. On the contrary, roots of both genera possess vessel members with simple perforation plates (Stern *et al.* 1993*a*), a derived condition in angiosperm phylogeny (as opposed to the ancestral state of scalariform plates), which makes less plausible the argument that *Apostasia* and *Neuwiedia* are close to the progenitors of all other orchids, particularly those in subfamilies Spiranthoideae and Orchidoideae (*sensu* Dressler 1993), which have scalariform perforation plates.

Rosso's (1966) study of leaves, axes (stem and rhizome), and roots of Cypripedioideae is still author-itative, although the sample size for *Paphiopedilum* in particular was too narrow. However, he suggested that on the basis of anatomy and distribution *Cypripedium irapeanum* was similar to *Selenipedium* and that *Paphiopedilum* and *Phragmipedium* share more vegetative characteristics than do *Selenipedium* and *Cypripedium* (except for *C. irapeanum*). Atwood (1984) updated and significantly expanded morphological, anatomical, and cytological data on Cypripedioideae and using the Wagner Groundplan-divergence Method (Wagner 1980) was the first since Pfitzer (1903) to treat the subfamily in a phylogenetic and biogeographic context.

Chesselet (1989) had little success in finding discriminating anatomical characters in leaves of the terrestrial subtribes Disinae and Coryciinae. Sclerenchyma bundle caps associated with a sclerified leaf margin is a synapomorphy for *Disa* sect. *Stenocarpa* and supports the inclusion of *Herschelianthe* in that same section. Kurzweil *et al.*

(1995) expanded Chesselet's anatomical survey to include the remainder of Diseae as well as other plant organs. They reported a similar scarcity of informative foliar characters in other subtribes but did find that presence of a dissected siphonostele in the tuber supported the inclusion of *Herschelianthe* in *Disa* sect. *Stenocarpa*.

From the standpoint of systematics, leaf characters in Diurideae are more useful than those relating to underground organs, probably because leaves are generally subject to more environmental selection pressures. Pridgeon (1994) reported that those characters showing the most variation among 25 species in 15 genera of Caladeniinae are presence and type of trichomes, cuticular sculpturing, anticlinal walls of epidermal cells, heterogeneity of chlorenchyma, and the distribution and length/width ratios of stomata.

Looking at the other tribe in Orchidoideae, Stern (1997*a*, *b*) surveyed the anatomy of leaf, stem, root, and tuber of taxa in Orchideae—21 species of four genera of Habenariinae and 35 species in 13 genera of Orchidinae. He confirmed the presence of velamen in root tubers in both subtribes, but due to the incomplete coverage of both he could not attach any systematic significance to the data.

In a more complete anatomical survey of subfamily Spiranthoideae, Stern *et al.* (1993*b*) demonstrated that Spiranthoideae *sensu* Dressler (1993) is polyphyletic and that tribes Tropidieae and Diceratosteleae are more properly placed in Epidendroideae, both conclusions supported by recent *rbcL* studies (Kores *et al.* 1997). Among the most useful characters were anticlinal epidermal cell walls in leaves, presence of glandular foliar hairs, development of subsidiary cells, presence of specialized amyloplasts termed spiranthosomes (Fig. III.6), presence of stegmata in leaf and stem, and wall thickness of the exodermal and endodermal cells of the root.

Epiphytic orchids generally show more anatomical variation than the terrestrials for the same reason that aerial organs (leaves, stems, aerial roots) are more variable than subterranean organs—selection pressures related to insolation, temperature, relative humidity, rainfall, etc. Using leaf characters related to the cuticle, stomata, hypodermis, helically thickened mesophyll cells, and extraxylary fibres, many of which related to the epiphytic habit, Baker (1972) grouped the genera of subtribe Laeliinae into five basic 'alliances': (1) *Cattleya–Laelia*; (2) *Brassavola–Domingoa*; (3) *Epidendrum*; (4) *Encyclia*; and (5)

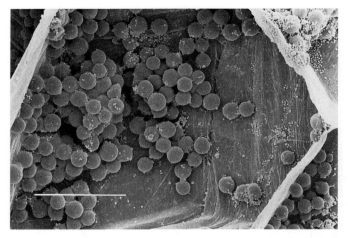

Fig. III.6. *Cranichis ciliata*. Cortical root cell showing globular spiranthosomes and smaller starch granules released from ruptured spiranthosomes. SEM photograph. Scale bar = 50 μm. (Reproduced, with permission of the author and of Springer-Verlag, Vienna, from Stern, W. L. *et al.* (1993*b*). *Botanical Journal of the Linnean Society* **113**, 161–97.)

Scaphyglottis. In a broad study of subtribe Pleurothallidinae, a Neotropical group of some 4000 species, Pridgeon (1982) examined the leaves, axes, and/or roots of 200 species in 22 genera and found that several characters associated with the leaf have diagnostic value, such as features of trichomes, cuticle, epidermis, hypodermis, spiral thickenings in hypodermal cells, and mesophyll idioblasts (Figs III.7, III.8), and number of vein series. Subtribal phylogenetic trends of reduction in number of pollinia and specialization of the perianth for pollination are generally correlated with particular anatomical trends, which are either also reductionary or involve specializations directly related to the water relations of the epiphytic habit.

Two significant studies of stomata in Orchidaceae have shown that not only presence, number, and arrangement of subsidiary cells but also their developmental patterns are useful at different levels of classification. Williams (1979) reported that these cells (adjoining guard cells of stomata) develop mesoperigenously from trapezoid cells which are themselves formed by oblique divisions in the files of cells adjacent to that containing the guard-cell mother-cell. The trapezoid cell usually divides unequally to form a subsidiary cell and a derivative cell. Polar subsidiary cells may also be formed. He observed subsidiary cells in 321 species in 145

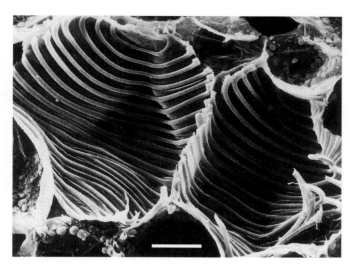

Fig. III.7. *Pleurothallis circumplexa.* Spirally thickened idioblasts in mesophyll of leaf viewed in transection. SEM photograph. Scale bar = 50 μm. (Reproduced, with permission, from Pridgeon, A. M. (1982). *American Journal of Botany* **69**, 921–38.)

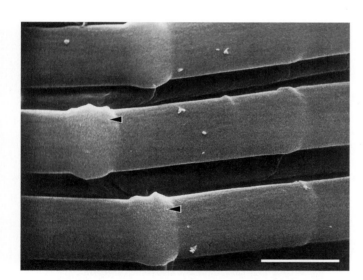

Fig. III.8. *Ophidion cymbula.* Helical thickenings of abaxial hypodermal cell as seen through the thin primary wall. SEM photograph. Scale bar = 5 μm. (Reproduced, with permission, from Pridgeon, A. M. (1982). *American Journal of Botany* **69**, 921–38.)

genera and in all tribes of Epidendroideae *s.l.* as well as in Cypripedioideae but not in tribe Orchideae of Orchidoideae nor in the primitive epidendroid tribe Gastrodieae. (Complementing this, Stern *et al.* (1993*a*) reported subsidiary cells in Apostasioideae.) Based on the absence of subsidiary

cells in primitive epidendroids and their presence in the most advanced epidendroids, Williams concluded that presence of subsidiary cells is the derived condition. Rasmussen (1981) expanded N. H. Williams' (1979) survey of stomatal development to include more representatives from Orchidoideae and found that both agenous and hemiperigenous types may occur within the same species. Although the ratio of the two types is relatively constant for a given species, there could be significant interspecific differences.

Several anatomical features of the root (Fig. III.9) have been used with profit in systematic work, particularly related to the velamen and exodermis (reviewed by Pridgeon 1987; Figs III.10–12). Porembski and Barthlott (1988) surveyed roots of 344 species in 262 genera and from these were able to describe twelve types of roots using the number of velamen layers, stratification of layers, presence of secondary wall thickenings ('helical thickenings') and pores, thickness of exodermal cell walls, and presence of '*Stabkörper*' (=tilosomes; see below) and 'tracheoidal idioblasts' in the cortex. Ten of the twelve types are mostly named after representative genera: *Calanthe, Bulbophyllum, Pleurothallis, Malaxis, Spiranthes, Coelogyne, Dendrobium, Epidendrum, Cymbidium, Vanda.* The other two types are 'unspecified', and 'velamen absent' (simple rhizodermis). The distribution of these characters and types agrees well with the classification of Dressler (1981). Although the authors did not polarize velamen syndromes, they suggested that the *Calanthe* type could be considered primitive because it lacks wall thickenings and is most frequent in putatively ancestral taxa. At the same time they consider the *Cymbidium* and *Vanda* types most derived. However, after studying variation in numbers of velamen layers within roots, among roots of the same plant, and among roots of different plants of the same species, Rüter and Stern (1994) advised caution in relying too heavily on number of velamen layers as a reliable character and that the level of sampling be standardized. They recommended that samples should be taken 40–45% distally from the root base to reach an approximation of the average number of velamen layers and that sections should also be made from different levels of the root and from several roots of the same plant to assess the degree of variation.

Expanding on the root studies of Leitgeb (1865) and Meinecke (1894), Pridgeon *et al.* (1983)

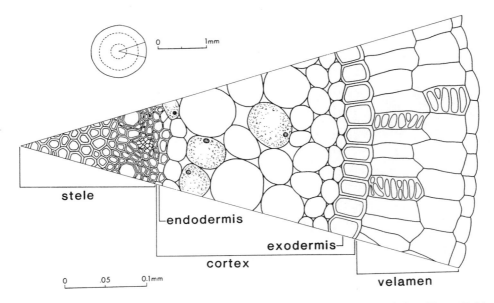

Fig. III.9. Root transection of *Restrepiella ophiocephala*, redrawn by Wendy B. Zomlefer. From Pridgeon, A. M. (1986). *Lindleyana* **1**, 90–101.

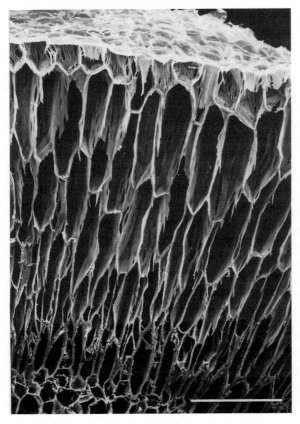

Fig. III.10. *Renanthera imschootiana*. Root transection showing multiseriate velamen. SEM photograph. Scale bar = 250 μm. From Pridgeon, A. M. (1986). *Lindleyana* **1**, 90–101.

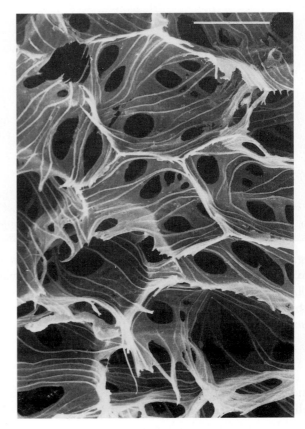

Fig. III.11. *Zygopetalum intermedium* var. *peruvianum*. Velamen with secondary wall thickenings and large perforations. SEM photograph. Scale bar = 25 μm. From Pridgeon, A. M. (1986). *Lindleyana* **1**, 90–101.

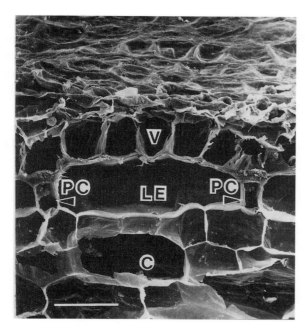

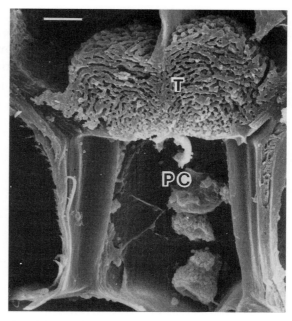

Fig. III.12. *Sobralia decora*. Radial section of root showing velamen (V), long exodermal cell (LE), passage cells (PC), and cortex (C). Note tilosomes above passage cells. SEM photograph. Scale bar = 50 μm. (Reproduced, with permission, from Pridgeon, A. M. *et al.* (1983). *American Journal of Botany* **70**, 1365–77.)

Fig. III.13. *Sobralia decora*. One passage cell and spongy tilosomes in root transection. SEM photograph. Scale bar = 5 μm. (Reproduced, with permission, from Pridgeon, A. M. *et al.* (1983). *American Journal of Botany* **70**, 1365–77.

surveyed 350 species in 175 genera for tilosomes, which are lignified excrescences from the walls of cells of the innermost velamen layer adjacent to thin-walled passage cells of the exodermis. Seven morphological types are recognizable: spongy (e.g. Sobraliinae; Fig. III.13), discoid (*Cryptophoranthus* (=*Zootrophion*)), lamellate (Laeliinae, Lycastinae), webbed (Polystachyeae), meshed (Maxillariinae), baculate (Maxillariinae), and plaited (*Trigonidium*). There is also a subtype of lamellate tilosomes, broadly lamellate, which, ranging up to the spongy type, is characteristic of *Bulbophyllum* (including *Cirrhopetalum*). Tilosomes are concentrated in Polystachyeae, Sobraliinae, Coelogyninae, Laeliinae, Pleurothallidinae, Bulbophyllinae, Lycastinae, and Maxillariinae. They are absent from Orchidoideae *s.l.* examined except for *Cryptostylis* (Stern *et al.* 1993*b*).

A useful character in estimating orchid phylogeny is presence of stegmata, bearing silica bodies, in longitudinal files lining fibre bundles and vascular bundle sheaths of all plant organs, though rarely (if ever) in the root (Møller and Rasmussen 1984). There are two different types of silica bodies, conical and spherical. Most orchid taxa surveyed either lack stegmata or have stegmata with conical silica bodies. Conical bodies characterize *Apostasia* and *Neuwiedia*, Tropidiinae, Arethuseae, Coelogyninae, Laeliinae, Pleurothallidinae, Glomerinae, and Cymbidieae. Spherical bodies are relatively rare but occur in Eriinae, Dendrobiinae, and Vandeae. Silica bodies are absent from examined species of Bulbophyllinae, *Sunipia*, and *Vanilla*. Both types occur in Podochilinae, conical in *Agrostophyllum* and spherical in *Appendicula* and *Podochilus*, suggesting either that the former is an early branch in the subtribe or that the subtribe as currently circumscribed is polyphyletic.

A recent phylogenetic analysis of Orchidaceae as a whole (Freudenstein and Rasmussen 1999) incorporated several of the anatomical characters mentioned above in addition to morphological characters: presence of root tubers and velamen, type of velamen secondary thickening, exodermal cell shape and thickening, presence of spiranthosomes, presence and type of stegmata, anticlinal walls of the abaxial leaf epidermis (straight or sinuous), presence of subsidiary cells, type of endothecial thickening, and presence of endocarpic trichomes. Results supported the monophyly of all subfamilies (including the concept of Vandoideae) except for Orchidoideae *s.l.*, which is paraphyletic.

Technological advances such as the scanning and transmission electron microscopes have opened up plant tissues to closer inspection and allowed us to develop new characters in assessing relationships. They have also given us new insights into features already described by the great German and French anatomists a century ago. More importantly, today we can apply those anatomical characters in a phylogenetic context along with molecular, cytological, and traditional morphological data—at different levels of classification—to produce complete analyses that have high levels of support.

ALEC M. PRIDGEON

Higher-level surveys

Features relating to pollinaria and pollinia are discussed in the introductory chapter on morphology, and pollen studies of individual genera will be treated separately with the discussions of those respective genera. Here we address only those characters relating to pollen and their application in systematic studies of higher taxa.

Reichenbach (1852) was among the first to report that orchid pollen is shed as either monads or tetrads. Erdtman (1944) observed that pollen of most species is inaperturate but that the grains of *Vanilla* and *Epistephium* are distinct in having three or four apertures.

Following Erdtman's (1952) classic monograph of angiosperm pollen including a brief treatment of Orchidaceae, two significant family-wide surveys of orchid pollen showed in much greater detail that certain sculptural features have systematic application. Schill and Pfeiffer (1977) used light microscopy, scanning electron microscopy (SEM), and transmission electron microscopy (TEM) in a study of 630 species representing almost 200 genera and all subfamilies. They found that although a tectate wall structure predominates in the family, semitectate types occur in Neottioideae and intectate forms in most Orchidoideae. The exine comprises only ektexine and lacks an endexine. Pollen grains are generally inaperturate, but grains of Cypripedioideae are colpate and those of some Neottioideae are porate. Ackerman and Williams (1980) later showed that grains of *Lecanorchis* in Neottieae have up to five pores.

Sculpturing varies from laevigate-scabrate to laevigate-porate, verrucose-hamulate, hamulate, reticulate-heterobrochate, reticulate-fragmentimurate and ornate to baculate-pilate (Schill and Pfeiffer 1977). Sculpturing in Cypripedioideae is generally laevigate-scabrate, whereas that of Neottioideae is reticulate-heterobrochate. The fine sculpture of grains in Epidendroideae and Vandoideae is homogeneous, but that of Orchidoideae is very heterogeneous and exhibits the whole spectrum. For example, in the genus *Orchis* alone there are species with laevigate-scabrate sculpturing (e.g. *O. coriophora*), laevigate-perforate (*O. mascula*), verrucose-hamulate (*O. laxiflora*), hamulate (*O. quadripunctata*), and baculate-pilate (*O. italica*). Sampling among genera was uneven. Out of an estimated 300 species of *Maxillaria*, 41 taxa were studied. However, out of about 2000 species of *Pleurothallis*, which is polyphyletic according to ongoing molecular work (Pridgeon and Chase, unpublished), pollen of only 15 species was examined. On balance, the study is still the most comprehensive of its kind for Orchidaceae and showed that characters related to sculpture are useful at least at the specific level but often the generic level and higher.

The other family-wide survey was by Williams and Broome (1976). The scope of their research was much less intensive (only SEM) and less extensive but included a few genera not studied by Schill and Pfeiffer (*Cleistes, Pogonia, Jacquiniella, Corallorhiza, Sievekingia*).

In the last 20 years published pollen studies have been limited to fewer taxa, either in subfamilies Apostasioideae and Cypripedioideae or lower taxa (tribes, subtribes, genera) in other subfamilies, but usually building on the foundation laid by Schill and Pfeiffer (1977). Schill (1978) supplemented the 1977 study with a treatment of nine species of *Apostasia* and *Neuwiedia* (Apostasioideae), and although he could not use pollen data to resolve the question of whether or not they belong to Orchidaceae, in the end he treated them as the separate family Apostasiaceae. In both genera pollen grains are shed as monads and are unique in Orchidaceae for having an operculate colpus. The exine is semitectate with reticulate-heterobrochate to microreticulate sculpturing.

Newton and Williams (1978) examined pollen of Apostasioideae and Cypripedioideae independent of Schill (1978) and Schill and Pfeiffer (1977). They confirmed the monosulcate condition and reticulate sculpturing of both *Apostasia* and *Neuwiedia* and also remarked on the operculate nature of the colpus of *Neuwiedia* (but not *Apostasia*). Newton and Williams also corroborated the monad, monosulcate condition of Cypripedioideae and the psilate (laevigate) sculpturing of grains of *Selenipedium*, *Cypripedium* (except foveolate in *C. parviflorum*), and *Phragmipedium* (except fossulate in *P. longifolium*). The greatest diversity in sculpture among the slipper orchids occurs in *Paphiopedilum*, in which the pollen of some species is psilate (e.g. *P. chamberlainianum*), foveolate (e.g. *P. barbatum*), fossulate (e.g. *P. concolor*) or scabrate (*P. bullenianum*). Finally, Newton and Williams observed two ridges on either side of the sulcus as well as in photographs by Schill and Pfeiffer (1977) and attributed them to wall collapse from air-drying rather than critical-point drying.

Burns-Balogh and Hesse (1988) used light microscopy (LM), SEM, and TEM to help resolve questions about the nature of the aperture in slipper orchid pollen. They found that there is, in fact, one sulcus, a circular field of thin sporoderm surrounded by a ring-like, thicker zone. More important, they were able to describe the exine of several taxa in *Cypripedium*, *Paphiopedilum*, *Phragmipedium*, and *Selenipedium*. The exine of *Selenipedium* comprises a large tectum, coherent columellae, and a compact foot-layer in the non-apertural region and a lamellated foot-layer and much-reduced tectum and columellae in the apertural region. Relative development of the exine is similar in *C. irapeanum* and *Phragmipedium*. However, in *Paphiopedilum* and other *Cypripedium* species, the foot-layer is absent, and columellae are present only in *Cypripedium* as fragmented globules embedded in the intine. As a result the exine of *Paphiopedilum* consists solely of sporopollenin 'lumps' on top of the intine, forming a smooth layer without sculpturing. Burns-Balogh and Hesse (1988) related the wall structure of *Selenipedium* pollen to that of other monocots and interpreted the reduced exine of *Paphiopedilum* and *Cypripedium* as derived, with that of *C. irapeanum* and *Phragmipedium* intermediate.

Among Diurideae of Orchidoideae, Ackerman and Williams (1981) studied pollen of Chloraeinae as well as that of the chiefly Australasian subtribes with SEM. The structure and sculpture of a good representation of Chloraeinae are similar—tectate-perforate to semitectate and foveolate to reticulate. The pollen unit of *Codonorchis* is an inaperturate monad, but in other genera (*Gavilea*, *Geoblasta*, *Bipinnula*, and *Chloraea*) it is an inaperturate tetrad. Pollen morphology in other subtribes is much more diverse, and there is no consistency in pollen unit within a subtribe or even the same sample. Generally, however, sculpturing is also diverse, from foveolate to reticulate. Within Caladeniinae, for example, some genera have finely reticulate exines, whereas those of *Drakaea* and *Spiculaea* have scabrate surfaces. Pollen of *Pterostylis* consists of loose monads with reticulate sculpturing and a porate aperture and is considered by the authors to be 'primitive'. Based on the fact that pollen structure of the Chloraeinae is most similar to that of Caladeniinae, Ackerman and Williams supported the placement of Chloraeinae in Diurideae.

Among tribe Diseae (Orchidoideae) Chesselet (1989) surveyed pollen morphology of subtribes Disinae and Coryciinae along with outgroups from Orchideae and Satyriinae and found that several characters provided some apomorphies at the subtribal level. In particular, there is a suite of synapomorphies for Coryciinae, including fasciculate massulae, elongated tetrads in a linear configuration, and a secondarily tectate exine. *Huttonaea* shares elongated tetrads with Coryciinae, which may indicate relationship, but the reticulate sculpturing of the genus is plesiomorphic among these taxa. Exine sculpturing in Disinae is either rugose or verrucose and fails to distinguish *Monadenia*, *Herschelianthe*, and *Disa*, which is consistent with rDNA ITS sequences (Douzery *et al.*, in press) nesting *Monadenia* and *Herschelianthe* within *Disa*. However, the molecular data contradict Chesselet's claim that section *Micranthae* of *Disa* may be segregated on the basis of distinctive rugose exines with free baculae. In fact, *Disa chrysantha* of that section is sister to *Herschelianthe*, and segregation of that clade would produce a paraphyletic *Disa*.

Sectile pollinia of tribe Cranichideae are calymmate massulae, the tetrads with predominantly reticulate exines (Williams and Broome 1976; Schill and Pfeiffer 1977; Freudenstein and Rasmussen 1997).

In Epidendroideae *s.l.* there have been few studies above the genus level other than those mentioned. Ackerman and Williams (1980) examined 41 species in all seven genera of Neottieae with SEM. In this group, pollen occurs as monads in the ancestral taxa and as tetrads in derived taxa. Grains are monoaperturate (except for *Lecanorchis*), porate or tenuate, and exine structure varies from tectate-perforate in *Lecanorchis* and *Cephalanthera* to semitectate in *Limodorum*, *Aphyllorchis*, *Epipactis*, *Neottia*, and

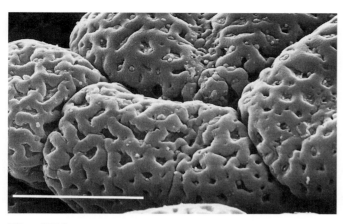

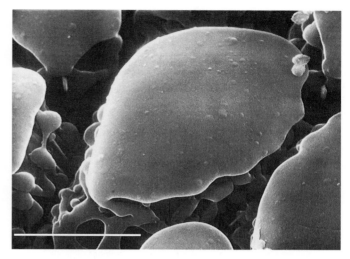

Fig. IV.1. *Platythelys querceticola.* Tetrads in a massula with reticulate sculpturing. Scale bar = 10 μm. With permission of J. V. Freudenstein.

Fig. IV.2. *Restrepiella ophiocephala.* Exine sculpturing of outer surfaces is laevigate and that of the inner surfaces reticulate. Scale bar = 5 μm. With permission of A. M. Pridgeon.

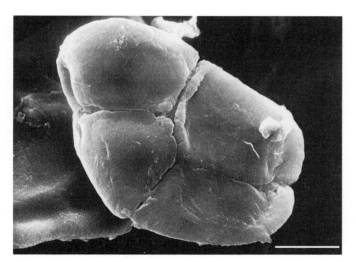

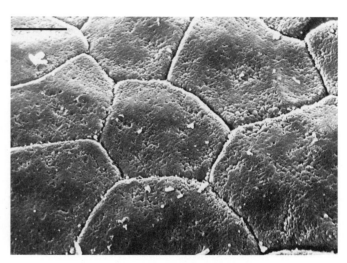

Fig. IV.3. *Pleurothallis mystax.* Tetrads are ± laevigate. Scale bar = 5 μm. With permission of A. M. Pridgeon.

Fig. IV.4. *Masdevallia* (=*Luerella*) *pelecaniceps.* Tetrads are tightly packed with fossulate sculpturing. Scale bar = 5 μm. With permission of A. M. Pridgeon.

Listera. The associated trend in sculpturing is from fossulate or foveolate to reticulate. The fact that *Lecanorchis* has 0–5 small pores (see above) leads the authors to argue that it is likely to be more closely related to Gastrodieae than Neottieae, although recent molecular and morphological evidence (Cameron 1996) shows that *Lecanorchis* is better placed in Vanilloideae (see below). Ackerman and Williams proposed that trends in Neottieae are:

(1) from tectate pollen in more ancestral genera (*Lecanorchis, Cephalanthera*) to semitectate in the derived genera (*Neottia and Listera*), even

though the tectate condition is most common in derived orchid tribes;

(2) from relatively large apertures to smaller ones; and

(3) from monads to easily separable tetrads to tightly coherent tetrads, related to progressively more efficient pollen transfer and more precise pollinating mechanisms.

Hesse *et al.* (1989) used both SEM and TEM to examine pollen wall ultrastructure of tribes Arethuseae, Gastrodieae, Triphoreae, and Vanilleae (now Vanilloideae) in an effort to characterize them for

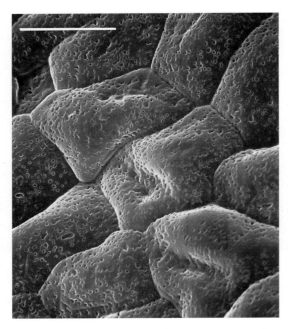

Fig. IV.5. *Leochilus oncidioides*. Tetrads with foveolate sculpturing. Scale bar = 15 μm. With permission of M. W. Chase.

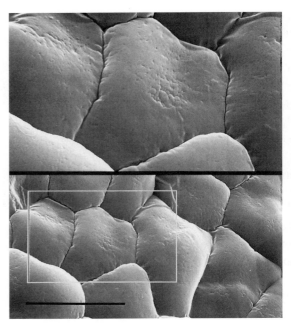

Fig. IV.6. *Oncidium reichenheimii*. Tightly packed tetrads with laevigate to foveolate sculpturing. Scale bar = 15 μm. With permission of M. W. Chase.

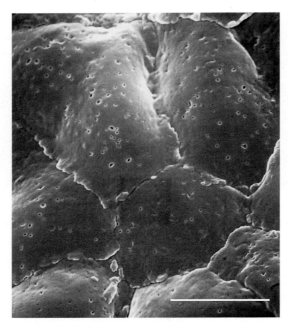

Fig. IV.7. *Rodriguezia venusta*. Exine sculpturing is laevigate-microfoveolate. Scale bar = 7.5 μm. With permission of M. W. Chase.

feature. The 'typical' exine of all these tribes is confined to the outermost walls of tetrads or polyads, whereas the exine of internal walls is reduced and may facilitate germination. Regrettably, the authors included subtribe Acianthinae in their analysis of the pollen of tribe Triphoreae; Acianthinae is now generally accepted (and supported by molecular results) as a subtribe of tribe Diurideae in Orchidoideae, not Epidendroideae.

Cameron (1996) surveyed pollen structure and sculpturing in Vanilloideae as part of a total evidence phylogeny of the subfamily and found trends mirroring those in other subfamilies. Within the Vanilloideae clade there is a trend from free monads (e.g. *Pogonia*, *Lecanorchis*) to tetrads (e.g. *Isotria*) to massulae (*Erythrorchis*, *Galeola*). Pollen of all taxa is inaperturate (except for *Lecanorchis*, which is polyporate) and has psilate sculpturing (except verrucose in *Epistephium*, *Eriaxis*, and *Vanilla*). Ultrastructural studies showed that the intine is channelled in *Pogonia* in addition to *Lecanorchis* and *Cleistes* as reported by Hesse *et al.* (1989).

Pollen and phylogeny

The first theory on the evolution of the exine in orchids is that of Burns-Balogh (1983), which has not

systematic purposes. A tectate-columellate exine and compact foot-layer characterizes some Vanilloideae, but the foot-layer is reduced or even absent in derived Vanilloideae, Arethuseae, and Gastrodieae. In Triphoreae there has been a reduction or loss of columellae and/or the tectum, regarded as an advanced been widely accepted. She argued that the tectate-perforate-columellate pollen of Apostasioideae is most primitive (in relation to unspecialized floral organs) and that there have been two major divergent evolutionary lines from that point. One of these terminates with Cypripedioideae and a tectate-imperforate exine with incipient columellae. The other line leads in two directions: (1) with loss of the foot-layer to walls that are semitectate (Orchidoideae, Epidendroideae) and then intectate (Orchidoideae), and (2) to walls that are tectate-perforate (Epidendroideae, Neottioideae, and Spiranthoideae), then tectate-imperforate (Epidendroideae, Orchidoideae) with loss of the foot-layer, and finally to walls that are tectate-imperforate-globular (Epidendroideae). The major problem with this theory, as Zavada (1990) points out, is that the sample size is too small and the number of characters too limited to make such sweeping generalizations about orchid phylogeny. Another problem with the scheme is that Burns-Balogh recognized the presence of the endexine throughout, and the endexine is not only absent in orchids (Schill and Pfeiffer 1977) but even in most monocots (Zavada 1983), the foot-layer being situated immediately above the intine.

Zavada (1990) identified general trends related to pollen unit, aperture, and exine structure across Orchidaceae:

(1) pollen unit. The trend is from monads to adherence of grains into polyads and/or mealy pollinia, and then into tightly packed, waxy pollinia;

(2) aperture. A well-defined sulcus is found in Apostasioideae, a poorly defined ulcerate pore in Cypripedioideae and Spiranthoideae, and finally the inaperturate condition in Orchidoideae and Epidendroideae;

(3) exine structure.

The pollen of Apostasioideae—tectate-columellate with a monosulcate aperture—characterizes primitive monocot families. Although this same condition occurs in other subfamilies, it is often associated with more derived pollen units and aperture types. The trend is toward tectate-granular, atectate, or exineless pollen with loss of the foot-layer and elaboration of the intine. Taking all of these trends into account, Zavada (1990) interpreted Vandeae as the most derived group in Orchidaceae.

Among the morphological and anatomical characters selected by Freudenstein and Rasmussen (1999) in their broad cladistic analysis of Orchidaceae, several characters associated with pollinaria, pollinia, and pollen were useful. Those related to pollen are: (1) unit (monads or tetrads); (2) tectum (reticulate or smooth); (3) aperture (colpate/sulcate, porate, inaperturate, or polyporate; (4) operculate colpus (absent or present (present only in *Apostasia* and *Neuwiedia*); (5) massulae (absent, orchidoid, epidendroid, arethusoid), and (6) pollinium texture (granular or solid). Results of the study are reported elsewhere in this volume, but briefly the study supported the monophyly of Apostasioideae, Cypripedioideae, Spiranthoideae, and Epidendroideae. Orchidoideae is paraphyletic, giving rise to spiranthoids. Vanilloid genera as treated by Cameron (1996) form a clade sister to all other epidendroids except Neottieae. The Neottieae form a polytomy (with no evidence for monophyly of the tribe) sister to vanilloids and epidendroids. Evidence from *rbcL* plastid sequences, however, supports the monophyly of Neottieae but still places them sister to all other epidendroids (Kores *et al.* 1997).

Future palynological studies should focus on the characters mentioned above and must include ultrastructural data to provide a complete picture of exine structure. This information, combined with other morphological and molecular data, is now producing the most complete picture of orchid relationships and phylogeny to date. Results will also help us to assess directionality of character-states and to explain the host of adaptations (such as reduction of the exine with concomitant elaboration of the intine) that have made Orchidaceae so successful worldwide.

ALEC M. PRIDGEON

V

EMBRYOLOGY

Embryology is here interpreted in the broader sense to include the processes involved in development of the male and female gametophyte, fertilization, and subsequent development of the embryo and ancillary structures through to maturation. Its phylogenetic significance will also be discussed in this chapter.

Historically the study of embryogenesis in Orchidaceae has been linked to the discovery of the process in higher plants. While studying various indigenous Italian plants including some orchids, Amici (1824, 1830, 1847) discovered the existence and development of pollen tubes on the stigma and subsequent growth into the ovary and penetration of the ovule. He also reported that ovules contained an egg cell prior to penetration of the pollen tube and that the embryo commenced formation only after fertilization of the egg cell. Soon thereafter the German botanist Hofmeister (1847) discovered the process of double fertilization while undertaking research on a range of plants including orchids. This discovery was later substantiated by Hildebrand (1863) based on his work with orchids. Since these discoveries numerous papers have been published providing detailed accounts of developmental processes in particular groups of plants including orchids. Most notable of these dealing with orchids are the works of Treub (1879), Swamy (1941–49), and Veyret (1955–74) who first reported and alluded to the comparative differences in embryological development in certain groups of orchids. The perceived phylogenetic significance of these embryological developmental patterns in orchids has recently been established through a detailed study of the so-called 'primitive' orchids with special emphasis on the mainly Australian tribe Diurideae and testing of the significance of these results through cladistic analysis (Clements 1995).

Microsporogenesis and microgametogenesis

Microsporogenesis in the Orchidaceae is essentially uniform (Heusser 1915; Swamy 1946*a,b*; 1949*a*; Sood 1986; Cocucci and Jensen 1969) and similar to that found in most other angiosperms (Johri 1984; Johri *et al.* 1992; McCormick 1993). The archesporal cells, which form in the anther, first divide to form many microspore mother cells. Through cytokinesis each mother cell divides by meiosis to form a microspore tetrad of haploid cells. Division is mostly by the simultaneous type in the Orchidaceae, although the successive type has been reported for some species. The exceptions are the apomictic species in which there is breakdown of the process during the early stages of development, resulting in infertile pollinia or their absence altogether in at least one genus, viz. *Cooktownia robertsii*. The actual mechanism causing the breakdown of microsporogenesis in these taxa is as yet unknown.

Microgametogenesis follows with the mitosis of the nucellus of each cell in the formation of a vegetative and a generative cell which is smaller, lens-shaped, and positioned toward the outside of the microspore. Thereafter the microspores either remain in tetrads or break up into monads. Both *Apostasia* and *Neuwiedia* produce pollen grains as monads by the breakup of microspores immediately after the second meiotic division and prior to maturation of the cell wall (exine and intine). Pollen grains produced in this way are typified by their uniform shape, something that has been observed previously in these two taxa (Schill 1978) and in *Vanilla planifolia* (Swamy 1947; Schill and Pfeiffer 1977). Structural differences in pollen morphology also exist in *Apostasia* and *Neuwiedia*, confirming earlier reports (Schill 1978; Newton and Williams 1978). These apparent differences in microspore morphologies appear significant in comparing them with other taxa from within and outside the family Orchidaceae. By comparison, the monad microspores reported to occur or produced in other taxa such as in the tribe Diurideae (e.g. *Thelymitra pauciflora* and *Orthoceras strictum*) are irregularly shaped and are often found in association with tetrahedral or occasionally linear-shaped microspores. Microspores of this type form through the breakup of pollen tetrads after the first mitotic cell division and maturation of the cell-wall. In addition, these irregularly shaped monads typically have thinner walls on part of the cell-wall margin. There is also a differential thickness of cell walls, particularly in those species in which microspores form massulae (Heslop-Harrison 1968; Vijayaraghavan *et al.* 1987; Yeung 1987).

Plasmodesmata and cytoplasmic channels have been reported in cell walls separating adjacent microspores such as those typically present in species with massulate pollinia (Yeung 1987). Delayed absorption of pollen grains and the synchronous germination of pollen grains have been attributed to possession of these cell structures. Confocal laser scanning microscopy study of pollen tetrads reveals that each microspore is completely surrounded by a layer of callose, of more or less even thickness, whereas in massulate pollinia the internal cell walls are much thinner (Clements 1995). The presence of these thin inter-microspore walls indicates that further development of the microspores has occurred during microgametogenesis, possibly facilitating near-synchrony of germination of all microspores, irrespective of their position within massulae. It has also been suggested that the close arrangement of pollen grains facilitates exchange of cytoplasm between cells (Vijayaraghavan *et al.* 1987), but there is no direct evidence to support this idea.

Possession of pollen in monads has been used in comparative studies of monocotyledons (Dahlgren and Clifford 1982) and as evidence to support the concept of the three separate families in the Orchidales, viz. Apostasiaceae, Cypripediaceae, and Orchidaceae (Vermeulen 1966; Rasmussen 1985). However, the situation regarding pollen configuration in the subfamily Cypripedioideae seems less clear. As early as 1813, Franz Bauer, a meticulous botanical artist, illustrated pollen in tetrads in *Cypripedium arietinum* (Stewart and Stearn 1993), whereas all recent studies reported possession of pollen in monads for the Cypripedioideae (Vermeulen 1966; Sood and Rao 1988; Rasmussen 1985). Apart from differences in pollen morphology and configuration, which is considerable within the Orchidaceae, no other major differences exist during microsporogenesis and megasporogenesis to suggest there are any features that would help elucidate phylogeny within the family.

Megasporogenesis

Megasporogenesis occurs within ovules which are usually initiated in response to pollination of the flower. Ovules which are formed on placental ridges in the ovary are of the anatropous type. At the apex of each ovule the megaspore mother cell undergoes two sets of meiotic divisions that result in the formation of four linear dyad cells. Typically the chalazal dyad functions as the megaspore and gives rise to the embryo sac, the remaining dyads later degenerating and taking no further part in development. This developmental pattern has been classified as the S-1 type and is present in most orchids. When the cell wall breaks down or is not formed between the two basal chalazal dyads, the resulting bisporic megaspore is classified as the S-2 type. The S-2 type condition is known to occur in certain species of *Cypripedium* and *Paphiopedilum* (Maheshwari 1955).

There has been reported diversity of developmental patterns in megasporogenesis within the Orchidaceae (Abe 1972; Tohda 1974), but this was not apparent in material examined in some recent studies (Fredrickson and Franksson 1988; Fredrickson 1990, 1991, 1992; Clements 1995).

Megagametogenesis

Megagametogenesis is the process whereby monosporic or bisporic megaspores give rise to the embryo sac. In the monosporic type, the first mitotic division of the functional megaspore gives rise to two nuclei that polarize toward the micropylar and chalazal ends of the cell. These nuclei then undergo two further divisions resulting in the formation of eight nuclei. Two nuclei at the micropylar end differentiate into synergids, one into the egg cell, and the remaining nucleus into one of the polar nuclei, which then migrates to the centre of the embryo sac. At the chalazal end, three nuclei remain and differentiate into antipodal cells while the fourth nucleus migrates to the centre of the embryo sac to become a second polar nucleus. The resulting structure is a G-1 or *Polygonum*-type embryo sac. Development of the embryo sac from the bisporic megaspore results in a similar configuration of nuclei and was classified as a G-2 or *Allium*-type embryo sac (Abe 1972; Johri 1984). In the Orchidaceae, most species studied have a *Polygonum*-type embryo sac, a condition considered by many authors to represent the ancestral state in monocotyledons (Maheshwari 1950; Davis 1966; Dahlgren and Clifford 1982; Johri 1984; Johri et al. 1992; Haig 1990). In comparison, *Neuwiedia* appears to possess a bisporic gametophyte, the S-2 type megasporogenesis, and G-2 type megagametogenesis (Clements 1995), which corresponds to that reported for some *Paphiopedilum* and *Cypripedium* species (Abe 1972).

Other features

Initiation of megasporogenesis prior to anthesis has been noted in the achlorophyllous *Epipogium* (Afzelius 1954) and was recently discovered in *Pterostylis*, *Chloraea*, *Spiranthes*, *Stenorrhynchos*, *Cyclopogon*, and *Sarcoglottis* (Clements 1995). Prepollination ovular initiation is not homologous to that found in cleistogamous species. If cleistogamy exists through auto-pollination, there is germination of pollen grains prior to the flower being fully mature and opening (in some species the flowers never fully open; Ridley 1888), initiating ovular development (Catling 1990).

The unitegmic condition is rare in orchids, having only been reported in several disparate taxa, viz. *Cypripedium insigne* [=*Paphiopedilum insigne*] (Afzelius 1916), *Epipogium* (Afzelius 1954; Davis 1966; Arkal and Mysore 1981), and *Spiculaea* (Clements 1995). The validity of the claimed occurrence of the unitegmic condition in the first of these has not been confirmed. Unitegmic ovules are also known to exist in at least one group of epiphytic orchids, the Podochileae, specifically in *Appendicula*, *Octarrhena*, *Phreatia*, *Rhynchophreatia*, and *Thelasis* (Clements, unpublished) as well as in the terrestrial *Epipogium*. The unitegmic condition is interpreted as a reduction event, occurring independently in apparently unrelated groups within the family.

A lack of vascularization of the funiculus was first noted by Brown (1831), a feature apparently overlooked by most subsequent authors in treatments of the Orchidaceae (except Vermeulen 1966). The funiculus was found to lack vascular tissue in *Apostasia*, *Neuwiedia*, *Paphiopedilum*, and *Phragmipedium* as well as representatives of all other major taxonomic groups within the Orchidaceae, but most outgroups such as *Hypoxis*, *Alstroemeria*, and *Tribonanthes* have a vascularized funiculus (Clements 1995). Funicular vascularization is presumably lost as a consequence of reduced ovule size. Its absence in even *Apostasia*, *Vanilla*, and *Disa*, which have among the largest seeds found in Orchidaceae, points to a genetically linked loss of character rather than an environmentally based loss.

The configuration in most orchids can be summarized as follows: ovules anatropous, bitegmic (rarely unitegmic), tenuinucellate with the inner and outer integuments or both forming a micropyle.

Embryogenesis

Embryogenesis has been extensively studied in higher plants (Maheshwari 1950; Natesh and Rau 1984; Johri 1984; Johri et al. 1992), but fewer than 700 species or approximately 3.5% of the total in Orchidaceae have been examined in any detail (Wirth and Withner 1959; Abe 1972; Veyret 1974; Savina 1978; Arditti 1992; Clements 1995).

There are two schools of thought regarding embryological classification in plants. The cell segmentation school, originated by Souéges (1934–39), centres around the concept that division of the zygote and the respective contribution each daughter cell (apical [ca] and basal [cb] cells) and derivatives make to the formation of the embryo (embryo proper including the suspensor if present), are a basis for recognition of embryological types. Major developmental patterns (called Series A, B, and C) are recognized, with respect to the arrangement of proembryo tetrads following division of the zygote. These series are based on the plane of cell division in the apical cell in a two-celled proembryo. For each series there are always two possibilities depending on the plane of division (transverse or longitudinal) of the proembryo basal cell. Embryological types are characterized by four 'laws': (i) origin, (ii) number, (iii) disposition or arrangement, and (iv) destinies of cells in the proembryo tetrad.

According to Veyret (1955a,b, 1956a–e, 1957, 1958a,b, 1961a,b, 1965, 1972, 1974), who adopted Souéges' system of classifying embryos, the Orchidaceae contains representatives of the A1, A2, C1, and C2 series type embryological tetrads. Further development of embryos was designated by labelling cells, using conventional abbreviations based on Souéges' system, in a series of stages from the tetrad to a mature embryo. As might be expected, this system of classifying embryos has proved somewhat cumbersome and has been used infrequently

		ZYGOTE	1ST DIV.	2ND DIV.	INTERM. STAGE	ADULT FORM
Asterad	GROUP A		BASAL CELL / TERMINAL CELL	SUSPENSOR INITIAL CELL / MIDDLE CELL		
Onagrad	GROUP B					

Fig. V.1. Diagrammatic representation of developmental patterns found in orchid embryos (after Swamy 1949*b*).

for orchids, although it does have its supporters. There have been several attempts at modifying the cell segmentation approach to embryological classification, but none of these has been used in orchid classification (Natesh and Rau 1984).

A second and developmental approach to embryogenesis and classification has been proposed (Natesh and Rau 1984; Johri *et al.* 1992). In this approach, it was recognized that the underlying plasticity in embryogenesis only becomes apparent if the process is studied as a whole, the latter stages of development being the most important. Rather than adopting the approach that specific cells are obliged to become cells of certain organs, these organs (cotyledons, epicotyls, hypocotyls, and hypophyses regions) only become apparent following cell divisions, growth, and organization of the embryo as a whole.

There have been some attempts at classifying the Orchidaceae by means of their developmental embryology. Ever since the work of Treub (1879), who published illustrations of the developmental sequences of a small but diverse group of orchids, the idea of using embryogenesis to classify orchids has been considered. Swamy (1949*b*), basing his work both on the suggestions of Treub and the laws of embryogenesis as outlined by Souéges (1934–1939, cited in Maheshwari (1950)), proposed a system for the classification of orchid embryos (Figs V.1, V.2). Six fundamental embryological types were identified based on two factors; (i) the division sequence of cells from a zygote through to the development of the proembryo, and (ii) the form and development of a suspensor. The first transverse division of the zygote results in a two-celled proembryo with basal (at the micropylar end) and terminal/apical cells (at the chalazal end). Thereafter cell division was recorded as being either regular (transverse and vertical only) or irregular (oblique, transverse, vertical). Those taxa of the irregular cellular divisional plane type were recog-

nized and exemplified by *Cymbidium* (type V embryos). The remaining taxa, all with regular type embryos, were further divided into two groups based on the development of a suspensor: (i) suspensor absent, all cells developing to form the embryo, asterad type, which was illustrated by Swamy as in Figure 5.1 but not typified in Figure 5.2, and (ii) suspensor present, onagrad type (Fig. V.1). Species with the onagrad-type development were further categorized into four additional embryo types based on the configuration of the

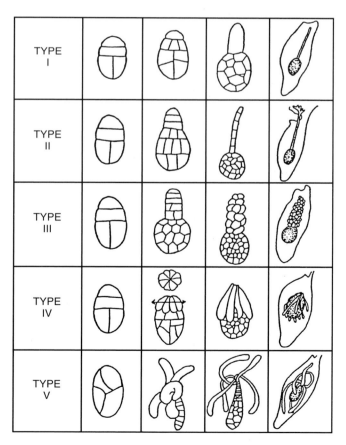

Fig. V.2. Embryo types found in the Orchidaceae as proposed by Swamy (1949*b*).

suspensor. Figure 5.2 summarizes Swamy's proposed system of classification based on embryogenesis.

Johansen (1950) modified Swamy's system, expanding the number of embryo types from five to 13. In Johansen's interpretation each embryo type (called a pattern) was typified by a particular species of orchid. Species without a suspensor or those with a single-celled suspensor were defined as type 1, the *Epipactis palustris* pattern. Taxa possessing relatively uniform suspensors were defined as type 2, the *Orchis latifolia* pattern. Table V.1 lists the remaining types in Johansen's scheme.

Unfortunately, the sample size used by both Treub and Swamy was very small (not all the major taxonomic groups were included) resulting in a fairly unrealistic alignment of taxa in both proposed classification systems. Not surprisingly, most authors who have studied orchid embryos have had difficulty applying either the Swamy or Johansen systems to the material in hand. Sood and Rao (1986), for example, found the embryo of *Microstylis wallichii* [=*Crepidium acuminatum*] was neither an onagrad nor asterad type but more or less intermediate between these. Despite this, they placed this species into the onagrad type. For these and similar reasons, neither Swamy's nor Johansen's system of classification has been considered as a viable alternative to traditional methods of classifying the Orchidaceae. Dressler (1974, 1979, 1981, 1983, 1986, 1989*b*,*c*, 1990*a*,*b*,*c*, 1993), for example, made no reference to either of these publications in his syntheses of the family.

Development of a suspensor

In many higher plants a suspensor arises from the development and division of the basal cells (cb) formed by the initial division of the zygote. In angiosperms, suspensors vary widely in their morphology (Yeung and Meinke 1993). Although the suspensor is a common feature in plant embryogenesis, it is absent from certain groups of plants including some orchids (Swamy 1949*b*; Veyret

Table V.1. Artificial key to embryological system of classifying the Orchidaceae (Johansen 1950)

A. Suspensor absent or present with a single cell . . . . .1. *Epipactis palustris* pattern
A. Suspensor or embryonal tubes or both present
    B. Embryonal tubes absent
        C. Suspensor linear or slightly curved, consisting of one row of cells only
            D. Suspensor cells relatively uniform . . . . .2. *Orchis latifolia* pattern
            D. Suspensor cells variously swollen or elongated
                E. Terminal cell of suspensor alone affected, excessively elongated . . . .3. *Goodyera discolor* pattern
                E. Terminal and subterminal cells of suspensor extended from micropyle, more or less swollen
                    F. Terminal cell only branched . . . . .4. *Serapias lingua* pattern
                    F. All extruded cells branched . . . . .5. *Herminium monorchis* pattern
        C. Suspensor massive
            G. Embryo lacking 'cotyledon' . . . . .6. *Epidendrum ciliare* pattern
            G. Embryo with presumptive 'cotyledon' .7. *Sobralia macrantha* pattern
    B. Embryonal tubes present
        H. Suspensor initial giving rise directly to embryonic tubes
            I. Embryonal tubes, eight or less in number
                J. Embryonal tubes growing in one direction only over the body
                of the embryo . . . . .8. *Vanda parviflora* pattern
                J. Embryonal tubes growing in all directions
                    K. Embryonal tubes essentially uniform in diameter
                        L. Tubes unbranched . . . . .9. *Cymbidium bicolor* pattern
                        L. Tubes each with two protuberances . . . . .10. *Phalaenopsis grandiflora* pattern
                  K. Embryonal tubes extended behind the embryo becoming
                    several times the diameter of those covering embryo . . .11. *Phalaenopsis schilleriana* pattern
            I. Embryonal tubes often more than eight in number . . . . .12. *Stanhopea oculata* pattern
        H. Suspensor initial developing into a linear primary suspensor with embryonal
        tubes arising from terminal cells of latter . . . . .13. *Eulophia epidendracea* pattern

1956a, 1974; Natesh and Rau 1984; Johri *et al.* 1992). Historically the suspensor was thought to function by positioning the developing proembryo amongst the nutritive endosperm (Maheshwari 1950). From his studies of orchid embryogenesis, Treub (1879) concluded that the embryo was differentiated into two parts: the suspensor for absorbing nutritive substances and the actual embryo which stored these substances as materials in reserve. He speculated that orchid embryos were equipped with special systems of absorption 'so they can pass the first stages of their development' and suggested that species without a suspensor stood less of a chance of establishment than those with a suspensor. Currently, following extensive research in cell biology, the suspensor is believed to function in actively assisting in the growth and development of the embryo proper (Yeung 1989; Yeung and Meinke 1993; Clements 1995). It also appears likely that absence of a suspensor is inconsequential to the process of establishment of an orchid seedling in the wild. Rather, successful establishment depends on infection and establishment of a mycorrhizal association with a compatible fungus (Warcup 1985; Clements 1988). Genera such as *Listera* and *Neottia*, which lack a suspensor, and *Pterostylis*, *Chloraea*, and *Spiranthes*, which apparently lack a suspensor (although there is a basal cell or group of cells present) are still dependent on establishing a mycorrhizal association for successful germination and establishment of plants in the wild (Bernard 1902; Burgeff 1932; Warcup 1985; Clements 1988). Examination of these basal cells using Nomarski optics has shown that they have a slightly different composition from those found in the remainder of the embryo. On the basis of these and other observations on seed germination (Arditti 1984), and their position in the embryo, it had been assumed these cells might be a rudimentary suspensor. However, when these cells were viewed using autofluorescence, it was discovered that they have cellular content identical to all other cells of the proembryo, which suggests that earlier interpretations about the classification of these basal cells are incorrect. It is more likely that these cells are the hypophysis region of the embryo rather than that of a true suspensor.

Evidence to support the first part of Treub's hypothesis is more forthcoming. Images of cleared material in the majority of taxa show that cells of the suspensor are uninucleate and highly vacuolate, abutting the hypophysis region in the embryo and attached to the embryo sac wall. Structurally the suspensor comprises a cell or cells, differentiated by their content from those of the true embryo. Suspensor cells are almost devoid of storage material such as oil droplets and starch grains. By comparison, proembryo cells are packed with storage material. Moreover, storage material is present in outer integument cells during ovule development, but these have mostly disappeared by the later stages of embryo growth. Nutrients from these outer integument cells are presumed to be transferred to cells of the developing embryo via the suspensor or through some other mechanism. There is no zone of activity in proembryo dermal cells to suggest there is any direct transfer of storage material to the proembryo other than via the suspensor (Clements 1995). This is the strongest evidence in support of the hypothesis that the suspensor is acting as a transfer cell or organ to facilitate the movement of nutrients from maternal ovular tissue to the developing proembryo.

For those species lacking a suspensor, the mechanism for the translocation of nutrients from cells of the outer integument to the proembryo is quite different from that described above. In suspensorless taxa, there appears to be direct transfer of nutrients either as the developing proembryo comes into direct contact with cells of the outer integument once it has broken through from the protection of the inner integument or directly through the cell walls. Confocal images show an intense autofluorescence zone immediately surrounding the dermal cells, possibly indicating greater metabolic activity in that area. Cells of the outer integument initially contained storage material such as starch grains and oil droplets, but these had disappeared by the later stages of development (Clements 1995).

Until recently it has been reported that the point of entry into the orchid embryo for mycorrhizal fungi is the suspensor (Bernard 1909; Burgeff 1932, 1959; Hadley 1982; Clements 1988). Where there is development of a suspensor, cells in this structure die as the embryo reaches maturity. In fact the initial point of entry of mycorrhizal fungus for all orchids is the hypophysis region rather than the suspensor. In *Epipogium roseum*, in which the mature embryo comprises 8–12 cells, or in embryos of *Listera* and *Neottia*, the hypophysis may comprise just one or two cells, each barely discernible from remaining cells of the embryo.

Lack of a suspensor has been seen as a key character in the classification of orchid embryos, but

Clements (1995) has demonstrated that this is an over-simplification of the actual situation. In *Listera* and *Neottia*, with the asterad type embryo development, the normal longitudinal division plane of the basal cell restricts the potential for formation of a suspensor. Studies of embryos infected with mycorrhizal fungi in *Neottia nidus-avis* revealed that one or two of these basal cells are the point of entry to the embryo by the fungus (Bernard 1902). In comparison, embryos lacking a suspensor in other genera such as *Pterostylis*, *Spiranthes*, and *Townsonia* have the onagrad type embryo and possess a basal cell or cells formed through the primary division of the zygote and subsequent transverse divisions of the basal cell. Developmental patterns in these three groups of taxa are therefore not homologous. Comparisons of this nature emphasize that an understanding of developmental patterns in orchids is necessary for accurate definition of the various states found within groups.

Endosperm

In all embryological studies in the Orchidaceae, there has been constant reference to the absence or almost total lack of any endosperm. Possession of an endosperm is linked to the double-fertilization process (Lopes and Larkins 1993). One hypothesis put forward was that endosperm originated from the production of a second embryo in ancestral angiosperms, later evolving into a storage organ (Sargant 1900,

cited in Lopes and Larkins 1993). Recent research on *Ephedra trifurca*, a close relative of angiosperms, has found evidence to support this hypothesis (Friedman 1990, 1992, cited in Lopes and Larkins 1993).

Table V.2 summarizes the current state of knowledge of endosperm in the Orchidaceae obtained from a survey of the literature. Evidence to support these authors' interpretations about the presence of endosperm is meager. Specifically, evidence that fusion of the second sperm nucleus and polar nuclei has taken place is often poorly supported or non-existent, mostly based on interpretations of cellular content of the embryo sac and illustrated by line drawings (Hagerup 1947; Swamy 1942, 1943*a*,*b*, 1944, 1945, 1946*a*,*b*, 1947, 1948*b*, 1949*a*,*b*). More recently, Fredrickson (1992) noted the occurrence of double fertilization in *Epipactis palustris*, providing as evidence images taken on a confocal laser scanning microscope to illustrate the phenomenon. Albert (1990) studied embryogenesis in *Cypripedium cordigerum* using modern fluorescent microscope techniques and computer imagery and found, contrary to previous records, that there is no double fertilization in this species. Instead, the second sperm nucleus merely moves toward and remains adjacent to the two polar nuclei. As a result, Albert questioned the interpretation of results on cellular development during the process of megasporogenesis in two earlier published reports involving *Cypripedium* (Pace 1907; Carlson 1954).

The presence or absence of endosperm during embryogenesis in the Orchidaceae is particularly

Table V.2. Reported cases of endosperm development in the Orchidaceae

Taxon	Number of endosperm nuclei	Reference
Cypripedium calceolus	4	Savina 1965
Cypripedium guttatum	4	Prosina 1930
Cypripedium parviflorum	2 or 4	Pace 1907
Paphiopedilum insigne	5–6	Poddubnaya-Arnoldi 1960, 1967
Paphiopedilum insigne	2	Afzelius 1916
Chamorchis alpina	2	Afzelius 1916
Vanilla planifolia	8–12	Swamy 1947
Galeola septentrionalis	16	Kimura 1971
Lecanorchis japonica	4	Tohda 1971a
Pogonia japonica	2	Abe 1968
Bletilla striata	8	Tohda 1968
Bromheadia finlaysoniana	2	Jeyanayaghy and Rao 1966

important since it is the only character Dahlgren and Clifford (1982) could identify to differentiate Orchidales from related taxa in the Liliiflorae.

Recent detailed studies on embryogenesis failed to find any direct evidence for the process of double fertilization having taken place in the Orchidaceae (Clements 1995). In the majority of species examined the polar nuclei are clearly visible during the early stage of embryogenesis. As proembryo development proceeds the polar nuclei are forced toward the chalazal end of the embryo sac. For most species there is no proliferation of polar nuclei following fertilization, the exceptions being in genera such as *Cephalanthera* and *Epipactis* in which four or possibly six cells surround the proembryo. Even in these cases there is no real evidence of these cells having arisen through fusion of the second sperm nucleus and the polar nuclei. The apparent cessation of division of cells interpreted as endosperm, on reaching the four- or six-cell stage, is most puzzling. Assuming these cells are the product of the second part of the double fertilization process, the mechanism that prevents their development beyond a few cell divisions remains a mystery. One possible explanation is that these cells might be products of division of polar nuclei rather than products of the triple fusion of the second sperm nucleus and two polar nuclei.

The lack of an endosperm in most orchids may have other developmental consequences. Lopes and Larkins (1993) pointed out that one possible function of endosperm is to influence epigenetic differences that direct development of the embryo. Presumably the relatively undifferentiated state of the mature embryo found in all orchids so far studied might be directly related to the lack of endosperm.

Embryological developmental patterns

From the recent study of the so-called 'primitive' members of the Orchidaceae (Clements 1995) and previously published results, the following embryological developmental patterns have been identified within the family and are summarized in Table V.3. These embryological patterns occur within major taxonomic groups within the Orchidaceae and therefore are thought to be of systematic significance (Clements 1995). Whereas almost all previous authors have concentrated on the development of the proembryo, here the development of all accessory structures is included in the description of each embryological developmental type. Each developmental type was named based on the principal taxon in which it occurs or was discovered, following Dressler's (1993) system of classification of the family.

Cypripedioid type (Figs V.3–5)

The bitegmic, anatropous ovule (*c.* $25 \times 50 \ \mu m$) is connected by a short funiculus, which lacks vascularization, and the outer integument is oblong-elongate with thin cell walls. The inner and outer integuments together form the micropyle in the mature ovule; the base of the inner integument is attenuate. The mature embryo sac is large (*c.* $20 \times 25 \ \mu m$) and is surrounded by a very thin, single-cell nucellus. Following fertilization of the egg cell and formation of the two-celled proembryo, the apical cell grows out into the embryo sac while the basal cell remains attached to the inner wall near the micropyle, enlarging to form a suspensor pad. Transverse, longitudinal, and anticlinal divisions of the apical cell result in the formation of a globular or spherical embryo proper. This general pattern of cell division and growth continues, the embryo proper eventually replacing and entirely filling the embryo sac. Development of both the embryo proper and the suspensor to maturity is confined to this area of the embryo sac. There is no development of an endosperm in this group of orchids, the polar nuclei being clearly visible during the early stages of embryogenesis. At maturity the suspensor is crushed against the inside wall of the embryo sac, and its cells die. All cells in the inner and outer integument also die, their cellular content having been lost during the development of the embryo. At maturity the actual embryo consists of approximately 20–40 cells encapsulated in a dark brown, slightly sclerified seed coat.

The embryology of *Apostasia* is fundamentally similar to that of the cypripedioid type but without the sclerification of cells in the outer layer of the outer integument. Rather, sclerification occurs in the outermost layer of cells in the inner integument, the membranous outer integument layer eventually sloughing off during seed dispersal. Sclerification of the inner rather than outer integument is thought to have developed in response to the mode of seed dispersal. In *Apostasia* seeds are encapsulated in a

Table V.3. Published records of embryogenesis in Orchidaceae

Taxon	Embryo type*	Number of suspensor cells	References
Aerides hookeriana	Vad	11	Rao and Chua 1978
Angraecum distichum	Vad?	0	Veyret 1958b
Arundina graminifolia	Cyp	5, 6	Rao 1967
Bletilla striata	Cym	4	Tohda 1968
Bromheadia finlaysoniana	Cyp?	2	Jeyanayaghy and Rao 1966
Bulbophyllum oreonastes	Cyp	3	Veyret 1957
Calanthe Veitchii	Cyp	1	Poddubnaya-Arnoldi 1967
Calypso bulbosa	Cal	6	Yeung and Law 1992
Coelogyne parishii	Cyp	6, 8	Veyret 1958a, 1974
Cymbidium bicolor	Cym	7	Swamy 1942, 1943a
Cypripedium cordigerum	Cyp	2, 3, 4	Sood and Rao 1988
Dactylorhiza aristata	Orch	8	Tohda 1971b
Dendrobium anosmum	Den	2	Pastrana and Santos 1931
Dendrobium nobile	Den	6	Poddubnaya-Arnoldi 1967
Dendrobium 'ccy' (sic)	Den	1	Rao and Chua 1978
Doritis pulcherrima	Vad	21	Rao and Chua 1978
Epidendrum ciliare	Epi	29, 32	Treub 1879
Epidendrum cochleatum	Epi	15	Sharp 1912
Epidendrum variegatum	Epi	?	Sharp 1912
Epidendrum verrucosum	Epi	17	Sharp 1912
Epipactis atrorubens	Cyp	0	Veyret 1974
Epipactis palustris	Cyp	0	Treub 1879
Epipogium aphyllum	Cyp	0	Afzelius 1954, Geitler 1956
Eulophia epidendracea	Cym	4, 5	Swamy 1943a
Galeola septentrionalis	Van	0	Kimura 1971
Geodorum citrinum	Cym	4	Rao and Chua 1978
Geodorum densiflorum	Cym	5	Swamy 1943a
Goodyera biflora	Goo	1	Sood 1984b
Goodyera discolor	Goo	1	Treub 1879
Goodyera repens	Goo	2	Veyret 1956c, Sood 1988
Gymnadenia camtschatica	Orc	7, 8	Tohda 1974b
Gymnadenia cucullata	Orc	7, 8	Tohda 1974a
Habenaria ciliaris	Orc	7, 8	Brown 1909
Habenaria integra	Orc	7, 8	Brown 1909
Herminium monorchis	Orc	6, 7	Treub 1879
Hetaeria nitida	Goo	1	Olsson 1967
Hetaeria shikokiana	Goo	1	Tohda 1967
Himantoglossum hircinum	Orc	6	Heusser 1915
Lecanorchis japonica	Van	2	Tohda 1971a
Limodorum abortivum	Cyp	0	Veyret 1974
Liparis pulverulenta	Lip	4	Veyret 1974
Liparis viridiflora	Lip	0	Rao and Rao 1983
Listera ovata	Neo	0	Hagerup 1947
Malaxis saprophyta	Lip	1	Sood 1992
Manniella gustavi	Goo	2	Veyret 1974
Masdevallia veitchiana	Ple	2	Veyret 1961b
Microstylis [Crepidium] wallichii	Lip	1, 2, 3	Sood and Rao 1986
Neottia listeroides	Neo	0	Sood 1984a
Neottia nidus-avis	Neo	0	Dume!e 1910
Orchis [Dactylorhiza] aristata	Orc	8	Tohda 1971b

Table V.3. Published records of embryogenesis in Orchidaceae (continued)

Taxon	Embryo type*	Number of suspensor cells	References
Orchis latifolia	Orc	12	Treub 1879
Orchis sp.	Orc	6	Amici 1847
Paphiopedilum insigne	Cyp	2	Poddubnaya-Arnoldi 1967
Paphiopedilum Leeanum	Cyp	2, 4	Francini 1931
Phaius grandifolium	Cyp	2	Sharp 1912
Phalaenopsis grandiflora	Vad	10, 15	Treub 1879
Phalaenopsis schilleriana	Vad	5	Treub 1879
Pogonia japonica	Neo	0	Tohda 1974c
Ponerorchis graminifolia	Orc	4	Nagashima 1989
Serapias lingua	Orc	6	Swamy 1943a
Serapias longipetala	Orc	6	Veyret 1974
Serapias pseudocordigera	Orc	4	Baranov 1915
Sobralia macrantha	Epi	13	Treub 1879
Spathoglottis aurea	Cyp	1	Rao and Chua 1978
Spathoglottis plicata	Cyp	1	Prakash and Lee-Lee 1972
Spathoglottis Premier	Cyp	1	Rao and Chua 1978
Spiranthes australis	Spi	0	Baranov 1915
Spiranthes carsei	Spi	0	Catling 1982
Spiranthes cernua	Spi	0	Catling 1982
Spiranthes lacera	Spi	0	Catling 1982
Spiranthes lacera var. *gracilis*	Spi	0	Catling 1982
Spiranthes laciniata	Spi	0	Catling 1982
Spiranthes magnicamporum	Spi	0	Catling 1982
Spiranthes ochroleuca	Spi	0	Catling 1982
Spiranthes romanzoffiana	Spi	0	Catling 1982
Spiranthes tuberosa	Spi	0	Catling 1982
Spiranthes vernalis	Spi	0	Catling 1982
Stanhopea oculata	Cym	7	Treub 1879
Vanda spathulata	Vad	8, 15	Swamy 1943a, Treub 1879
Vanilla fragrans	Van	7?	Veyret 1955a
Vanilla planifolia [as *V. fragrans*]	Van	0	Swamy 1947
Vanilla roscheri	Van	0	Kurpko *et al.* 1954
Zeuxine sulcata	Goo	0	Joshi 1933, Seshagiriah 1941 Swamy 1946b

* Codes for embryo patterns and types: Cal = *Calypso*; Cym = *Cymbidium*; Cyp = Cypripedioid (plesiomorphic type); Den = Dendrobieae; Epi = Epidendreae; Epg = Epipogoniinae; Goo = Goodyerinae; Lip = *Liparis*; Neo = Neottieae; Orc = Orchideae; Ple = *Pleurothallis*; Pod = Podochileae; Spi = Spiranthinae; Vad = Vandeae; Van = Vanilloid. Binomials are as originally published in references cited above.

fleshy, succulent, indehiscent fruit that is both aromatic and yellow at maturity, some 12 to 15 months after it commenced development. The seed dispersal mechanism is unknown, but it is possible that the fruit might be consumed by an animal, passing through its gut and being deposited in faeces some distance away from the source plant.

Despite the unusual nature of the seed development of *Apostasia*, both it and *Paphiopedilum*, which has wind-blown seeds, fundamentally have the same embryological developmental pattern. This developmental pattern is therefore interpreted as the plesiomorphic state for the Orchidaceae. With few modifications, this embryological pattern is found in a number of orchid groups that have traditionally been placed in separate subfamilies.

Present in *Aphyllorchis*, *Apostasia*, *Arundina*, *Bromheadia*, *Calanthe* (Veitchii), *Coelogyne*,

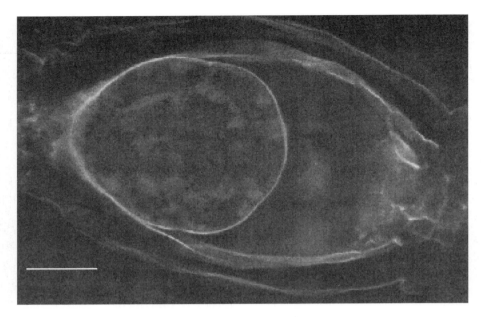

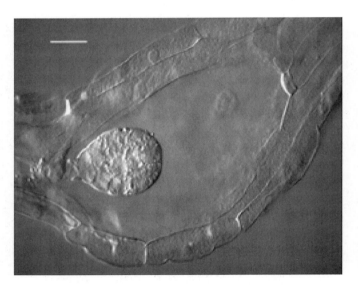

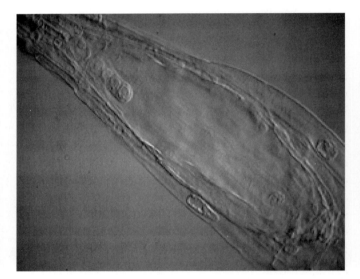

Fig. V.3. *Aphyllorchis queenslandica.* Photomicrograph using autofluorescence showing developing proembryo in embryo sac, surrounded by inner integument. Scale bar = 10 μm. With permission of M. A. Clements.

Fig. V.4. *Gastrodia sesamoides*, an example of the cypripedioid- or plesiomorphic-type developmental pattern showing a partially developed proembryo attached to the wall of the embryo sac by a stalked suspensor, with suspensor pad, and with polar nuclei clearly visible in the main body of the embryo sac, with the second sperm-nucleus adjacent. Scale bar = 10 μm. With permission of M. A. Clements.

Fig. V.5. *Phragmipedium longifolium.* A mature embryo sac at the early stage of proembryo development, with polar nucleus visible at antipodal end. Scale bar = 10 μm. With permission of M. A. Clements.

Corymborkis, Cypripedium, Epipactis, Gastrodia, Limodorum, Paphiopedilum, Phaius, Phragmipedium, Spathoglottis, and *Tropidia.*

Neuwiedia type (Fig. V.6)

In *Neuwiedia* details on embryogenesis are incomplete as the material available for study covered only the earlier and mature stages of development. Despite this limitation, there are embryological features in *Neuwiedia* that separate it from those of the cypripedioid type. The anatropous ovule is bitegmic with both layers forming the micropyle prior to any penetration by the pollen tube. Both integuments elongate as the ovule reaches maturity, resulting in the formation of an attenuate inner integument and

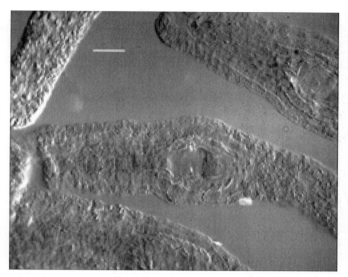

Fig. V.6. *Neuwiedia veratrifolia*, at the early developmental stage at or soon after fertilization. Scale bar = 10 μm. With permission of M. A. Clements.

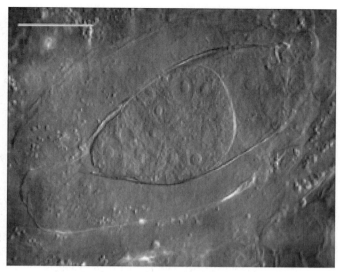

Fig. V.7. *Aporostylis bifolia*, an example of a diurid-type embryo sac development with polar nuclei clearly visible between the proembryo and antipodal cells. Scale bar = 10 μm. With permission of M. A. Clements.

micropylar cavity. The embryo sac is relatively small (*c.* 10 × 18 μm) compared to the remainder of the ovule (*c.* 30 × 150 μm), and there is a well-developed hypostase. The funiculus lacks any vascularization in the form of phloem and xylem cells with reinforced walls. The outer wall of the outer integument is relatively thin, and there is no secondary thickening of other cell walls in the integuments. In mature embryos reconstituted in ammonia solution and subsequently cleared in lacto-phenol, there is evidence of the restricted growth and development of the inner integuments and that the proembryo cells were packed with storage material. There was also no development of any endosperm.

This embryological developmental pattern is not replicated exactly in any other orchid group, so it has been retained as distinct pending the results of future studies. The embryological pattern does, however, bear similarities to the patterns found in other orchid groups, some *Calanthe* species for example, the significance of which has yet to be determined.

Present in *Neuwiedia* and *Calanthe*(?)

Diurideae type (Figs V.7–9)

A distinct embryological developmental sequence was identified among most genera traditionally placed in the tribe Diurideae (Clements 1995). Initial embryological development takes place within the confines of the elongate to obovoid embryo sac

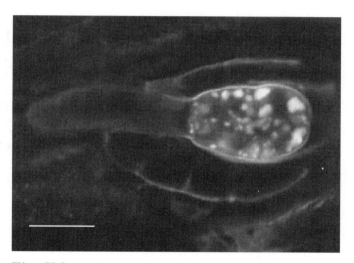

Fig. V.8. *Arthrochilus dockrillii*. Photomicrograph using autofluorescence, showing near-fully developed embryo encapsulated in the inner integument and a single-celled suspensor protruding through to the micropyle. Storage bodies are present in all cells of the embryo but absent in the suspensor. Scale bar = 10 μm. With permission of M. A. Clements.

(measuring *c.* 5–10 × 15–30 μm). At this stage of development the embryo sac is small and surrounded by a thin nucellus which is often barely discernible from cell walls of the inner integument. The inner integument consists of two cell layers, in turn surrounded by a two-cell-layered outer integument. Both basal and terminal or apical cells undergo division, generating a daughter basal cell

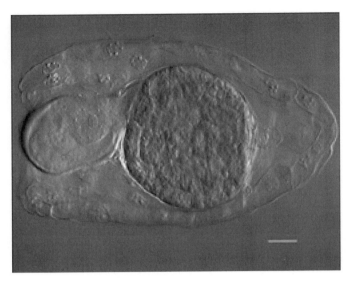

Fig. V.9. *Genoplesium apostasioides*, an example of a diurid-type embryo developmental pattern, with a single large suspensor cell protruding into the micropylar space. Scale bar = 10 μm. With permission of M. A. Clements.

and daughter terminal cell. The two inner cells then become respectively the lower daughter cell of the basal cell and the upper daughter cell of the apical cell. These two basal cells do not appear to undergo further division; rather, the outermost cell grows toward the micropyle and develops into a suspensor. Concurrently the apical daughter cells continue to grow and divide in a near random state in both vertical and transverse planes. This group of cells is the proembryo and eventually forms the actual embryo. During this phase of development the proembryo increases in size while still remaining encapsulated within the inner integuments and nucellus, although the latter becomes compressed and almost unrecognizable in most samples examined. Concurrent with proembryo development, the suspensor grows out through the inner micropyle into the micropylar space formed by the outer integuments. In some species the suspensor may grow further and emerge from the outer micropyle into an ovarian cavity, but this can be a variable condition even within the same sample or capsule. In most taxa the inner basal cell forms part of the proembryo, as the hypophysis, and in a few species this cell appears to form part of the suspensor. Its cellular composition distinguishes it from cells in the proembryo.

Autofluorescent images of fresh ovules of representatives of this type taken using the confocal microscope confirmed that suspensor cells have a different composition than cells of the embryo proper. Fluorescent storage bodies are all but absent from cell cytoplasm in the embryo proper and suspensor during the early stages of development. Cells of the inner integuments are also devoid of many of these storage bodies. Correspondingly, cells of the outer integument contain many storage bodies, presumably fat bodies and starch grains, the latter being distinguishable by their characteristic crystalline shape. Through the course of development, cells in the actual embryo become packed with storage bodies, mostly fat bodies, while the cell(s) of the suspensor becomes highly vacuolate and contains few detectable storage bodies. Corresponding to the increased incidence of storage bodies in cells of the embryo is a marked decrease and vacuolation of the outer integument cells, the cytoplasm being present mainly around the nuclei and cell periphery. At maturity the embryo comprises 30 to 50 cells, packed with storage bodies and encapsulated in cells of the inner integument, suspended by the skeletal remains of cells from the outer integument.

In all taxa an endosperm fails to develop, the polar nuclei remain undivided, eventually becoming squashed against the inner wall of the embryo sac and then being obliterated by the developing embryo. Taxa with this embryological type include *Diuris*, the type genus for the tribe Diurideae.

Townsonia deviates from this general Diurideae type pattern by the non-development of a suspensor. In this taxon both apical and basal cells contribute to proembryo development although unequally so. Most of the development is derived from the apical cell (ca). The embryo does not develop a suspensor, and it remains surrounded by the inner integument suspended by cells of the outer integument, the cells of which are much enlarged. As the proembryo develops it simply grows and becomes larger, filling up much of the space within the inner integument. Both integument layers are fully developed at the commencement of embryogenesis. Examination of the available material revealed a similar cell composition to that described above for Diurideae, a concentration of storage bodies in the embryo and highly vacuolate cells in inner and outer integuments. The fully developed embryo contains approximately 12–20 cells.

Present in *Acianthus*, *Adenochilus*, *Aporostylis*, *Arthrochilus*, *Burnettia*, *Caladenia*, *Caleana*, *Calochilus*, *Chiloglottis*, *Coilochilus*, *Corybas*, *Cryptostylis*, *Cyrtostylis*, *Diuris*, *Drakaea*, *Elythranthera*, *Epiblema*,

Eriochilus, Genoplesium, Glossodia, Leporella, Lepto-ceras, Lyperanthus, Megastylis, Microtis, Orthoceras, Prasophyllum, Pyrorchis, Rhizanthella, Rimacola, Spiculaea, Stigmatodactylus, Thelymitra, Townsonia, and *Waireia.*

Orchideae type (Fig. V.10)

In subtribes Orchidinae and Habenariinae embryogenesis is almost identical to that described above for the Diurideae with one notable difference. The group is characterized by the basal cell (cb) undergoing two or more divisions in the development of a suspensor, to form a linear chain of cells comprising at least four cells and up to sixteen cells long. In many of these samples the suspensor protruded through the micropyle into the ovarian cavity, sometimes reaching the placental tissue. Samples of taxa representative of the tribes Orchideae and Diseae provided a near-uniform pattern of embryological development. Apart from the number of cells in the suspensor chain and the overall size of the embryo, no other differences in embryogenesis have been found in the Diseae from that described for the Orchidinae and Habenariinae. *Disa uniflora*, for example, has the largest embryos (measuring *c*. 100 × 140 μm), whereas species of *Peristylus* and *Herminium* possess minute seeds with embryos measuring *c*. 30 × 50 μm. Embryogenesis in tribes

Orchideae and Diseae is very similar to that found in tribe Diurideae, differing only by the configuration of the suspensor (4–16 cells in the Orchideae and Diseae versus one (rarely two) cells in the Diurideae).

Present in *Cooktownia, Cynorkis, Dactylorhiza, Disa, Gymnadenia, Habenaria, Herminium, Himantoglossum, Monadenia, Neotinea, Neottianthe, Ophrys, Orchis, Peristylus, Ponerorchis, Serapias,* and *Stenoglottis.*

Spiranthinae type (Fig. V.11, V.12)

Members of the Spiranthinae, such as *Spiranthes*, as well as *Pterostylis, Chloraea,* and *Geoblasta* possess this embryological developmental pattern. After the

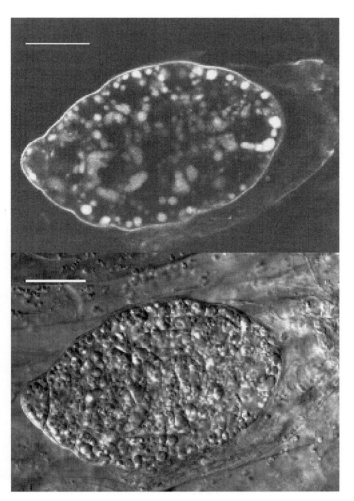

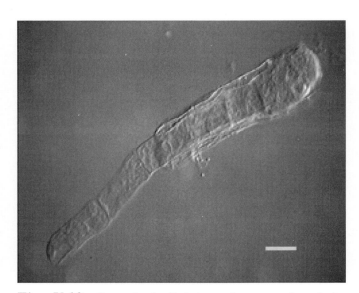

Fig. V.10. *Stenoglottis fimbriata.* An example of the orchidoid-type developmental pattern with outer integument removed, the proembryo and basal cells of the linear suspensor encapsulated in the inner integument. Scale bar = 10 μm. With permission of M. A. Clements.

Fig. V.11. *Pterostylis × ingens.* DIC (above) and autofluorescence (below) of a mature embryo of the spiranthoid-type developmental pattern, all cells including the basal cell with storage bodies. Scale bar = 10 μm. With permission of M. A. Clements.

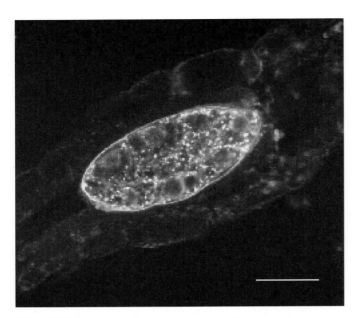

Fig. V.12. *Spiranthes australis.* An image taken using autofluorescence of an example of a spiranthoid-type developmental pattern, the partially developed proembryo and integuments both with storage bodies. Scale bar = 10 μm. With permission of M. A. Clements.

initial division of the zygote, the basal cell (cb) either divides once more or fails to develop further. Thereafter, all development of the proembryo originates from the apical cell (ca). The embryo sac is typically small (*c.* 10–15 × 15–20 μm) at the commencement of embryogenesis. At this stage of development the micropyle is composed only of the inner integuments. The developing proembryo then quickly increases in size to occupy the entire embryo sac. During this phase of development the proembryo apex comes in contact with the chalazal end of the embryo sac supported by the hypostase. Continued growth of the proembryo then proceeds in the opposite direction toward the micropyle forcing the expansion of the inner integuments and nucellus, which results in a break-out of the base of the developing embryo either through the inner micropyle into the outer micropylar spaces (which have formed by this time) or laterally through rupture of the inner integument usually near the apex. The developing proembryo continues to expand and grow basally, remaining within and in direct contact with the outer integument cells.

Confocal scanning images of fresh material of the spiranthoid species show that the composition of all cells, including the basal cells, is similar throughout the embryo. Storage bodies are present in all embryo cells. The contact zone on the entire outer surface of dermal cells of the proembryo is particularly autofluorescent. By comparison, cells of the inner and outer integuments are highly vacuolate and contain fewer storage bodies. Endosperm also fails to develop in this group of taxa, the polar nuclei being pushed toward the chalazal end and eventually obliterated by the developing proembryo as above.

Present in *Bipinnula*, *Chloraea*, *Cyclopogon*, *Gavilea*, *Geoblasta*, *Odontorrhychos*, *Pachyplectron*, *Prescottia*, *Pterostylis*, *Sarcoglottis*, *Spiranthes*, and *Stenorrhynchos*.

Goodyerinae type

Like the preceding group the outer integument becomes attenuate during ovular development. The anatropous bitegmic ovule develops quickly after pollination. Pollen tubes enter via the micropyle made up of both integuments. Thereafter the outer integument develops and elongates rapidly. Following division of the zygote, most subsequent divisions occur from apically generated cells, but the basal cell also contributes to form a tail, consisting of one or two cells. As the embryo increases in size it first fills the embryo sac area (at which time growth toward the chalazal end is halted by the hypostase) and then grows in a micropylar direction pushing the inner integuments apart or possibly tearing them away from connection with the hypostase. At maturity the embryo is an elongate to ovate structure, with or without a tail. Endosperm is not formed during embryogenesis. Embryos in species of *Zeuxine* are often only four cells thick at their widest point and, in total, comprise less than 20 cells. The outer and inner integuments have thin cell walls. Cellular content is mostly uniform throughout the embryo except for the basal cells which have less dense cytoplasm with little storage material evident in these cells.

Present in *Chamaegastrodia*, *Cheirostylis*, *Gonatostylis*, *Goodyera*, *Hetaeria*, *Hylophila*, *Ludisia*, *Macodes*, *Manniella*, *Pristiglottis*, and *Zeuxine*.

Neottieae type (Fig. V.13)

Both the basal and apical cells, produced by the transverse division of the zygote, contribute equally to the formation of the embryo. Vertical divisions of these cells produce a proembryo tetrad in which all cells are of near equal size. Vertical divisions of

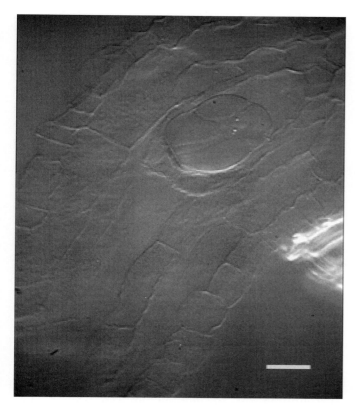

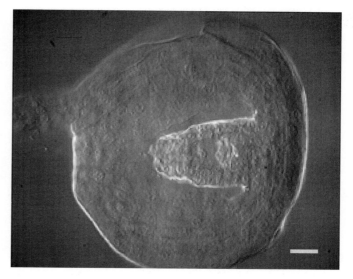

Fig. V.14. *Vanilla planifolia*. An example of the vanilloid-type developmental pattern. Scale bar = 10 μm. With permission of M. A. Clements.

Fig. V.13. *Listera ovata*, an example of a Neottieae-type developmental pattern at the eight-celled stage of pro-embryo development. Scale bar = 10 μm. With permission of M. A. Clements.

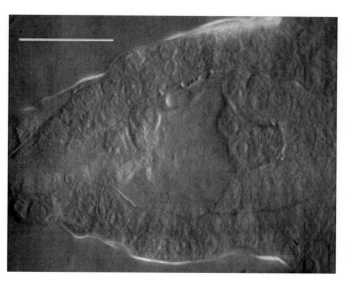

Fig. V.15. *Vanilla planifolia*. A close-up of the embryo sac showing the development of accessory cell layer at the antipodal end, the polar nuclei attached to the antipodal cells, and the zygote (largest cell at the micropylar end). Scale bar = 10 μm. With permission of M. A. Clements.

cells, followed by other transverse and vertical divisions plus growth, result in the formation of a near-spherical embryo. There is no suspensor, and the developing embryo remains within the inner integuments as the whole structure increases in size to reach *c.* 70 × 90 μm at maturity. Cells in the proembryo are packed with lipids and starch grains. Starch grains are plentiful in outer integument cells. These starch grains disappear toward maturity of the embryo, leaving an embryo suspended in an integumental skeleton that is slightly sclerified on the external wall. Endosperm fails to develop in these taxa.

Present in *Listera, Neottia,* and *Pogonia*.

Vanilloideae type (Figs V.14, V.15)

At the time of fertilization, the embryo sac is suspended in the inner integument positioned basally and isolated in the well developed outer integument. Following fertilization the hypostase and inner integument cells develop rapidly to form a unique layer of secondarily thickened cells surrounding the developing proembryo. The proembryo develops to a spherical structure, apparently without a suspensor, eventually filling and replacing the area of the embryo sac. Cells of the outer integument thicken and or develop secondarily into elaborate winged extensions before hardening to form sclerified winged or spherical seeds. In *Vanilla* the mature seed measures *c.* 180 × 180 μm.

Present in *Cyrtosia, Eriaxis, Erythrorchis, Lecanorchis, Pseudovanilla*, and *Vanilla*.

Liparis type

At the commencement of development the embryo sac and integuments are much reduced. Proembryo development follows that described for the plesiomorphic type, except the suspensor is proportionally larger and inflated, expanding into the inner micropylar space. The whole structure expands as the proembryo develops to replace the volume of the embryo sac, growing to the limits of the outer wall of the outer integument.

Present in *Crepidium, Dienia, Liparis*, and *Oberonia*.

Epidendreae type

Embryogenesis has been studied in *Sobralia* (Treub 1879) matching that found in *Cattleya* and *Encyclia*. Following fertilization the proembryo elongates through a series of transverse divisions and growth of cells. Terminal cells also undergo longitudinal divisions, resulting in the eventual formation of a many-celled, obovoid embryo. A series of divisions in the basal cells results in the formation of an elongate elaborate suspensor, typically comprising 20–50 cells, in a chain of single or paired cells. The apex extends beyond the micropyle but without attaching the placental ridge. A central band of cells in the suspensor fluoresces when viewed with confocal microscopy, indicating different cell content. At maturity an embryo measures *c.* 30×130 μm, the suspensor being *c.* 80 μm of that total length. There is no development of endosperm.

Present in *Cattleya, Encyclia, Epidendrum*, and *Sobralia*.

Cymbidium type (Fig. V.16)

At fertilization the bitegmic, anatropous ovule contains a relatively large embryo sac. The proembryo develops initially through transverse divisions of cells and then longitudinal divisions of terminal cells to form an elongate, slightly oblong structure. The apical cell of the suspensor first grows into the micropylar channel formed by the inner integument and then undergoes several longitudinal divisions. Cells created through division of the apical suspensor cell elongate and enlarge to many times their original size, growing into intercellular spaces in the outer and inner integument, eventually and partially

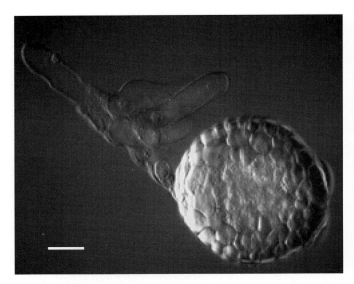

Fig. V.16. *Cymbidium canaliculatum.* A mature embryo removed from the embryo sac showing the complexed suspensor. Scale bar = 10 μm. With permission of M. A. Clements.

surrounding the developing embryo proper. The embryo continues to develop, eventually filling and replacing the embryo-sac space. At maturity the embryo measures *c.* 40×80 μm. There is no development of endosperm.

Present in *Bletilla, Cymbidium, Dipodium, Eulophia, Geodorum, Grammatophyllum, Oeceoclades*, and *Stanhopea*.

Nervilieae type

Embryogenesis in *Nervilia*, the only genus in the subtribe, resembles the general pattern reported for the Cypripedioideae, *Cephalanthera*, and *Tropidia*, for example, but in a reduced slightly elongate form. Mature embryos measure *c.* 20×50 μm, and there is no development of endosperm.

Present in *Nervilia*.

Epipogoniinae type

This subtribe contains two saprophytic genera, including *Epipogium*. The developmental cycle in representatives of this genus takes only 10–12 days from emergence of the inflorescence from the ground surface to dehiscence of a seed capsule. The unitegmic, anatropous ovule contains a small embryo sac. Following fertilization the zygote undergoes an initial transverse division, and then both terminal and basal cells undergo one or two more longitudinal, oblique or transverse divisions forming an 8–12-celled mature embryo measuring *c.* 10×17 μm. Concurrently there is rapid development of integument cells to form large vacuolate cells in the mature seed. There is no development of endosperm.

Present in *Epipogium*.

Calypso type

At fertilization the bitegmic, anatropus, ovate ovule has a small embryo sac. The outer integuments expand and elongate basally. The proembryo first grows to replace the volume of the embryo sac and then expands to about twice the size and out into the inner micropylar space, pushing the suspensor further toward the micropyle. The suspensor develops by the transverse division of the primary basal cell (cb), both cells expanding and elongating linearly, the inner cell longer than the apical cell. The two suspensor cells and proembryo are comparatively similar. There is no development of endosperm.

Present in *Calypso* and *Corallorhiza*.

Podochileae type

In this group, embryogenesis commences in a unitegmic or bitegmic, anatropous, ovate to obovate ovule, the embryo sac occupying most of the volume of the structure. The obovate proembryo develops with a single globular suspensor attached to the inner cell wall at the micropylar end and eventually replaces the volume of the embryo sac with little, if any, expansion of the integuments.

Present in *Appendicula*, *Ceratostylis*, *Eria*, *Mediocalcar*, *Octarrhena*, *Phreatia*, *Rhynchophreatia*, and *Thelasis*.

Pleurothallis type

In *Pleurothallis* the proembryo elongates following fertilization through a series of transverse divisions and growth of cells. Terminal cells undergo longitudinal divisions, resulting in the eventual formation of a small few-celled, obovoid embryo proper. The basal cells divide twice to form a narrow, elongate suspensor stalk and smaller, terminal, globular suspensor. At maturity an embryo measures *c.* 10–15 μm, the suspensor being *c.* 5–7 μm of the total length.

Present in *Masdevallia* and *Pleurothallis*.

Dendrobieae type

As interpreted by Dressler (1993), this tribe comprises the subtribes Dendrobiinae and Bulbophyllinae made up of six and 14 genera respectively, but in total containing over 2000 species. Embryogenesis was investigated in a range of species representative of five of these genera. Substantial variation in embryo developmental patterns exists in the taxa investigated, especially among taxa within *Dendrobium*. At present eight patterns have been identified. There are almost certainly more, but it is difficult to know their comparative significance.

Overall, in the Dendrobiinae embryogenesis commences with a small narrow embryo sac, the proembryo producing a single or two-celled suspensor that is retained within or grows out of the embryo sac/inner integument. There is also expansion of the embryo sac/inner integument as the proembryo grows, resulting usually in an ovate, obovate to elongate, sometimes globose mature embryo, with or without prominent basal cells as a transition to the suspensor. The suspensor is typically ovate to rectangular and insignificant but may be inflated or even elongate to linear.

Within the Bulbophyllinae the overall pattern appears to be the same in the few species examined, represented by the development of a small narrow, almost linear embryo sac at the commencement of embryogenesis, followed by internal development of the proembryo and expansion of the embryo sac-inner integument, a prominent slightly inflated suspensor, with or without basal connective cell or hypophysis. The resulting embryo is small, narrow, obovate to elongate, and surrounded by the slightly larger outer integument.

Present in *Bulbophyllum*, *Cadetia*, *Cannaeorchis*, *Dendrobium*, *Diplocaulobium*, *Eriopexis*, *Flickingeria*, *Grastidium*, *Hapalochilus*, *Pedilonum*, *Sayeria*, and *Winika*.

Vandeae type

The tribe comprises three subtribes – Aeridinae, Angraecinae, and Aerangidinae – containing 157 genera. These are among the most recognizable and distinct groups of orchids in the family and include horticulturally popular genera such as *Phalaenopsis* and *Vanda*. Material from 23 species representing all three subtribes was examined for embryological developmental patterns and found to be of a uniform type which has previously been reported and characterized by Swamy (1949*b*).

Embryogenesis commences with development of the proembryo in a small narrow embryo sac. Terminal cells divide transversely and longitudinally to form a narrow, elongate–obovate embryo. Basal cells initially divide twice transversely, forming a row of cells that grow out of the embryo sac to the inner micropylar space and beyond. These basal cells form a narrow tail to the main body of the embryo. The outer cell undergoes three or four vertical divisions, the cells then growing from their base to form narrow, almost thread-like, structures and reflexed to the main body of the embryo.

Present in *Angraecum*, *Drymoanthus*, *Phalaenopsis*, *Plectorrhiza*, *Taeniophyllum*, and *Vanda*.

Phylogenetic significance

Developmental studies, especially embryogenesis, of nearly four hundred species of orchids, coupled with a cladistic analysis based on vegetative and floral morphological features, have proved highly significant in enunciating the underlying phylogeny of the groups at higher evolutionary levels (Clements 1995, 1996). They provide additional verifiable characters and states that are both conservative and easily recognized with the appropriate methods of study and analysis. With the incorporation of developmental data, the historical arguments as to which characters may or may not determine the limits of the family or taxa within are no longer valid. There are now ample data which provide unequivocal support for the monophyly of the Orchidaceae *s.l.* irrespective of what rank is given to the taxa therein.

The concept Orchidaceae, which includes *Apostasia* and *Neuwiedia*, is well supported and isolated from the outgroup taxa by five synapomorphies:

(i)   lack of development of the endosperm;

(ii) possession of a protocorm;

(iii) lack of vascularization of the funiculus;

(iv) complete or partial fusion of the style and anther filaments to form a column (gynostemium); and

(v)   the reduction of fertile anthers from six to one (one outer dorsal), two (two inner lateral) or three (one outer dorsal, two inner lateral).

Without the first three characters the recognition of the Orchidaceae would depend on acceptance of the

importance of the formation of the gynostemium and reduction of anther number from six to two or three to distinguish it from other members of the Lilianae, a situation that has been debated frequently during the past two centuries (Lindley 1852–1859; Pfitzer 1889; Rolfe 1890; Garay 1960, 1972; Vermeulen 1966; de Vogel 1969; Cronquist 1981; Kurnar and Manilal 1988; Stern *et al.* 1993*b*; Judd *et al.* 1993). These results agree in principle with those generated by Burns-Balogh and Funk (1986), who provided three synapomorphies for the monophyly of the Orchidaceae (including *Apostasia* and *Neuwiedia*). Their three synapomorphies (partial fusion of style and filaments, lack of endosperm, one outer and two inner lateral stamens) are equivalent to three of five characters identified as supporting the Orchidaceae.

Setting aside the characters involving partial fusion of the style and filaments and anther number, lack of development of an endosperm appears to be the most significant character defining the ingroup. As pointed out recently by Friedman (1994), endosperm appears to have arisen as a supernumerary embryo early on in the evolution of all flowering plants through a divergence of functions. Presumably, there has been a loss of function for the development of endosperm in orchids since it is present, in various forms, in all outgroup taxa. Both possession of a protocorm and nonvascularization of the funiculus can almost certainly be attributed to lack of development of an endosperm, but they are not dependent characters. The Orchidaceae is therefore defined as those plants in the Lilianae with (i) zygomorphic flowers, particularly having fully fused stamens and anther filaments forming a gynostemium, (ii) no vascularization of the funiculus, (iii) a lack of development of endosperm (incomplete double fertilization), (iv) mycorrhizal dependence, and (v) production of a protocorm as an interim stage prior to differentiation of a plumule and radical.

Within the Orchidaceae there is evidence to support the group as a whole or to recognize several major taxa as distinct families. However, developmental studies indicate that many of the historical interpretations of subfamiliar taxa, based on traditional interpretations of anther and pollen morphology, cannot be sustained. Other arrangements are more significant. For example, there is little difference in embryogenesis among *Apostasia*, *Neuwiedia*, and Cypripedioideae, compared with various members of Epidendroideae, e.g. *Tropidia*,

Gastrodia, and *Coelogyne*, suggesting a strong evolutionary link between these groups. By comparison, within the Epidendroideae *sensu* Dressler (1993), there are considerable developmental differences between groups such as Cymbidieae and Vandeae compared to Cypripedioideae, suggesting significant evolutionary change has taken place and that these are perhaps better treated as separate subfamilies. Similarly, the Orchidoideae is also distinct.

Embryological studies also provide new evidence that supports the concept of the Cypripedioideae as the next most divergent taxon to *Neuwiedia* and *Apostasia*. Despite the possession of autapomorphic floral characters in the Cypripedioideae, specifically the possession of a slipper-shaped labellum and a prominent discoid dorsal staminode, the vegetative and developmental embryo patterns of these plants (the cypripedioid pattern) have much in common with those found in *Apostasia* and *Neuwiedia*.

Overall the topologies of the trees generated in phylogenetic studies using embryological data (Clements 1995, 1996) bear a strong resemblance to those published from recent studies of plastid *rbcL* sequence data (Duvall *et al.* 1993; Cameron *et al.* 1999; Albert 1994; Albert and Pettersson 1994).

Embryological studies also provide support for the hypothesis that the so-called 'primitive' orchids comprise several major divergent lineages, one involving plants that are predominantly perennial in habit (Vanilleae, Arethuseae, Tropidieae), the remainder involving taxa with an annually deciduous growth habit and possession of storage organs such as root tubers, corms, and protocorm-like tubers. Only the presence of apomorphic embryo pattern characters provides evidence to support the recognition of distinct lineages within this group of taxa.

In tribe Neottieae which has traditionally been treated either as ancestral to or closely related to the Diurideae (Schlechter 1926) or ancestral to Spiranthoideae (Garay 1960, 1972; Rasmussen 1984) because of similarities of floral morphology, the inclusion of developmental and embryological characters in these studies renders the Neottieae paraphyletic to the Orchidoideae clade if included therein. Embryological studies provide evidence that the Neottieae are more closely related to the Vanilleae, Tropidieae, and Arethuseae than they are to the Orchidoideae or Spiranthoideae *sensu* Dressler (1993).

Groups of putatively related taxa (based on similarities of floral morphology) within a single genus generally possess the same embryological developmental pattern. For example, in the tribe Diurideae the *Pterostylis curta* group characteristically produces obovoid embryos, whereas those in the *P. ophioglossa* and *P. nana* groups (traditionally treated in the same section) have elongate obovoid embryos with an extended basal region. Similarly the morphologically distinct *Caladenia gemmata* and its allies characteristically possess embryos with a short, stout suspensor compared with those of species in other groups within the genus, which have an elongate, slightly inflated suspensor. These similarities and differences in embryo morphology possibly indicate phylogenetic relationships.

At the generic level, general developmental patterns conform to type, allowing easy detection of irregularities. For example, in *Microtis* the morphologically distinct species *M. orbicularis* and *M. atrata* both possess embryos with a slightly obovate, inflated suspensor as opposed to the ovate embryo with a short narrow suspensor in remaining species in the genus. Embryos in both *Lyperanthus suaveolens* and *L. serratus* are nearly identical, each possessing a long and attenuate suspensor, whereas those seen in the morphologically distinct species *L. nigricans* have a large embryo proper with a short narrow suspensor.

The embryos of *Acianthus* characteristically comprise a small, few-celled, ovate, embryo proper with a single-celled, attenuate suspensor that is often approximately the same width as the embryo. This same embryo form is present in *Chiloglottis*, *Arthrochilus*, and *Caladenia s.l.* suggesting also that there might be a relationship among these taxa. However, Dressler (1993) treats *Acianthus* in the subtribe Acianthinae, *Caladenia* in the subtribe Caladeniinae, and *Arthrochilus* in the subtribe Drakaeinae. Likewise, *Diuris* and *Orthoceras* in subtribe Diuridinae and *Epiblema*, *Calochilus*, and *Thelymitra* in subtribe Thelymitrinae (Dressler 1993) or in a separate subfamily Thelymitroideae (Szlachetko 1991) in general have the same embryo morphology.

In conclusion, the development of orchid embryos from the plesiomorphic state of a relatively large embryo sac with minimal internal development of the suspensor (e.g. *Neuwiedia*, Cypripedioideae) to the more advanced types in which the embryo sac is smaller with a more complex external suspensor (e.g. Diurideae, Epidendreae, and Vandeae) correlates with the general interpretation of the phylogeny of the family based on other criteria. However, additional research is required for many other taxa, especially some of the more obscure groups, to elucidate their embryological developmental patterns and possible significance when compared to the embryological types found in the remainder of the Orchidaceae.

MARK A. CLEMENTS

VI

SEED MORPHOLOGY

Introduction

Orchid seeds have been called 'dust seeds' due to their minute size (0.15–6.0 mm) and light weight, sometimes no more than a microgram (Ziegler 1981). They generally lack endosperm, and at maturity the uniseriate epidermis of the outer integument forms a loose sheath or tunic around the embryo. Their shapes vary from small ovoid or ellipsoid seeds with a proportionately large embryo that may have relatively poor aerodynamic qualities (more commonly found in terrestrial orchids) to balloon-shaped, winged or filiform seeds designed to be carried over great distances on air currents. The presence of orchids on most remote oceanic islands attests to the long-distance dispersal of these seeds, although published studies on orchid seed dispersal are rarely performed. Although easily dispersed by wind, orchid seeds have little desiccation tolerance, and it is doubtful whether they can sustain long periods of suspension in the atmosphere without damage to the embryo.

Orchid seeds form in vast numbers, usually within dehiscent capsules. One of the better-known exceptions is the indehiscent fruit of *Vanilla*. The lowest number of seeds per capsule is around 6000, found in some terrestrials. Most have tens or hundreds of thousands of seeds, and the highest numbers, four million, are known in epiphytes, *Cycnoches* and *Anguloa* (Ziegler 1981).

At maturity most orchid seeds have a thin, papery coat formed by a single layer of dead cells derived from the outer integument (Ziegler 1981; Clements 1995). Ziegler (1981) noted that the loss of cell contents in this layer is the last stage before maturity and that the previously hard capsule becomes softer and more pliable once this occurs. He also mentioned that seeds in several taxa have hygroscopic elaters that prevent a simultaneous dumping of all seeds upon dehiscence and also eject the seed far from the capsule. The elaters have independent origins in different taxa, and therefore have little obvious phylogenetic utility.

Seed-coat characters

Despite their apparent macroscopic similarity, orchid seeds are highly variable, as can be seen in a magnificent drawing by Beer (1863), the earliest worker to study these structures (Fig. VI.1). Recent work (Clifford and Smith 1969; Rauh *et al.* 1975; Barthlott 1976a; Arditti *et al.* 1979, 1980; Ziegler 1981; Tohda 1983, 1985, 1986; Chase and Pippen 1988, 1990; Chase and Hills 1992; Molvray and Kores 1995) has examined orchid seed-characters both to find phylogenetically useful characters and determine which characters are diagnostic for given taxa, usually at suprageneric ranks.

Morphological variability in orchid seeds encompasses every available character: size, shape, and hardness of the whole seed; size, shape, number, and arrangement of cells making up the coat; features of the cell walls, their adhesion zones, and sculpturing composed both of wall material and cuticular or waxy deposits. The most comprehensive work on seed coat morphology to date is an unpublished dissertation by Ziegler (1981), parts of which have been published in a short article by Barthlott and Ziegler (1981) and summarized in Dressler (1993).

Hypotheses can easily be made concerning the functional significance of some of the variability in morphology of orchid seeds. Seed shape, in particular, as well as projecting walls and other characters that affect the ability to drift on air, adhere to surfaces, or absorb moisture are likely to be subject to selection. The few orchid seeds with thick, hard coats (e.g. *Vanilla*, *Selenipedium*, *Palmorchis*) are likely to be bird- or insect-dispersed. Seed-coat cells sometimes have gaps between adjoining cells which may facilitate entry of the symbiotic fungi necessary to sustain growth once germination begins. All these examples relate to functions essential to the continued survival of the plant and hence may have evolved more than once. Convergent evolution is always a possibility in characters with potential functional significance, which therefore need careful

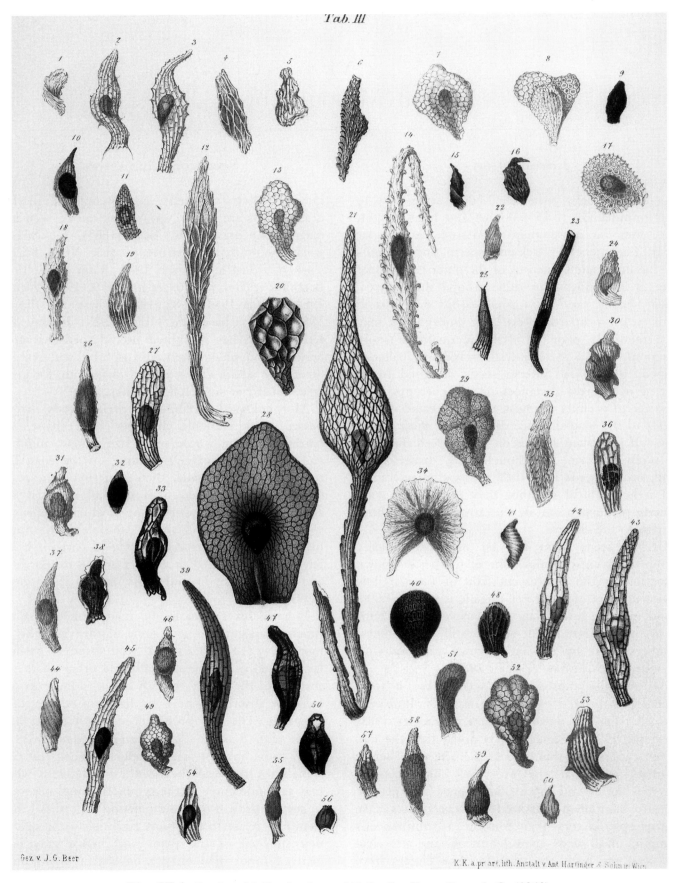

Fig. VI.1. Seed variability in the orchid family. From Beer, J. G. (1863).

study before phylogenetic conclusions can be drawn, and additional data from other sources should always be used in conjunction with seed morphological data.

Some coat characters can be better understood by considering their development. In most orchids, the ovule itself is undeveloped until a pollinium lands on the stigmatic surface. While the pollen tubes are growing toward the ovules, the two integuments of the ovules develop. In monandrous orchids (except Vanilloideae) the inner integument is no longer evident in the mature seed. Once fertilization occurs, cell division in the integuments appears to cease or slow down in some tribes, such that all or most of subsequent growth must come from elongation of existing cells. In the most extreme forms found in Orchideae and Diseae, this results in a seed with a few small basal and apical cells and the body of the seed composed of single cells stretching all the way from base to apex (Fig. VI.2). In others, cell division continues, and the mature seed has numerous small cells. Extreme examples of the latter condition can be found in some Goodyerinae (Fig. VI.3) and Epidendroideae. Thus cessation of cell division in the outer integument explains two observed associations: seed coats composed of only few cells also have cells variable in length with the medial cells highly elongate, whereas seeds with many cells all tend to be equivalent in size and of square to rectangular shape.

Characteristics of the outer periclinal and anticlinal cell walls can be diagnostic at high taxonomic ranks, although their developmental trajectories are not yet known. In particular, in those Vanilloideae and terrestrial Epidendroideae examined to date there is a characteristic layer covering the outer surface of the seed and thereby obscuring the cell-to-cell adhesion zones (Fig. VI.4a). This cuticular layer is a synapomorphy for monandrous orchids, although it may be reduced or lost in more derived members of the clade (e.g. Oncidiinae, Fig. VI.4b; Chase and Pippen 1988). Cypripedioideae do not appear to share this layer, a condition which agrees with recent molecular data indicating that Cypripedioideae are sister to all monandrous orchids including Vanilloideae (Kores *et al.* 1997).

At a lower taxonomic level, gaps in the cell-to-cell adhesion zones are diagnostic of the spiranthoids in the same sense as they were delimited in molecular work (Kores *et al.* 1997), i.e. including Diurideae (Molvray and Kores 1995). Variation in the pattern of gaps is largely correlated with tribal affinities. Diurideae tend to have most of their gaps along the longitudinal anticlinals (Fig. VI.5), whereas Cranichideae tend to have them at the cell vertices (Fig. VI.3). Orchideae and Diseae lack these 'intercellulars', as Ziegler (1981) termed them. It should be noted that intercellular gaps do occur in epidendroids, but these are not homologous to those in the spiranthoid clade.

A number of features become evident only after cell contents are lost and cannot be seen before maturity. This is important when studying herbarium specimens in which cell death may have occurred prior to maturity, leading to a deceptive absence of certain characters, especially various forms of sculpturing (Chase and Pippen 1990). Some seeds have a visible middle lamella between cells, which is obscured when the cells are still turgid. Ridging can be invisible until cells lose their turgidity. Orchid seeds display great variety of thickenings: ridges may be present only on inner or outer periclinal walls, or only on anticlinals, or everywhere; they may be longitudinal, diagonal, transverse, undulate or even helical; they may be no more than faint striations, or they may be elevated into marked prominences and protuberances. The great variety limits the usefulness of ridge patterns

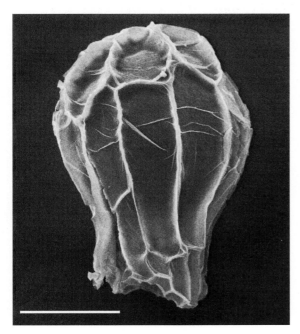

Fig. VI.2. *Disa rungweensis.* Note medial cells extending nearly from chalazal to micropylar ends and the small and irregular shapes of the cells confined to the ends. Scale bar = 100 μm. With permission of M. Molvray.

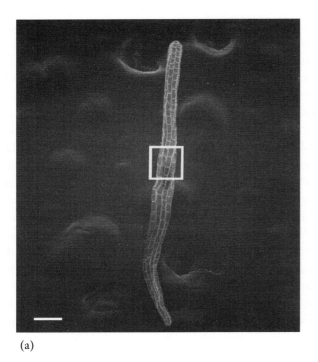

(a)

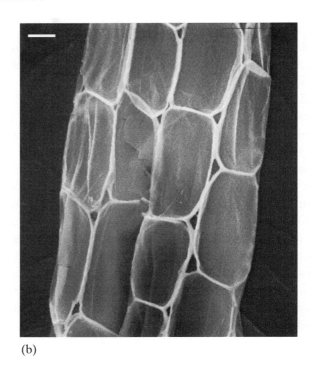

(b)

Fig. VI.3. *Zeuxine elongata.* (a) Note numerous, regularly rectangular, small cells. Scale bar = 100 μm. (b) Close-up of a section of *Z. elongata* testa cells. Note the intercellular gaps at cell vertices. With permission of M. Molvray. Scale bar = 10 μm.

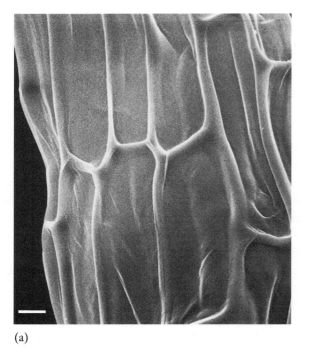

(a)

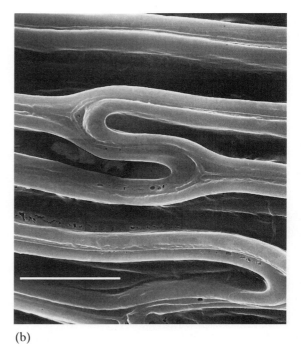

(b)

Fig. VI.4. Epidendroid seeds with varying degrees of covering layer obscuring cell-to-cell adhesion zones. (a) *Cephalanthera erecta*, with adhesions zone completely covered. Scale bar = 10 μm. With permission of M. Molvray. (b) *Oncidium maculatum*, with adhesion zones only partially covered. Scale bar = 12 μm. With permission of M. W. Chase.

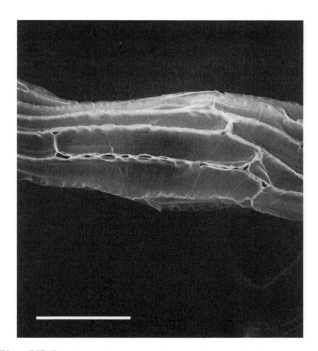

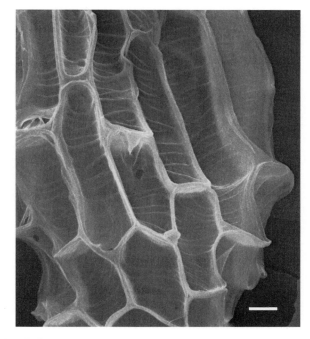

Fig. VI.5. *Acianthus macroglossus*. Intercellular gaps are visible between many of the longitudinal anticlinal walls. Scale bar = 100 μm. With permission of M. Molvray.

Fig. VI.6. Elevation and arching of transverse anticlinal walls in *Spiranthes vernalis*. Scale bar = 10 μm. With permission of M. Molvray.

at higher taxonomic levels, although they may have utility at generic or specific levels.

Not all features contributing to seed surface relief appear late in development. One anticlinal wall character evident early in development is elevation and arching of transverse anticlinals in *Spiranthes* (Fig. VI.6). Some epidendroids have similar arches, e.g. *Sobralia macrantha*, *Bletia purpurea*, *Calanthe regnieri*, *Gomesa verbonii* (Ziegler 1981), that may also develop early. As one can see from the taxa listed, however, there is no easily discernible phylogenetic significance to this character despite its appearance early in development.

There are a number of features generally found in Epidendroideae, but their phylogenetic utility within that subfamily is unclear. Testa cell-shape can be rounded at the ends in a characteristic pattern found predominantly in Maxillarieae (Fig. VI.7), but also appearing in *Coelogyne* and *Dendrobium* (Ziegler 1981) which molecular results indicate are only distantly related within the subfamily. Verrucose sculpturing of the outer periclinal walls is found most commonly in epidendroids, but again its appearance is sufficiently sporadic that phylogenetic significance is not obvious. Pitted surfaces are found in the subfamily but have not been observed outside of it. One of the most striking features of the epidendroid

seed coat can be wall sculpturing. This can be formed from wall material or cuticular deposits and varies from long protrusions and 'grappling hooks' (Fig. VI.8) to small bumps, verrucosities (Fig. VI.9),

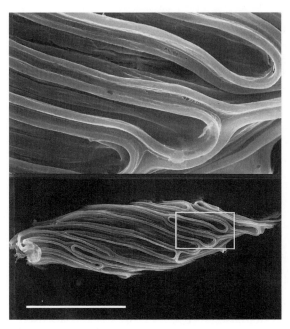

Fig. VI.7. *Trichocentrum pfavii*. Note sinuous transverse anticlinal walls. Scale bar = 75 μm. With permission of M. W. Chase.

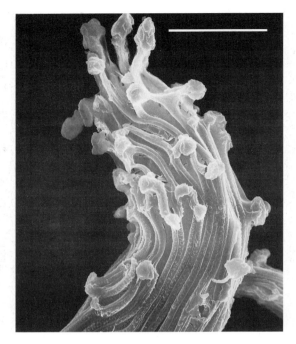

Fig. VI.8. *Tolumnia variegata.* This obligate twig epiphyte exhibits the protuberances at cell vertices that are typical of plants living under these conditions. Scale bar = 30 μm. With permission of M. W. Chase.

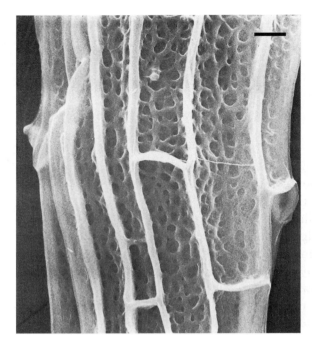

Fig. VI.10. *Tipularia discolor.* Reticulate thickening of the outer periclinal walls. Scale bar = 10 μm. With permission of M. Molvray.

Summary of subfamilial distribution of characters

Apostasioideae

Unlike the monandrous orchids, *Apostasia* and *Neuwiedia* have two integuments present in the mature seed, as do outgroups to the family and most other angiosperms. A hard and relatively thick seed coat is another primitive character present in this subfamily. However, *Neuwiedia* has balloon-like extensions in either or both basal and apical directions. The seeds are small, as in all orchids, and generally ellipsoid, if the extensions in *Neuwiedia* are disregarded.

Cypripedioideae

Seeds with two integuments evident at maturity are present in this subfamily as well. Hard seeds with thick seed coats are found in *Selenipedium*, although this does not appear to be directly related to the condition in Apostasioideae. It is more likely that the character was independently derived to survive bird dispersal, which is common in the genus. Other slipper orchids have ellipsoid or fusiform seeds, sometimes rather large for orchids, without notable extremes in cell size or shape and without sculpturing.

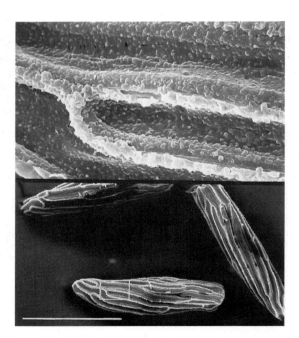

Fig. VI.9. *Dryadella edwallii.* Verrucosities of the outer periclinal wall. Scale bar = 150 μm. With permission of M. W. Chase.

and reticulations (Fig. VI.10). When formed from extensions of the wall, the protrusions tend to be found at cell vertices (Fig. VI.8) or along the anticlinals, especially transverse anticlinals.

Orchidoideae (Figs VI.2, VI.3, VI.6)

Fusiform and ovoid seeds composed of few cells and having highly elongated medial cells are typical of Orchidoideae in the strict sense. In Diseae and Orchideae gaps of any kind are rare, and sculpturing is limited to relatively simple ridges found in a few genera. Some members of *Disa* have the smallest seeds found in the family, and others have relatively hard seeds for this subfamily, although not approaching the truly hard seeds found in, for instance, *Vanilla*. In Cranichideae, however, intercellular gaps are common, as are reticulations and ridges in the subtribe Spiranthinae (Fig. VI.6). In some Spiranthinae and Prescottinae, adjoining anticlinal cell-walls tend to roll inward at their edges, away from each other. More study of the anticlinals is needed to define that character precisely and determine how consistently it is present in members of the two subtribes.

Vanilloideae (Figs VI.11 and VI.12)

Unusual seeds are the rule in this subfamily (Cameron and Chase 1998). *Vanilla* has hard, lenticular seeds adapted to bird dispersal. Other members of the subfamily also have relatively hard seed bodies, but the central kernel is surrounded by extensive wings (e.g. *Epistephium*, *Eriaxis*, *Galeola*). There appear to be two cell layers at maturity in these seeds, as in the other primitive subfamilies, although the inner layer is reduced to the point that cell structure may not be evident (Ziegler 1981).

The other clades in this subfamily have such widely divergent seeds that the genera would not be grouped on the basis of seed characters. For instance, *Lecanorchis* has filiform seeds with a ballooning around the centrally placed embryo. *Cleistes* has rounded, balloon-shaped seeds with coats composed of numerous cells. *Pogonia* has fusiform seeds with severely elongate single cells stretching all the way from base to apex. *Duckeella* and *Lecanorchis* both have some intercellular gaps and *Isotria* has 'ordinary' fusiform seeds with rectangular cells (Cameron and Chase 1998). This clade appears to have one integument in mature seeds.

Epidendroideae (Figs VI.4, VI.7–VI.10)

Epidendroids span the range from plesiomorphic monandrous orchids to highly derived species, and the subfamily includes the vast majority of taxa in

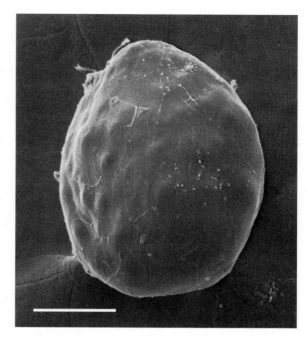

Fig. VI.11. *Vanilla* sp. with thickened, hard outer seed coat. Scale bar = 100 μm. With permission of M. Molvray.

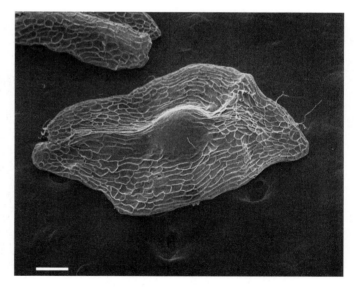

Fig. VI.12. *Epistephium laxiflorum*. A relatively hard central kernel surrounded by a wing. Scale bar = 100 μm. With permission of M. Molvray.

the family. It is to be expected therefore that the range of seed variability will also be great. All have one integument at maturity, but otherwise they display every kind of seed size and shape, including hard seeds in *Palmorchis*. Seed shape is most commonly fusiform or oblong as in other subfamilies; however, filiform seeds also occur frequently, and seeds with inflated, balloon-like sections can be found. The wall-adhesion zones, as mentioned earlier, tend to be covered with a membranous layer. However, given that this layer is also found in Vanillinae and is thus plesiomorphic within Epidendroideae, it has little phylogenetic utility below the rank of subfamily. Sculpturing, as mentioned earlier, is most highly developed in Epidendroideae, with ridges, reticulations, perforations, protrusions, prominences or verrucosities of the wall, as well as sometimes bizarre waxy deposits on the surface of the seed (Chase and Pippen 1990).

Conclusions

As can be seen from the subfamilial distribution of characters, there is considerable convergence in characteristics of the seed coat. Seed shape, for instance, shows numerous convergences at the subfamily level but can be consistent at the tribal, subtribal or generic levels. Intercellular gaps are another example. Although they form a synapomorphy for Cranichideae, their presence in some epidendroids is unrelated; hence they are not useful as single defining characters at the subfamily level. Some characters, such as seed hardness, appear to be so subject to selective pressures that their phylogenetic utility would be limited to intergeneric levels. Similarly, Chase and Pippen (1988) note that seed morphology in Oncidiinae shows convergence with Old World vandoid genera, although at the subtribal level seed coat characters appear phylogenetically informative. When it is not possible to estimate *a priori* which is the optimum taxonomic level for a given character, the utility of seed coat characters for discovering phylogenetic relationships can be limited.

On the other hand, seed characters give indications of phylogenetic affinities in some difficult taxa, relationships that were subsequently confirmed by molecular data. The cuticular covering present in *Tropidia*, for instance, indicated epidendroid affinities at a time when the tribe was placed in Spiranthoideae (in the DNA sequence studies, they are members of Epidendroideae). Similarly, intercellular gaps indicated that Diurideae were more closely allied to 'Spiranthoideae' than to Orchidoideae in which they were placed at the time (Dressler 1993). Both of these characters are associated with cell walls in adhesion zones. Details of cell walls may be one of the more promising areas to look for phylogenetically informative characters. Micrographs from transmission electron microscopy, including both fractures and sections, as well as from high-magnification SEM are needed. Use of types (e.g., *Maxillaria*-type: Chase and Pippen 1988; Ziegler 1981; Molvray and Kores 1995) is not recommended because it obscures what the variable characters actually are; types are simply amalgams of several characters, and it is the characters, not the types, that are useful in phylogenetic analyses. Further research to discover new phylogenetically informative characters and the judicious use of many existing characters can help to elucidate orchid phylogeny at many levels, from family down to genus and species. In addition an understanding of the ecological adaptations of these traits would be most desirable.

MIA MOLVRAY
MARK W. CHASE

VII

Cytogenetics

The orchids have been subjected to considerable numbers of cytological investigations over many years, perhaps reflecting the overall biological/horticultural appeal of the family rather than the intrinsic interest of their chromosomes which, with only few exceptions (see Karasawa 1979, 1981; Martínez 1985), are small and uniform in size. A brief survey of orchid chromosomes will be presented here, which will consider mainly the wide differences in chromosome numbers (basic numbers and levels of polyploidy) that occur within and between the genera. It will not deal with differences between species within a genus, since the scale of such a study would be larger than the range of detail chosen for this work, and except in one or two cases it will not consider hybrids, which are often chromosomally irregular.

In comparison with chromosomes of other petaloid monocotyledons, most of which are large and easily prepared, orchid chromosomes present several problems that must be overcome before any worthwhile project on them can be completed. These will be addressed below, and some ways to minimize them will be suggested.

Techniques for preparation of orchid chromosomes

Somatic chromosomes

Initially, there are problems of chromosome availability and preparation. Since mitotic plant chromosomes can be seen only in the actively dividing cells that are found in a meristem, the natural choice of many workers is the root tip. In the aerial roots of epiphytic orchids, the number of nuclei in mitosis versus that of non-dividing nuclei at any particular time (= the mitotic index) can often be very low, and only some of the dividing nuclei are at metaphase with chromosomes that can be seen clearly. One or more of a range of chemicals can be applied to living roots to increase the mitotic index by arresting mitosis at metaphase and to contract the

chromosomes, making them more easily separable in the squash preparation. The presence of a velamen can delay or prevent the penetration of these pretreating agents into the meristem, thereby reducing the quality of the preparations. These difficulties of pretreatment can be overcome to some extent by adding dilute sucrose (1%) and aerating or agitating when carrying out the process. Frequently-used pretreating agents are alpha-bromonaphthalene (saturated aqueous solution for 24 h at 4°C), 8-hydroxyquinoline (0.002 M aqueous for 4–8 h at room temperature) or colchicine (0.02%–0.05% aqueous for 4–8 h at room temperature). Partial removal of the velamen or longitudinal splitting of the living root tips can aid penetration of the pretreating agents, resulting in greater contraction of the chromosomes and better preparations.

The ovary is not so widely available as root tips but is another excellent source of mitotic material that avoids the problem of the velamen. In developing ovaries of young buds, and again during the period of ovary growth after pollination, ovary wall cells and to a much lesser extent ovule cells can be found in nuclear and cell division. Ovaries should be split longitudinally to improve penetration of the pretreatment agent and processed as above.

Following pretreatment, root or ovary material should be fixed in freshly prepared 1:3 (v/v) glacial acetic acid and absolute ethyl alcohol and subsequently stored in 70% ethyl alcohol. It should be hydrolysed in 1 M HCl at 60°C for 8 min and stained either in Feulgen or 1% orcein in 45% acetic acid. Permanent squash preparations made with the latter have the advantage of being more stable in long-term storage.

Meiotic chromosomes

When they are available, buds can be used as sources of meiotic material, both in the developing anthers and in the ovules, which are abundant in the ovaries of almost every orchid species. Generally, meiosis occurs in the embryo mother cells of the ovules in buds that are considerably older than buds undergoing meiosis in the anther. Buds should be dissected to reveal the anther(s), and the ovaries split open. Fixation and staining are as above, although if material is available at the right time and adjacent to a laboratory, the best meiotic preparations can be made by squashing fresh unfixed material directly on a slide in 1% orcein in 45% acetic acid.

Chromosome numbers can sometimes be difficult to obtain from meiosis, especially when it is not completely regular and bivalent-forming. Meiosis gives very little information on chromosome morphology, but in the orchids it is invaluable in clarifying problems of ploidy level in species and also in the determination of chromosome homology in interspecific and intergeneric hybrids. The last of these problems is of particular significance in that large numbers of bigeneric and multigeneric orchid hybrids have been produced in cultivation and some also appear in the wild, whereas they are rare or unknown in most other angiosperm families. That many bigeneric orchid hybrids are fertile enough to be used by breeders in the production of even more complex ones suggests the presence of a reasonably efficient meiosis in them to produce this fertility. This in turn suggests high levels of chromosomal homology between these relatively easily hybridized orchid genera (e.g. *Laelia* and *Cattleya*), to a level which might even be used to question the validity of maintaining these genera as separate entities. This is a complex taxonomic problem that is not pursued further here.

Chromosome morphology

With the exception of a few genera, notably *Paphiopedilum* and its relatives (see below), most orchids have small chromosomes that are rather undistinguished in general morphology. The majority have large amounts of heterochromatin which remains condensed in the interphase nucleus as groups of dense bodies (chromocentres). Together with the usually light staining of the remainder of the nucleus by the most common staining techniques such as Feulgen or aceto-orcein, these strongly heterochromatic interphase nuclei resemble chromosome spreads so closely that true chromosome spreads at metaphase of mitosis can often be difficult to see.

Variation in chromosome number

The annotated list below gives records of published somatic chromosome numbers found in orchid genera. The data were obtained from the general plant chromosome indices prepared by Ornduff (1968, 1969), Fedorov (1969), Moore (1970, 1971, 1972, 1973, 1974, 1977), Goldblatt (1981, 1984, 1985, 1988), Goldblatt and Johnson (1990, 1991, 1994, 1996) and the orchid chromosome number index of Tanaka and Kamemoto (1984). These indices range from a comprehensive survey of the early plant chromosome work (Fedorov 1969) to one of all plant chromosome papers published in 1993 (Goldblatt and Johnson 1996), which is the latest available at the time of writing. Some records were also obtained from Chase and Palmer (1992). To avoid possible confusion, all the published records of gametic numbers (n) have been doubled to create a single series ($2n$). The names of the genera used here, and the chromosome records that accompany them, are as given by the authors of the original papers, and the supplied generic authorities follow Brummitt and Powell (1992). Genera given in the original papers but not recognized by Brummitt and Powell (1992) are retained for the purpose of the information that is provided, even when it is clear that the records refer to species later transferred to other genera (e.g. *Oncidium pusillum*, now *Psygmorchis pusilla*) or that they refer to genera that are now widely recognized as synonymous (e.g. *Dactylorchis* Vermeulen and *Dactylorhiza* Neck. ex Nevski). The authorities of these genera were checked in Willis (1973).

The relationships of *Orchis*, *Dactylorhiza*, and several other genera of the Orchidoideae have recently been analysed extensively, based on their nuclear internal transcribed spacer (ITS) DNA sequences, by Bateman *et al.* (1997) and Pridgeon *et al.* (1997). As a result of this investigation, several well-known European species are transferred into other genera (Bateman *et al.* 1997).

In many orchid genera the different records of somatic chromosome number seem to present erratic series with no clear pattern or single basic number visible (the basic number being the gametic chromosome number of a diploid plant). There are many causes of this irregularity, and an important one that must be pointed out is the questionable technique of many investigators, especially the early ones, which often resulted in the publication of erroneous counts. Another source of error is the occasional publication of n (i.e. gametic) chromosome numbers as $2n$ (i.e. sporophytic) numbers or vice versa. It is often difficult to detect these errors, so the published records must be taken at their face values and discussed accordingly. Another source of confusion is the possible misidentification of the plants, although, as the data given below are presented at generic level, this source of error should not be as great as it would be in a similar study that compared records at the species level.

Some clear patterns of chromosome numbers appear from the data, with polyploid series being frequent and seemingly arising from a wide range of basic numbers. Chase and Olmstead (1988) have suggested that the variation of chromosome numbers in orchids can be attributable more frequently to aneuploidy than to polyploidy. Aneuploidy is clearly present in many orchid genera, e.g. *Oncidium*, in which at least thirty different somatic chromosome numbers have been recorded, but where it seems appropriate from the *prima facie* evidence provided by the chromosome counts to do so, the data presented below will be interpreted in terms of polyploidy.

The polyploid series are sometimes obscured by aneuploidy and dysploidy. Several cases are noted below (e.g. in *Cymbidium* and several other genera) in which one basic number (i.e. $x = 10$) forms a polyploid series in a genus, but in which there is also a second number derived from the first by chromosome doubling followed by loss of one or more chromosomes (i.e. to the neobasic $x = 19$), which forms the start of another polyploid series. In other cases the neobasic series is produced by the doubling of a basic number and the duplication or fission of one chromosome, e.g. the neobasic $x = 21$ derived from $x = 10$ in *Dactylorhiza* and *Orchis*.

Dysploidy is a stabilized change in the basic chromosome number by steps that are smaller than in the above example. It is usually an increase from a low primitive state, resulting from chromosome gain or fission, but it can also be a reduction arising from chromosome fusion. Some genera are recognized below as having a dysploid variation in basic chromosome number, with two or more basic numbers differing by one or only a few chromosomes and sometimes forming the origins of independent polyploid series. Examples are *Hetaeria*

with $x = 10, 11, 12$, *Odontoglossum* with $x = 13, 14, 15, 16, 17$ etc., *Oncidium* with $x = 13, 14$ etc., and many others.

Aneuploidy is common in polyploid plants, and individuals with one or more chromosomes above or below the eu-polyploid level frequently survive in nature (Brandham and Johnson 1977), so some of the variants from the polyploid series identified below could be accurate records. Aneuploidy at the diploid level is normally very rare but does seem to occur in many orchid genera. Trisomics (diploid plus 1) are sometimes recorded in other plant families, but monosomics (diploid minus 1) are usually non-viable.

B-chromosomes occur in many genera in addition to the normal or A-chromosome complement. These have very limited genetic function, carry hardly any genes, and fluctuate in numbers from one individual to another within a single population (Jones and Rees 1982). Thus their presence or absence, though noted below, has no significance to any systematic investigation involving chromosome studies.

One general point that is relevant is that odd-number polyploids ($3x$, $5x$ etc.) always have unstable meiosis with irregular segregation of chromosomes. They are therefore largely sterile (especially triploids, but pentaploids less so) and would not be expected to persist in the wild unless they are apomictic or reproduce solely by vegetative means. Records that seem to be triploid (e.g. in *Ancistrorhynchus* and a number of other genera) have been scrutinized closely for the possibility that they are actually hexaploid, derived from a basic number that is half of the one that is immediately apparent. Triploids nevertheless occur sporadically in populations of diploids (Brandham and Johnson 1982) and are often more vigorous plants with flowers that are larger than those of diploids. Consequently, they are liable to be selected preferentially by collectors, especially in decorative groups such as the orchids, and can thus enter chromosome surveys of material in cultivation. In the following analysis the basic chromosome number is given as x and diploids, triploids, tetraploids etc. are referred to as $2x$, $3x$, $4x$, etc. Aneuploidy is common in cultivated hybrids and especially in polyploid ones, because of the reduction of the selection pressures against them, so records that are clearly of hybrid plants will not be considered below, except where they are of particular interest (e.g. high polyploids etc.).

Chromosome number variations in orchid genera are considered below, and comments are made on some of them with particular reference to basic numbers and ploidy levels, although genera listed with only one or very few records will not be treated individually. Inevitably, more questions are raised by the observations than are answered by them. If there are sufficient records in a genus, the most frequently occurring numbers are given in italics. In cases of abundant records of more than one number in a genus, both are italicized.

Acampe Lindl. $2n = 30, 36, 38$
A complex genus with no clear relationship between the chromosome numbers reported.

Acanthephippium Blume $2n = 40, 42, 48$
A complex genus with no clear relationship between the chromosome numbers reported.

Aceras R.Br. $2n = 42$
Acineta Lindl. $2n = 40, 42$
Acriopsis Blume $2n = 40$
Ada Lindl. $2n = 60$
The count of $2n = 60$ was recognized by Chase and Olmstead (1988) as a diploid based on $x = 30$. Analysis of meiosis in the genus might confirm this or might establish lower basic numbers of 15, 10 or even 6, depending on the levels of multivalent formation, if any.

Aerangis Rchb.f. $2n = 42, 46, 50, 51, 52,$
                                               $54, 200$
$x = 25$, with many records of $2x = 50$. This genus reaches $8x = 200$. The rather high basic number suggests that it might be allopolyploid in origin, derived from hybridization of plants with two or more lower ones, such as $x = 10$ and $x = 15$ ($2n = 20$ and 30 respectively; these numbers are chosen solely as an example of allopolyploid formation, and there is no evidence of their existence in *Aerangis*), with doubling of the chromosome number to produce a neobasic $x = 25$ ($2n = 50$ etc.). The occurrence of aneuploids around the $2n = 50$ value supports an allopolyploid state for species with this number, but dysploid variation also seems to be present.

Aerides Lour. $2n = 36, 38, 40, 76$
$x = 19$. Many records with $2x = 38$; fewer with $4x = 76$. There is dysploid variation also.

Agrostophyllum Blume $2n = 38, 40, 46$
Amblostoma Scheidw. $2n = 40$
Amerorchis Hultén $2n = 42$
Amitostigma Schltr. $2n = 42, 43, 44$
Anacamptis Rich. $2n = 20, 36, 42, 54, 63,$
                                               72
$x = 18$. Many records with $2x = 36$; fewer with $4x = 72$. Triploids with $3x = 54$ have been recorded.

Ancistrorhynchus Finet $2n = 48, c. 50, 72, 96$
$x = 24?$, with records of $2x = 48$, $3x = 72$, $4x = 96$. The problem of the occurrence of sterile triploids is avoided if the basic number is taken to be $x = 12$. The records then become $4x = 48$, $6x = 72$ and $8x = 96$, all of which would be fertile. If the latter alternative is the true one there would be no diploid records in the genus.

Angraecopsis Kraenzl. $2n = 48, 50$
Angraecum Bory $2n = 38, 40, 42, 42–48,$
                                               $46, 50, 63, c. 76, 92$
$x = 19$. There are many records of diploid $2x = 38$, with a possible $4x = 76$, but there are many other numbers, including the highest record of $2n = 92$, that do not fit the series, so dysploid variation must be present.

Ania Lindl. $2n = 40, 76$
Anoectochilus Blume $2n = 15+Bs, 19, 24, 30,$
                                               40
Most of the counts are of $2n = 40$, but the existence of lower ones suggests that the $2n = 40$ plants might be tetraploid ($x = 10?$). It is also possible that the basic number could be as low as $x = 5$, giving $8x = 40$, since all the other recorded numbers ($2n = 15, 19, 24, 30$) are sufficiently close to being multiples of 5 to be interpreted in terms of polyploids of that number modified by aneuploid loss of one chromosome in two of them. Further investigation should seek a $2n = 10$ diploid, not recorded so far.

Ansellia Lindl. $2n = 42$
Anthogonium Wall. ex Lindl. $2n = 38, 40, 42, 54$
There is no clear pattern in the four recorded counts and the basic numbers could be almost any three of $x = 7, 10, 19, 20, 21$. It seems that polyploids exist based on $x = 18$ ($?3x = 54$), but the diploid ($2x = 36$) that gave rise to them remains unreported.

Apatostelis Garay $2n = 32$
Aphyllorchis Blume $2n = 36, 40$
Apostasia Blume $2n = 48$
Appendicula Blume $2n = 40$
Arachnanthe Blume $2n = 28, 38$
Arachnis Blume $2n = 38, 42$
Arethusa L. $2n = 40$
Armodorum Breda $2n = 38$
Arundina Blume $2n = 32, 38, 40, 42$
The majority of records are of $2n = 40$, but the presence of aneuploid records near this value suggests that it is not diploid but tetraploid ($x = 10?$). Doubling followed by loss of one chromosome gives a neobasic $x = 19$, the origin of the record of $2n = 38$. This process occurs in *Bulbophyllum* and several other genera.

Ascocentrum Schltr. ex J.J.Sm. $2n = 38$
Ascotainia Ridl. $2n = 36$
Aspasia Lindl. $2n = 56, 58, 60$
Barlia Parl. $2n = 36$

Beadlea Small	$2n = 32$
Beloglottis Schltr.	$2n = 60$
Biermannia King & Pantl.	$2n = 36$
Bifrenaria Lindl.	$2n = 38, 40$
Blephariglottis Raf.	$2n = 42$
Bletia Ruíz & Pav.	$2n = 60$
Bletilla Rchb.f.	$2n = 66, 32, 36, c. 38,$ $64, 76$

Two polyploid series seem to be present in this genus, with a dysploid difference in basic number: $x = 8$ giving $2x = 16$, $4x = 32$ and $8x = 64$; also a neobasic $x = 19$, derived probably from a doubling of $x = 8$ with the addition of three chromosomes, giving $2x = 38$ and $4x = 76$.

Bolusiella Schltr.	$2n = 50$
Braasiella Braem, Lückel & Rüssmann	$2n = 42$
Brachycorythis Lindl.	$2n = 42, 46$
Brachypeza Garay	$2n = 36–38$
Brachystele Schltr.	$2n = 46, 92$

$x = 23$, giving $2x = 46$ and $4x = 92$.

Brassavola R.Br.	$2n = 40$
Brassia R.Br.	$2n = 18, 50, 52–58, 60$

A problematic genus with most records of $2n = 60$. It was suggested by Chase and Olmstead (1988) that these are diploid ($x = 30$), but the presence of lower counts suggests that the basic number might be as low as $x = 10$, making the $2n = 60$ plants hexaploid and the $2n = 50$ plants pentaploid. The range of aneuploids between $2n = 52$ and 58 would be expected to survive in the progeny of a pentaploid. The record of $2n = 18$ does not fit the series.

Broughtonia R.Br.	$2n = 40$
Bulbophyllum Thouars	$2n = 36, 38, 38+Bs, 39,$ $40, 40+Bs, 42, 57,$ $58, 60, 80, 95$

$x = 19$. There are many records in this genus, with Lim and Jones (1982) recording $2x = 38$, $3x = 57$ and $5x = 95$ but oddly no tetraploids ($4x = 76$). Another series based on $x = 20$ is also present, giving $2x = 40$, $3x = 60$ and $4x = 80$, although if based on $x = 10$ the latter series would become $4x = 40$, $6x = 60$ and $8x = 80$ (see under *Ancistrorhynchus*, above). It is possible that $x = 10$ is the original basic number in this genus, with $x = 19$ being a neobasic number derived from it by doubling followed by dysploid loss of one chromosome.

Calanthe Ker Gawl.	$2n = 20, 28, 38, 40,$ $40+Bs, 42, 44, 45,$ $46, 48, 52, 58, 60$

Two records of $2n = 20$ by Hsu (1972), suggest a low basic number of $x = 10$ that gave rise the common $4x = 40$ and rarer $6x = 60$. Many counts of $2n = 42$ suggest the appearance of a neobasic number of $x = 21$ by doubling of $x = 10$ and addition of one chromosome. There is also much aneuploidy.

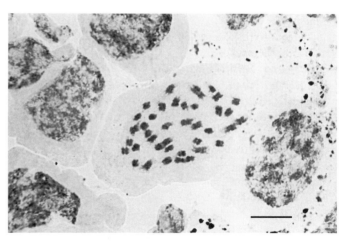

Fig. VII.1. Somatic chromosomes of *Calopogon tuberosus* var. *simpsonii* ($2n = 40$). Scale bar = 10 μm. With permission of D. H. Goldman.

Calopogon R.Br.	$2n = 26, 40$ (Fig. VII.1), 42

No clear relationship among the recorded counts in this genus.

Calymmanthera Schltr.	$2n = 38$
Calypso Salisb.	$2n = 28, 32$

No clear relationship between the two recorded counts in this genus.

Calyptrochilum Kraenzl.	$2n = 38$
Camarotis Lindl.	$2n = 38$
Campylocentrum Benth.	$2n = 44$
Catasetum Rich. ex Kunth	$2n = 54, 56, 108, 162$

$x = 27$, giving $2x = 54$, $4x = 108$, $6x = 162$. The high basic number is probably neobasic, derived from two lower ones.

Cattleya Lindl.	$2n = 40, 41, 42, 43, 46,$ $55, 59, 60, 62,$ $c. 67, 75, 76, 80,$ $81, 82, 83, 84,$ $85–103, 100, 102$

$x = 20$?, giving $2x = 40$, $3x = 60$, $4x = 80$ and $5x = 100$, but if the basic number is $x = 10$ the ploidy values would become $4x$, $6x$, $8x$, and $10x$ respectively, avoiding the problem of sterility or irregular meiosis in odd-number polyploids (see also *Ancistrorhynchus*).

Caularthron Raf.	$2n = 40$
Cephalanthera Rich.	$2n = 32, 33, 34, 36, 42,$ $44, 48, 54, 64, 68$

$x = 18$, giving $2x = 36$ as the most frequent record. Schwarzacher *et al.* (1980) have shown that plants with $2n = 54$ are triploids. In the same genus, $x = 16$ gives $2x = 32$ and $4x = 64$. Much aneuploidy and/or dysploidy is also present.

Cephalantheropsis Guillaumin	$2n = 40$
Ceratostylis Blume	$2n = 40$

Chamaeangis Schltr. $2n = 50, c. 100$

x = 25, giving $2x = 50$ and $4x = 100$. Again, this could be neobasic.

Chamaeanthus Schltr. ex J.J.Sm. $2n = 38$

Chamaeorchis Koch $2n = 42$

Chamorchis Rich. $2n = 24, 42$

It is difficult to reconcile these markedly different counts (x = 12, 21), unless they are based on x = 6 and 7, respectively.

Changnienia Chien $2n = 46$

Cheirostylis Blume $2n = 26$

Chiloschista Lindl. $2n = 38$

Chloraea Lindl. $2n = 16$

Chrysoglossum Blume $2n = 36$

Cirrhopetalum Lindl. $2n = 20, 38, 38–40, 48$

x = 10, giving $2x = 20$, $4x = c. 40$. Doubling this basic number with dysploid loss of one chromosome gives the neobasic x = 19, with $2x = 38$, a pattern that occurs in many orchid genera. The count of $2n = 48$ does not fit the series.

Cleisocentron Brühl $2n = 38$

Cleisostoma Blume $2n = 26, 38, 40, 42, 72$

A complex genus with no clear relationship between the five chromosome numbers reported.

Cleistes Rich. ex Lindl. $2n = 18$

Coeloglossum Hartm. $2n = 40, 41, 42, 80$

x = 20. giving $2x = 40$, $4x = 80$. The counts of 41 and 42 might indicate the presence of B-chromosomes, raising the apparent diploid count by 1–2.

Coelogyne Lindl. $2n = 38, 38+Bs, 40, 42,$
                                     $40–44, 44, 76, 80$

x = 19, 20. Two polyploid series present, with x = 19 giving $2x = 38$, $4x = 76$ and x = 20 giving $2x = 40$ (the majority of records; Fig. VII.2), $4x = 80$. As is also possible in *Arundina* and *Bulbophyllum*, the original basic number could have been x = 10 in this

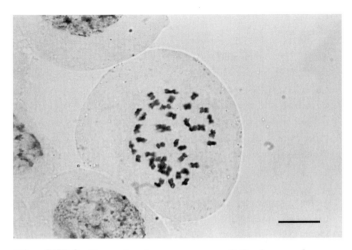

Fig. VII.2. Somatic chromosomes of *Coelogyne fragrans* ($2n = 40$). Scale bar = 10 μm. With permission of K. Y. Lim.

genus, giving $4x = 40$, $8x = 80$, with all diploids extinct or still undiscovered. This can give rise to a neobasic number of x = 19 by doubling and the stabilized dysploid loss of one chromosome.

Comparettia Poepp. & Endl. $2n = 42, 44$

Corallorhiza Gagnebin $2n = 38, 40, 42, 84$

x = 21, giving $2x = 42$ (the most frequent record), $4x = 84$, with dysploid variation at the diploid level that might indicate a series increasing from x = 19 to 20 to 21.

Coryanthes Hook. $2n = 40$

Corybas Salisb. $2n = 40$

Corymborkis Thouars $2n = 40, 58$

Cottonia Wight $2n = 38, 40$

In this genus, the frequent dysploid series of x = 10–20–19 seems to be again present.

Cremastra Lindl. $2n = 42, 48, 50, 52$

A complex genus with no clear relationship between the chromosome numbers reported.

Cryptochilus Wall. $2n = 38$

Cryptopus Lindl. $2n = 76, 95$

x = 19. No diploids recorded, but $4x = 76$ and $5x = 95$.

Cryptostylis R.Br. $2n = 42$

Cyclopogon C.Presl $2n = 28, 30, 32, 36, 45,$
                                     $56, 66$

x = 14, 15, 16? x = 14 gives $2x = 28$, $4x = 56$, but several other counts do not fit. Perhaps there are other series based on x = 15 giving $2x = 30$ and $3x = 45$ and x = 16 giving $2x = 32$. This genus is a clear case of close dysploid variation.

Cycnoches Lindl. $2n = 64, 68$

Cymbidiella Rolfe $2n = 52, 54$

Cymbidium Sw. $2n = 32, 36, 38, 39, 40,$
                                     $41, 42, 43, 44, 46,$
                                     $52, 57, 58, 60, 64,$
                                     $65, 66, 68, 70, 75,$
                                     $80, 81, 82, 85, 100$

A very complex series of records with high levels of aneuploidy. Probably x = 20 gives $2x = 40$, $3x = 60$, $4x = 80$, although to avoid triploid sterility, these counts could be based on x = 10, producing $4x = 40$, $6x = 60$, $8x = 80$, $10x = 100$. Counts of $2n = 38, 57$ could also be related to these ($2x$ and $3x$ based on x = 19) by the doubling of x = 10 and dysploid reduction by one chromosome to give x = 19, as in *Coelogyne*, *Bulbophyllum* etc.

Cynorkis Thouars $2n = 14$

Cyperorchis Blume $2n = 40$

Cypripedium L. $2n = 20, 21, 24, 22, 26,$
                                     $30, 32, 38, 42$

x = 10 gives $2x = 20$ (the majority) and $3x = 30$, but there are also records that are aneuploid or dysploid.

Cyrtopodium R.Br. $2n = 46$

Cyrtorchis Schltr. $2n = 44–46, 46, 50, 92,$
                                     $138, c. 150$

$x = 23$, giving $2x = 46$ (the majority), $4x = 92$ and $6x = 138$. Other counts do not fit the series.

Dactylorchis (Klinge) Verm. $2n = 20, 32, 40, 42, 60,$
                                      $80, 100, 120$

Dactylorhiza Neck. ex Nevski $2n = 16-28, 40, 41, 42,$
                                      $56, 59, 60, 61, 78,$
                                      $80, 82, 100, 120,$
                                      122

In this genus (including *Dactylorchis*) $x = 10, 20$? $x = 20$ gives $2x = 40, 3x = 60, 4x = 80, 5x = 100, 6x = 120$. As in several other genera noted above, the problem of sterility in triploids (and also pentaploids in this case) is avoided if the basic number is taken as $x = 10$, giving $4x = 40, 6x = 60, 8x = 80, 10x = 100, 12x = 120$. This alternative is supported if the few records of $2n = 2x = 20$ in the genus by Vermeulen (1947) and others are accepted. This establishes tetraploidy ($4x = 40$) and octoploidy ($8x = 80$) as the common ploidy levels in this genus, with the diploid being very rare. Counts of $2n = 42$ could indicate the formation of a neobasic number from $x = 10$ by doubling and the addition or fission of one chromosome to give $x = 21$ (see also *Orchis*).

Dactylostalix Rchb.f. $2n = 42$

Dendrobium Sw. $2n = 18, 30, 32-35, 36,$
                               $36+Bs, 38, 38+Bs,$
                               $39, 40, 40+Bs, 41,$
                               $42, 43, 57, 76, 80$

This genus was the subject of a very large survey conducted by Jones *et al.* (1982). There is a complex group of records, but $x = 19$ seems to be the main basic number, giving $2x = 38, 3x = 57, 4x = 76$. The count of $2n = 40$ is also frequent in the genus, with $2n = 80$ less so. These records suggest the presence of another basic number ($x = 20$ or 10), so once again there is the possibility of a basic number of $x = 10$ giving rise to $x = 19$ by doubling and dysploid reduction, as is suggested in *Bulbophyllum*, *Coelogyne*, *Cymbidium*, etc. The record of $2n = 18$ in *D. fimbriato labellum* by Hsu (1972) must be treated with circumspection.

Dendrochilum Blume $2n = 30, 38, 40$
This genus could be another case of a basic number of $x = 10$ that has doubled and produced a neobasic $x = 19$ by loss of one chromosome.

Dendrophylax Rchb.f. $2n = 42$

Diaphananthe Schltr. $2n = 50, 100$
A high basic number of $x = 25$, giving $2x = 50$, $4x = 100$.

Dichaea Lindl. $2n = 52$

Dilomilis Raf. $2n = 42$

Dimerandra Schltr. $2n = 40$

Diplocaulobium (Rchb.f.) Kraenzl. $2n = 38$

Diplocentrum Lindl. $2n = 38$

Diplomeris D.Don $2n = 42$

Diploprora Hook.f. $2n = 16, 38$

Dipodium R.Br. $2n = 46$

Disa Bergius $2n = 36, 36-38, 38, 40,$
                               $48, 52, 54, 57, 68$
A complex series with no clear pattern, although $x = 18$ is possible, giving $2x = 36$ and $3x = 54$.

Doritis Lindl. $2n = 38, 40, 57, 76$
$x = 19$, giving $2x = 38, 3x = 57, 4x = 76$. The record of $2n = 40$ suggests a basic number of $x = 10$ which gave rise to $x = 19$ by doubling and loss, as in many other genera.

Eggelingia Summerh. $2n = 46$

Eleorchis F.Maek. $2n = 40$

Eltroplectris Raf. $2n = 26, 46$
Two very dissimilar basic numbers $x = 13$, giving $2x = 26$ and $x = 23$, giving $2x = 46$ (the most frequent record). Both of these complements are strongly bimodal, with one pair of chromosomes that is much larger than the remainder (Martínez 1985). Bimodality of the karyotype is uncommon in the orchids, although it is widespread in other families, for example the Aloaceae (Brandham and Johnson 1977, 1982).

Encyclia Hook. $2n = 40$

Ephemerantha P.F.Hunt & Summerh. $2n = 40$

Ephippianthus Rchb.f. $2n = 36, 40, 42$

Epidendrum L. $2n = 20, 24, 28, 36, 38,$
                               $40, 44, 46, 48, 56,$
                               $57, 60, 62, 70, 80,$
                               $80-90, 160.$

$x = 10$, giving $2x = 20, 4x = 40$ (the most frequent record), $6x = 60, 7x = 70, 8x = 80, 9x = 90, 16x = 160$. There are several aneuploid records in this genus, or perhaps another basic number, i.e. $x = 19$, giving $2x = 38, 3x = 57$. If this is so, the genus represents yet another case of doubling of $x = 10$ and dysploid reduction by one chromosome to produce a neobasic number of, $x = 19$ (see *Bulbophyllum*, *Coelogyne*, *Cymbidium*, and *Dendrobium*).

Epigeneium Gagnep. $2n = 40$

Epipactis Zinn $2n = 20, 24, 32, 34+Bs,$
                               $36, 38, 38+Bs, 40,$
                               $40+Bs, 44, 46, 48,$
                               $60, 60+Bs, 80$

$x = 10$, giving $2x = 20, 4x = 40, 6x = 60, 8x = 80$. Perhaps also $x = 12$, giving $2x = 24, 3x = 36, 4x = 48, 5x = 60$. The relation between $x = 10$ and 12 is not clear, but possibly $x = 6$ instead of 12 gives the series $4x = 24, 6x = 36, 8x = 48, 10x = 60$. This avoids the problems of sterility in odd-number polyploids (see *Ancistrorhynchus*, above), and $x = 6$ can again form a neobasic number of $x = 10$ by doubling and dysploid reduction, this time by two chromosomes. Other counts fit neither series ($2n = 32, 34, 38, 44, 46$), the whole making this genus a good candidate for further investigation.

Epipogium Borkh. $2n = 68$

Eria Lindl. $2n = 18, 24+Bs, 34, 36,$
                                 $38, 40, c.42, 44, 66,$
                                 76

$x = 19$, giving $2x = 38$. The tetraploid $4x = 76$ is much less common (Lim 1985). There are several other confusing records, including the very low one of $2n = 18$ in *E. muscicola* by Vij and Shekhar (1985). $x = 22$ (or 11) commonly gives $2n = 44$.

Eriopsis Lindl. $2n = 40$
Erycina Lindl. $2n = 52, 56$
Eulophia R.Br. ex Lindl. $2n = 28, 32, 34, 38, 40,$
                                 $41, 42, 44, 46, 48,$
                                 $50, 52, 54, 56, 60,$
                                 $62, 66, 68, 70, 72,$
                                 $74, 76, 80, 82, 84,$
                                 $94, 96, 100, 112,$
                                 120

This large pantropical genus is chromosomally very complex. The most common chromosome number is $2n = 32$, suggesting $x = 16$, although more of the records fit into a series based on $x = 8$. Thus there are no recorded diploids based on $x = 8$, but there is a nine-step polyploid series: $4x = 32, 5x = 40, 6x = 48, 7x = 56, 9x = 72, 10x = 80, 12x = 96, 13x = 112, 15x = 120$. Some other records do not fit this series, and there are probably groups of species with other basic numbers.

Eulophidium Pfitzer $2n = 48, 58$

A complex genus with no clear relationship between the chromosome numbers reported.

Eulophiella Rolfe $2n = 52$
Eurychone Schltr. $2n = 50$
Flickingeria A.D.Hawkes $2n = 38$
Galeandra Lindl. $2n = 56$
Galearis Raf. $2n = 42$
Galeola Lour. $2n = 28, 30$
Gastrochilus D.Don $2n = 30, 34, 38, 40$
Gastrodia R.Br. $2n = 16, 18, 22, 24, 30,$
                                 $36, 40$

Another confusing genus, with perhaps $x = 8$ giving $2x = 16, 3x = 24, 5x = 40$, but the tetraploid $4x = 32$ has not been recorded. Other counts ($2n = 18, 22, 30, 36$) do not fit the series, but $x = 9$ produces diploids and tetraploids.

Gennaria Parl. $2n = 34, 36$
Genyorchis Schltr. $2n = 38$
Geodorum G.Jacks. $2n = 36, 52, 54$

$x = 18$?, giving $2x = 36, 3x = 54$, but perhaps $x = 9$, giving $4x$ and $6x$ records.

Glossorhyncha Ridl. $2n = 40$
Gomesa R.Br. $2n = 56$
Gongora Ruíz & Pav. $2n = 38, 40, 42$

An original $x = 10$ in this genus (no diploids known) has seemingly doubled, with loss and gain of one chromosome to derive the neobasic $x = 19$ and 21.

Goodyera R.Br. $2n = 22, 26, 28, 30, 30+Bs,$
                                 $32, 32+Bs, 38, 40,$
                                 $42, 43–45, 44, 56,$
                                 $58, 59–61, 60$

There is no clear series in this genus, except perhaps $x = 14$ (7?), giving $2x = 28, 3x = 42, 4x = 56$. Aneuploidy and chromosome fission or fusion are probably widespread in it.

Grammangis Rchb.f. $2n = 54$
Grammatophyllum Blume $2n = 38, 40$
Graphorkis Thouars $2n = 52, 54$
Grobya Lindl. $2n = 56$
Gymnadenia R.Br. $2n = 20, 30, 36, 38, 40,$
                                 $42, 80, 100, 120$

A very clear series based on $x = 10$, giving $2x = 20, 3x = 30, 4x = 40$ (the most common), $8x = 80, 10x = 100, 12x = 120$, with small amounts of aneuploidy at the tetraploid level.

Habenaria Willd. $2n = 16, 26, 28, 30, 32,$
                                 $37, 38, 40, 42,$
                                 $42+Bs, 44, 44–50,$
                                 $46, 48, 62, 63, 64,$
                                 $66, 84, 88, 108,$
                                 $c. 112, 122,$
                                 $122–126, 134, 168$

There are many different records in this genus, but the most frequent one of $2n = 42$ suggests $x = 21$, giving $2x = 42, 3x = 63, 4x = 84, 6x = 126, 8x = 168$. There are clearly several other basic numbers in this complex genus, with one of them perhaps $x = 16$, giving $2x = 32, 3x = 48, 4x = 64$.

Haemaria Lindl. $2n = 44$
Hammarbya Kuntze $2n = 28$
Hemipilia Lindl. $2n = 42, 44$
Herminium L. $2n = 24–26, 34, 36, 38,$
                                 $40, 42, 72, 76, 80,$
                                 114

Common chromosome numbers in this genus are $2n = 38$ and 40, suggesting $x = 19$ and 20, with tetraploids $4x = 76$ and 80 being recorded, also the hexaploid $6x = 114$ from $x = 19$. Once more, these plants could be ancient high polyploids based originally on $x = 10$, with doubling and dysploid reduction to produce the neobasic $x = 19$.

Hetaeria Blume $2n = 20, 22, 24, 42$

$x = 10, 11, 12$. The numbers suggest diploidy ($2n = 20, 22, 24$) and tetraploidy ($2n = 42$). It is possible that the last is an allotetraploid derived from hybridization between species with $2n = 20$ and 22, followed by the doubling of the chromosome number.

Hexisea Lindl. $2n = 38$
Himantoglossum K.Koch $2n = 24, 36, 37$

$x = 6$, giving $4x = 24, 6x = 36$, with some aneuploidy. The frequency of $2n = 36$ records precludes triploids of $x = 12$.

Hispaniella Braem $2n = 40$

Holcoglossum Schltr. $2n = 38, 57$

     $x = 19$, giving $2x = 38$, $3x = 57$.

Hormidium (Lindl.) Heynh. $2n = 40, c. 52$

Ione Lindl. $2n = 40$

Ionopsis Kunth $2n = 46$

Isotria Raf. $2n = 18$

Jumellea Schltr. $2n = 38, 38–40$

Katharinea A.D.Hawkes $2n = 40, c. 76$

     $x = 20$?, giving $2x = 40$ and $4x = 80$ (reported as *c.* 76), or possibly $x = 10$ with no diploid $2n = 20$ yet recorded.

Kingidium P.F.Hunt $2n = 36, 38$

Kingiella Rolfe $2n = 38$

Koellensteinia Rchb.f. $2n = 48$

Laelia Lindl. $2n = 40, 41, 42, 60,$ *c.* $63, 80$

     $x = 20$?, giving $2x = 40$, $3x = 60$, $4x = 80$, or perhaps $x = 10$, giving $4x = 40$, $6x = 60$, $8x = 80$, with no diploid yet recorded.

×*Laeliocattleya* Rolfe $2n = 40, 60, 61, 62, 80,$ $100, 102, c. 108,$ 114

The chromosome counts reported in this widely-grown group of bigeneric hybrids confirm the basic numbers which are the same in both parents ($x = 10$ or 20). They also show the genetic similarity of *Laelia* and *Cattleya* and explain the relative ease of hybridization among their species to produce fertile hybrids.

Lanium (Lindl.) Benth. $2n = 38, 40$

Lemboglossum Halbinger $2n = 56$

Leochilus Knowles & Westc. $2n = 42, 48$

These two numbers could both be based on $x = 6$, although the occurrence only of $7x$ and $8x$ plants is not likely. It is more probable that they are less closely related, e.g. $x = 7$ or 21 and 8, 12 or 24.

Leptotes Lindl. $2n = 40$

Leucolaena (DC.) Benth. $2n = c. 36, c. 150$

Leucorchis Blume $2n = 40, 42$

Limnorchis Rydb. $2n = 42, 80, 84$

     $x = 21$, giving $2x = 42$, $4x = 84$, with aneuploidy.

Limodorum Boehm. $2n = 56, 60, 64$

Liparis Rich. $2n = 20, 22+Bs, 26, 28,$ $30, 32, 36, 38, 40,$ $42, 56, 60, 66–80,$ $76, 80$

A low basic number of $x = 5$, giving $4x = 20$, $6x = 30$, $8x = 40$, $12x = 60$, $16x = 80$, with some aneuploidy. The number $2n = 30$ is commonly reported in wild species, reducing the likelihood that the basic number is $x = 10$, which would make them sterile triploids. As in several other genera, the derived neobasic number of $4x - 1 = 19$ gives $2n = 38$ (neo-diploid) and 76 (neotetraploid). Another neobasic

$4x + 1 = 21$ gives $2x = 42$, which has also been reported frequently.

Listera R.Br. $2n = 20, 20+Bs. 32, 34,$ $34+Bs, 35, 36,$ $36–38, 37, 38,$ $38+Bs, 39, 40, 42,$ $44, 46, 51+Bs, 56,$ $56+B, 57$

A very complex dysploid series with no clear single basic number, although there are several records indicating a basic number of 17, giving $2x = 34$, $3x = 51$ (Simon 1968).

Listrostachys Rchb.f. $2n = 46$

Lockhartia Hook. $2n = 14, 56$

     $x = 7$, giving $2x = 14$, $8x = 56$, with no intermediate records

Loroglossum Rich. $2n = 24$

Luisia Gaudich. $2n = 38, 40$

Lycaste Lindl. $2n = 40, c. 48$

Lysias Salisb. ex Rydb. $2n = 42$

Lysiella Rydb. $2n = 42, 82–84, c. 120$

     $x = 21$, giving $2x = 42$, $4x = 84$, $6x = c. 120$ (126?). The basic number could also be $x = 7$, with a corresponding trebling of the ploidy levels.

Macodes (Blume) Lindl. $2n = 42$

Macradenia R.Br. $2n = 48, 52$

Macroclinium Barb. Rodr. $2n = 42$

Malaxis Sol. ex Sw. $2n = 28, 30, 30 +Bs,$ $30–34, 36, 38,$ *c.* $40, 42, 44, 60$

A very complex dysploid series with no single basic number that is more frequent than the rest. $x = 7$ gives $4x = 28$ and $6x = 42$. Derived by doubling of $x = 7$ and the addition of one chromosome, the neobasic $x = 15$ gives $2x = 30$ and $4x = 60$.

Manniella Rchb.f. $2n = 84$

Masdevallia Ruíz & Pav. $2n = 36, 44, 64$

Maxillaria Ruíz & Pav. $2n = 40, 42$

Mediocalcar J.J.Sm. $2n = 38, 85$

Megastylis Schltr. $2n = 44$

Mesadenella Pabst & Garay $2n = 46$

Mexipedium V.A.Albert & M.W.Chase $2n = 26$

                         (Fig. VII.3–4)

$x = 13$. The cytology of this genus places it closer to *Phragmipedium* than to *Paphiopedilum* (Cox *et al.* 1997*b*). Because its haploid karyotype contains three telocentrics, it has 23 chromosome arms, not the *nombre fondamental* of $x = 26$ arms typical of *Paphiopedilum*.

Microcoelia Lindl. $2n = 47, 48$

Micropera Lindl. $2n = 38$

Microstylis Nutt. $2n = 26, 30, 36, 38–42,$ $42, 44$

A very complex dysploid series with no clear single basic number.

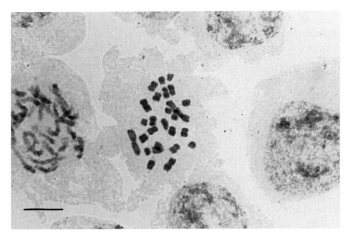

Fig. VII.3. Somatic chromosomes of *Mexipedium xerophyticum* (2n = 26). Scale bar = 10 μm. From Cox, A. V. et al. (1997b). *Lindleyana* **12**, 162–5.

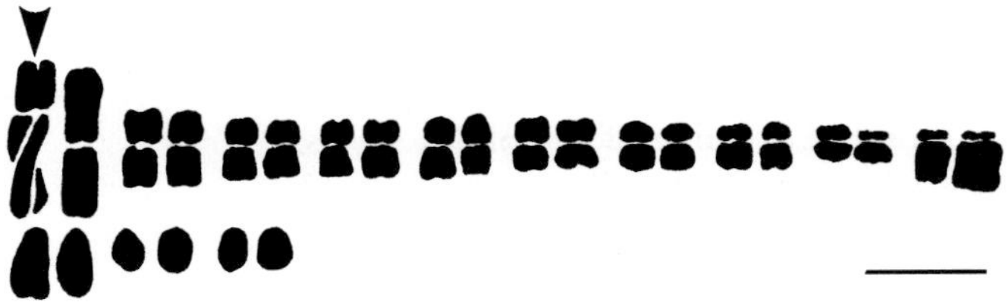

Fig. VII.4. Karyotype of *Mexipedium xerophyticum*. Telocentric chromosomes are shown in the second row. Arrowhead indicates a pericentric inversion in one large metacentric chromosome. Scale bar = 5 μm. From Cox, A. V. et al. (1997b). *Lindleyana* **12**, 162–5.

Microtis R.Br.   2n = *44*, 88
  x = 22, giving 2x = 44, 4x = 88. There is also a possibility that the basic number might be x = 11, which would double the above ploidy levels.
Miltonia Lindl.   2n = 48, 56, 59, *60*, 86, 98, 112
  x = 15?, giving the frequent 4x = 60 and aneuploids at 3x and 5x levels, together with further aneuploidy in high polyploid hybrids. There seem to be no discovered diploids in this genus.
Miltonioides Brieger & Lückel   2n = 56
Mormodes Lindl.   2n = *54*
Myrmechis (Lindl.) Blume   2n = 28, 56
  x = 14?, giving 2x = 28, 4x = 56, but perhaps x = 7 giving 4x = 28, 8x = 56.
Mystacidium Lindl.   2n = 48
Neofinetia Hu   2n = *38*
Neolehmannia Kraenzl.   2n = 36, 40
Neolindleya Kraenzl.   2n = 36–38
Neotinea Rchb.f.   2n = *40*, 42

Neottia Guett.   2n = 20, 32, *36*, 40, 42, 46
  The frequently-recorded number of 2n = 36 suggests a basic number of x = 18, but the other reports do not agree with this suggestion. Another basic number of x = 10 gives 2x = 20, 4x = 40.
Neottianthe (Rchb.) Schltr.   2n = 42
Nephelaphyllum Blume   2n = 18+Bs, 36+Bs
  x = 9, giving 2x = 18, 4x = 36.
Nervilia Comm. ex Gaud.   2n = 20, 54, 72, 108, 144
  x = 9, giving 6x = 54, 8x = 72, 12x = 108, 16x = 144. No diploid with 2x = 18 or tetraploid with 4x = 36 has been reported. The count of 2n = 20 suggests a primitive basic number of x = 10, from which the more frequent series based on x = 9 was derived by dysploid reduction.
Neuwiedia Blume   2n = 96, *c.* 144
  The high chromosome numbers recorded in this genus suggest a basic number of x = 24, with 4x = 96 and 6x = c. 144, but they could also suggest x = 12, with 8x = 96 and 12x = c. 144, or even x = 8, with 12x = 96 and 18x = 144, although the latter seems unlikely. The low ploidy levels have not been found.
Nigritella Rich.   2n = 32, 38, 40, 60, 64, 80
  x = 20?, giving 2x = 40, 3x = 60, 4x = 80 with some dysploidy. The basic number could also be x =10, giving 4x = 40, 6x = 60, 8x = 80.
Notylia Lindl.   2n = 32, *42*
Oberonia Lindl.   2n = 26, *30*, 60, *c.* 72
  x = 15, giving 2x = 30 (the most frequent), 4x = 60.
Odontochilus Blume   2n = 28, 60, *c.* 120
  x = 15, giving 4x = 60, 8x = c. 120, with no diploid being recorded. The count of 2n = 28 is either a dysploid difference at the diploid level (x = 15 – 1 = 14) or a misinterpretation of 2x = 30.
Odontoglossum Kunth   2n = 26, 28, 30, 32, 34, 37, 38, 40, 42, *44*, 44–48, 50–60, *56*, 60, 84, 112
  A very complex dysploid series with no clear single basic number, although Wimber (1989) reported 2n = 56 to be common in the genus. Thus x = 14, 2x = 28, 3x = 42, 4x = 56, 6x = 84, 8x = 112. It is also possible that the basic number is x = 7, which would double the above ploidy levels. Other counts in the genus, notably the several reports of 2n = 44 (x = 11 or 22), do not fit into this series, so the genus is another which is chromosomally complex. The complexity supports the suggestion by Chase and Palmer (1992) that *Odontoglossum* is an artificial genus.
Oeceoclades Lindl.   2n = 58
Oeoniella Schltr.   2n = 38
Olgasis Raf.   2n = 42

Oncidium Sw. $2n = 10, 14, 26, 28, 30, 32, 34, 36, 37, 38, 39, 40, 42, 44, 48, 49, 52, 54, 56, 57, 60, 61, 63, 72, 84, 105–111, 112, 126, 133, 140–150, 168$

Another complex genus with several basic chromosome numbers, although $2n = 56$ is particularly common. This suggests $x = 7$, giving $2x = 14$ (very rarely—see below), $4x = 28$, $6x = 42$, $8x = 56$, $12x = 84$, $16x = 112$, $18x = 126$, $24x = 168$. It is unlikely that the basic number is $x = 14$, since there are several species with $2n = 42$. These would be sterile if they were triploid, so it is more probable that they are hexaploid. Several other polyploid series clearly exist in this genus, complicated by dysploidy and aneuploidy. One of these is $x = 13$, giving $2x = 26$, $3x = 39$, $4x = 52$. It is a neobasic number derived from $x = 7$ by doubling, followed by loss of one chromosome. These differing series support the contention of Chase and Palmer (1992) that the genus is an artificial one, encompassing several entities that are related only loosely. The very low number of $2n = 10$ reported by Dodson (1957) and confirmed by other authors (see Tanaka and Kamemoto 1984) is unusual in the family and raises the possibility that the plant might be a rarely occurring haploid sporophyte. This would, of course, be meiotically sterile and would need to reproduce vegetatively or apomictically in order to survive. Haploid sporophytes are usually small in size compared to related diploids, and the species name quoted by Dodson (1957), *O. pusillum* (although the taxon is now known as *Psygmorchis pusilla*), supports this idea. On the other hand, Chase and Olmstead (1988) have suggested that the species is a diploid based on $x = 5$. An alternative count of $2n = 14$ in the same species by Kugust (1966) could be the only record of a diploid based on $x = 7$ in this largely polyploid genus.

Ophrys L. $2n = 22–24, 36, 36+Bs, 37, 38, 40, 54, 72, 72+Bs, c. 73, 76$

$x = 18$, giving $2x = 36$, $3x = 54$, $4x = 72$, although for reasons stated under *Ancistrorhynchus* and several other genera above, the basic number could be $x = 9$, giving $4x = 36$, $6x = 54$, $8x = 72$, in which case diploids remain unreported.

Orchis L. $2n = 14, 20, 21, 32, 34, 36, 38, 40, 40+Bs, 42, 42+Bs, 54, 63, 79, 80, 82, 84, 120, 122$

$x = 10$, giving $2x = 20$, $4x = 40$ (the most frequent), $8x = 80$ (also very common), $12x = 120$. The frequent count of $2n = 42$ in this genus suggests the existence

of another basic number $x = 7$, giving records of $2x = 14$, $3x = 21$, $6x = 42$. The same count ($2n = 42$) might also suggest the formation of a neobasic number of $x = 21$ by doubling of $x = 10$ and the addition or fission of a single chromosome. This forms the basis of another polyploid series with $2n = 42, 63, 84$. Other frequent counts (e.g. $2n = 36, 38$) do not fit any of the above series, so clearly the genus is extremely complex as far as its chromosomes are concerned, with extensive dysploid variation, with $x = 10$ doubling and reducing to the neobasic $x = 19$ and 18. Pridgeon *et al.* (1997) and Bateman *et al.* (1997) have shown that *Orchis* is triphyletic based on rDNA ITS sequences and that the three clades have different karyotypes: (1) the type *O. militaris* clade with $2n = 42$; (2) the *O. tridentata* clade (including *Neotinea*) with $2n = 42$ (+ Bs); and (3) the *O. morio* clade (including *Anacamptis*) with $2n = 32, 36$.

Oreorchis Lindl. $2n = 42, 48, 50, 84$

$x = 21$, giving $2x = 42$, $4x = 84$. This high basic number could be neobasic, derived from $x = 10$, or it could indicate allopolyploidy, but plants with the original numbers have not been reported. Alternatively, the basic number could be $x = 7$, giving $6x = 42$, $12x = 84$, with no lower ploidy levels known. This genus merits further investigation.

Ornithidium Salisb. $2n = 48$
Ornithochilus (Lindl.) Benth. $2n = 38$
Ornithophora Barb.Rodr. $2n = 56$
Ossiculum P.J.Cribb & Laan $2n = 34$
Otochilus Lindl. $2n = 36, 40$
Pachystoma Blume $2n = 40$
Panisea (Lindl.) Lindl. $2n = 40$
Paphiopedilum Pfitzer $2n = 21, 24, 25, 26, 26–29, 27, 28, 30, 32, 33, 34, 34+Bs, 35, 35–36, 36, 37, 38, 39, 39–42, 40, 40+Bs, 40–45, 41, 42, 48, 50, 53–56, 57, 58, 70$

This genus is unusual in the orchids in having large metacentric chromosomes that commonly undergo Robertsonian fission to produce stable telocentrics, thus changing the basic number. The lowest basic number is $x = 13$ metacentrics, giving the common $2x = 26$. A single fission gives $x = 14$, with 12 metacentrics and 2 telocentrics. Further fissions increase the basic numbers up to a maximum of $x = 26$ telocentrics, but the *nombre fondamental* of $x = 26$ chromosome arms remains constant (Karasawa 1979, 1981; Cox *et al.* 1997*b*). The different basic numbers combined with polyploidy explain the many records between $2n = 26$ and 70 but do not explain those below $2n = 26$.

Papilionanthe Schltr. $2n = 38$
Paraphalaenopsis A.D.Hawkes $2n = 38$
Pecteilis Raf. $2n = 32, 42, 48, 64$

$x = 16$?, giving $2x = 32$, $3x = 48$, $4x = 64$, although for reasons stated under *Ancistrorhynchus* and several other genera, the basic number could be $x = 8$, giving $4x = 32$, $6x = 48$, $8x = 64$, with no diploids reported. The report of $2n = 42$ in the genus does not fit the above series.

Pelatantheria Ridl.   *$2n = 38$*
Pelexia Poit. ex Lindl.   *$2n = 46$*
Peristeria Hook.   *$2n = 40$*
Peristylus Blume $2n = 28, 36,$ *42,* *c.* $44,$
                                       $46, 46+Bs, 47, 88$

A complex genus with no clear single basic number, although $2n = 42$ is frequent ($x = 7$ or 21). The record of $2n = 28$ would constitute a tetraploid based on $x = 7$.

Perularia Lindl.   *$2n = 42$*
Phaius Lour. $2n = 28, 38, 40,$ *42,* $44,$
                                       $44+Bs, 48, 50, 62$

A complex genus with no clear single basic number although $2n = 42$ is frequent ($x = 7$ or 21), as in *Peristylus*.

Phalaenopsis Blume $2n = 38, 55, 57, 69+Bs,$
                                       c. $74, 76, 92, 100,$
                                       $114, 152 \pm 2$

$x = 19$ is common, giving $2x = 38$ (the most frequent), $3x = 57$, $4x = 76$, $6x = 114$, $8x = c. 152$.

Pholidota Lindl. ex Hook. $2n = 38,$ *40,* $42, 44$

A complex genus with no clear single basic number, although $x = 20$ (?10) is the basis of the most frequent record of $2n = 40$. The others could be derived by dysploid addition or reduction to $x = 19, 21, 22$.

Phragmipedium Rolfe $2n = 18, 20, 20+Bs, 21,$
                                       $22, 23, 24, 28, 30,$
                                       $32, 36, 40$

A complex genus with no clear single basic number. There are some indications of a series based on $x = 9$ giving $2x = 18$, $4x = 36$ and another of $x = 10$ giving $2x = 20$, $3x = 30$, $4x = 40$, but other numbers are reported that do not fit these series.

Phreatia Lindl. $2n = 32$
Phyllomphax Schltr. $2n = 38, 42, 46$
Phymatidium Lindl. $2n = 46–56$
Physosiphon Lindl. $2n = 32, 34, 36, 38, 42$

A complex genus with no clear single basic number.

Physurus Rich. $2n = 22$
Platanthera Rich. $2n = 16, 20, 21, 36, 38,$
                                       $40,$ *42,* $44, 63, 84,$
                                       c. 126

The common basic number is $x = 21$, giving $2x = 42$, $3x = 63$, $4x = 84$, $6x = c. 126$. The single record of $2n = 21$ (Tanaka and Kamemoto 1984) is either from a haploid plant, or indicates a lower basic number of

$x = 7$. Other records do not fit this series, indicating dysploid variation.

Platyclinis Benth. $2n = 38, 40$
Plectrelminthus Raf. $2n = 46$
Pleione D.Don $2n = 20+Bs, 38, 39,$ *40,*
                                       $40+Bs, 42, 43, 44,$
                                       $56–61, 60, 80, 120$

$x = 10$, giving $2x = 20$, $4x = 40$ (the most frequent), $6x = 60$, $8x = 80$, $12x = 120$. Aneuploids also occur.

Pleurothallis R.Br. $2n = 20, 32, 34, 36, 38,$
                                       $40, 42, 43, 44, 45,$
                                       $64, 68, 72, 76, 77,$
                                       $78, 80, 84$

In this chromosomally complex genus several basic numbers are present and form polyploid series. $x = 10$ gives $2x = 20$, $4x = 40$, $8x = 80$. $x = 16$ gives $2x = 32$, $4x = 64$, $5x = 80$. $x = 17$ gives $2x = 34$, $4x = 68$. $x = 21$ gives $2x = 42$, $4x = 84$. The genus was the subject of an extensive investigation by Nakata and Hashimoto (1983).

Plocoglottis Blume   *$2n = 38$*
Podangis Schltr.   *$2n = 46$*
Podochilus Blume   *$2n = 38$*
Pogonia Juss. $2n =$ *18,* $19, 20, 21, 23,$
                                       $36, 62, 68$

$x = 9$, giving $2x = 18$ (the most frequent), $4x = 36$, but other records do not fit into the series.

Polystachya Hook. $2n = 38, 39,$ *40,* $41, 80,$
                                       $81, 120$

The frequent count of $2n = 40$ (Jones 1966) indicates a basic number of $x = 20$, but the aneuploid records around $2n = 40$ ($2n = 38, 39, 41$) suggest that it is more probably tetraploid. Thus in this genus $x = 10$, with $4x = 40$, $8x = 80$, $12x = 120$ and no recorded diploids.

Pomatocalpa Breda, Kuhl & Hasselt   *$2n = 38$*
Ponerorchis Rchb.f.   *$2n = 42$*
Ponthieva R.Br. $2n = 46$
Porpax Lindl. $2n = 24, 42$

A complex genus with no clear relationship between the chromosome numbers reported except perhaps $x = 6$, although $4x$ and $7x$ records are unlikely in the absence of any others.

Pristiglottis Cretz. & J.J.Sm.   *$2n = 26$*
Promenaea Lindl. $2n = 46$
Pseudorchis Ség.   *$2n = 42$*
Psychopsis Raf. $2n = 38$
Psygmorchis Dodson & Dressler $2n = 10, 14$

For discussion of this genus see *Oncidium*.

Rangaeris (Schltr.) Summerh. $2n = 46, 50, 92,$ *c.* $100,$
                                       c. 108

Perhaps there are two basic numbers in this genus: $x = 23$ giving $2x = 46$ and $4x = 92$; and also $x = 25$ giving $2x = 50$ and $4x = 100$. The count of $2n = c. 108$ fits neither series.

Renanthera Lour. $2n = 38, 76, 114, 115$

$x = 19$, giving $2x = 38$ (the most frequent), $4x = 76$, $6x = 114$. This is one of the few cases of a single series with no reported major deviations (see also *Satyrium*), except at the hexaploid level, which is to be expected.

Rhynchostylis Blume   *2n = 38*
Ritaia King & Pantling $2n = 38$
Robiquetia Gaudich.   *2n = 38* (Fig. VII.5)
Rodriguezia Ruíz & Pav. $2n = 28, 29, 42$

As the count of $2n = 42$ is common in this genus (Sinotô 1962), it is not likely to be triploid. Thus $x = 7$, giving $4x = 28$, $6x = 42$. There are no diploid records.

Rossioglossum (Schltr.) Garay & G. C. Kenn. $2n = 44$
Saccoglossum Schltr. $2n = c. 40$
Saccolabium Blume $2n = 30, 36-44, 38, 76$

$x = 19$, giving $2x = 38$ (the most frequent), $4x = 76$, plus other records that do not fit the series.

Sacoila Raf. $2n = 46$
Sarcanthus Lindl. $2n = 36, 38$
Sarcochilus R.Br. $2n = 36, 38, 76$

$x = 19$, giving $2x = 38$ (the most frequent), $4x = 76$, plus another record that does not fit the series.

Sarcoglottis C.Presl   *2n = 46*
Sarcorhynchus Schltr. $2n = 50$
Satyrium Sw. $2n = 41, 82, 123, 164$

$x = 41$, giving $2x = 82$, $3x = 123$, $4x = 164$. This very high basic number, occurring in *S. nepalense* (Jorapur 1980; Vij *et al.* 1981; Kashyap and Mehra 1983), is almost certainly a neobasic number resulting from the combination of two or more lower ones. Thus the 'diploid' plants with $2n = 82$ are probably allo-tetraploids or even higher polyploids. The record of $2n = 41$ in the same species by Swamy (1944) possibly represents the original diploid hybrid plant.

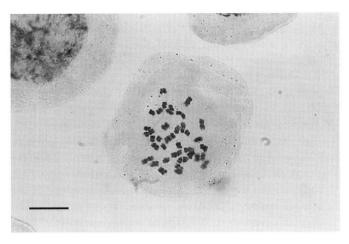

Fig. VII.5. Somatic chromosomes of *Robiquetia moore-ana* ($2n = 38$). Scale bar = 10 μm. With permission of K. Y. Lim.

Sauroglossum Lindl. $2n = 44, 46$
Scaphyglottis Poepp. & Endl. $2n = 40$
Schoenorchis Blume $2n = 36$
Schomburgkia Lindl. $2n = 40$
Serapias L. $2n = 24, 36, 72$

$x = 18$, giving $2x = 36$ (the most frequent), $4x = 72$. The numbers also fit other series based on $x = 6$ or 12. These would be supported by the report of $2n = 24$ mentioned by Fedorov (1969), but as the original record dates from 1915 it must remain doubtful.

Sigmatostalix Rchb.f.   *2n = 60*
Smitinandia Holttum $2n = 38$
Sobralia Ruíz & Pav. $2n = 54$
Solengangis Schltr. $2n = 50$
Sophronitis Lindl. $2n = 40$
Spathoglottis Blume $2n = 18, 38, 40, 42, 60$

A complex genus with no clear single basic number. $x = 9$ is its possible primitive basic number, giving the record of $2n = 18$. This could have doubled to give $x = 18$ (for which there are no records), but small dysploid changes to the latter give the neobasic $x = 19, 20$ and 21 to account for all of the other records.

Sphyrarhynchus Mansf. $2n = 50$
Spiranthes Rich. $2n = 20, 24, 26, 28, 30,$
                                     $30+2B, 32, 35, 40,$
                                     $44, 45, 50, 56, 60,$
                                     $61, 74$

Because $2n = 30$ is frequently recorded in this species, it is not likely to be triploid ($x = 10$). Thus the main polyploid series is probably based on $x = 15$, giving $2x = 30$, $3x = 45$, $4x = 60$. Many other counts do not fit this series and bear no clear relationship to each other.

Stanhopea Frost ex Hook. $2n = 40, 41, 42, 80$

The counts of $2n = 41, 42$ in this genus are aneuploids of $2n = 40$. Because aneuploidy at the diploid level is uncommon, it is more likely that the basic number is $x = 10$, giving $4x = 40$, $8x = 80$.

Staurochilus Ridl. ex Pfitzer   *2n = 38*
Stauropsis Rchb.f.   *2n = 38*
Stelis Sw. $2n = 28, 30, 32$

There are relatively few counts available in this large genus. Of these, the majority of species have $2n = 32$, suggesting a basic number of $x = 8$ or 16, but the other counts of $2n = 28$ and 30 do not support this. Clearly much more chromosome work needs to be done on *Stelis*.

Stenoglottis Lindl. $2n = 36$
Stenorrhynchos Rich. ex Spreng. $2n = 46$
Stolzia Schltr. $2n = 34$
Summerhayesia P.J.Cribb $2n = 44$
Taeniophyllum Blume $2n = 24, 36, 38, 40, 56$

A complex genus with perhaps two basic numbers. One is $x = 6$ or 12, giving records of $2n = 24, 36$ and the other is $x = 19$, giving records of $2x = 38$, $4x = 56$. The records of $2n = 40$ fit neither series.

Tainia Blume $2n = 20, 30+Bs, 32, 36,$
$36+Bs, 40, c. 72$

Two basic numbers are present: $x = 10$, giving $2x = 20$, $3x = 30$, $4x = 40$; also $x = 18$, giving $2x = 36$, $4x = 72$. It is also possible that one or both of these basic numbers could be reduced, to $x = 5$ in the first instance, or $x = 6$ or 9 in the second one. These would raise the ploidy levels of the recorded counts. A possible evolutionary sequence from $x = 5$ to 10 to 9 agrees with the pattern seen in some other genera (e.g. *Cymbidium* and several others), although the initial basic number is not the same. The record of $2n = 32$ fits neither series, unless it represents a further dysploid reduction from $x = 9$ to $x = 8$.

Thelasis Blume $2n = 30, 32$

Thelymitra J.R.Forst. & G.Forst. $2n = 26$

Thrixspermum Lour. $2n = 38$

Thunia Rchb.f. $2n = 36, 38, 40, 42, 44,$
$c. 80$

There are five different chromosome numbers recorded between $2n = 36$ and 44. These might be diploid or polyploid, but there is no other evidence available. Another count of $c. 80$ is not precise enough to represent a doubling of any of these, so it cannot be determined which of the recorded basic numbers is the basis for a polyploid series.

Tipularia Nutt. $2n = 38$

Tolumnia Raf. $2n = 40, 42, 84, 126$

$x = 21$, giving $2x = 42$, $4x = 84$, $6x = 126$. Alternatively, the series could be based on $x = 7$, giving $6x = 42$, $12x = 84$, $18x = 126$, although as 18-ploid plants are rarely recorded this would seem unlikely.

Trachoma Garay $2n = 38$

Traunsteinera Rchb. $2n = 42$

Trias Lindl. $2n = 38$

Trichocentrum Poepp. & Engl. $2n = 24, 28$

The two numbers are quoted with equal frequency, suggesting $x = 6$ and 7, with only tetraploids being found so far.

Trichoceros Kunth $2n = 48$

Trichoglottis Blume $2n = 38$

Trichopilia Lindl. $2n = 56$

Two species are recorded with this high number (Tanaka and Kamemoto 1984). They are probably octoploids based on $x = 7$.

Trichosma Lindl. $2n = 36$

Tridactyle Schltr. $2n = 44, 46, c. 50, c. 100$

$x = 23$, giving $2x = 46$ (the most frequent) and $c. 50$, $4x = c. 100$. Because the basic number is high when compared with those of most other orchids, it could well be a combination of two or more lower ones (e.g. $10 + 13$, or $7 + 8 + 8$), making the $2n = 46$ plants allopolyploid.

Triphora Nutt. $2n = 44$

Tropidia Lindl. $2n = 56, 57, 60$

These records are probably dysploid variants of a highly polyploid genus ($x = 7$ or 10).

Tulotis Raf. $2n = 42$

Tylostylis Blume $2n = 20, 40$

$x = 5$ or 10, giving $4x = 20$, $8x = 40$ in the first alternative and $2x = 20$, $4x = 40$ in the second one.

Uncifera Lindl. $2n = 38$

Uraria Desv. $2n = 20$

Vanda Jones ex R.Br. $2n = 18, 20, 28, 38,$
$38–42, 40, 42, 57,$
$72, 76, 95, 114, 115$

$x = 19$, giving $2x = 38$ (the most frequent), $3x = 57$, $4x = 76$, $5x = 95$, $6x = 114$. There is some dysploidy and the low records of $2n = 18, 20, 28$ do not fit the series. It is possible that basic numbers $x = 9$ and 10, which are indicated to exist by two of these low records, have combined to give a neobasic number of $x = 19$, making the $2n = 38$ plants allotetraploid and similarly doubling the ploidy level of the other four counts recorded in the $x = 19$ polyploid series.

Vandopsis Pfitzer $2n = 38$

Vanilla Mill. $2n = 28–32, 32, 36, 64$

$x = 16?$, giving $2x = 32$ (the most frequent), $4x = 64$. The basic number could also be $x = 8$, giving $4x = 32$, $8x = 64$, which might better account for the observed aneuploidy around $2n = 32$.

Vermeulenia A.Löve & D.Löve $2n = 32$

Vexillabium F.Maek. $2n = 26, 40$

The common count of $2n = 26$ suggests $x = 13$, with the record of $2n = 40$ possibly being an aneuploid or an inaccurately-counted triploid.

Warrea Lindl. $2n = 52$

Xylobium Lindl. $2n = 40$

Yoania Maxim. $2n = 24$

Zeuxine Lindl. $2n = 20, 22, 40, 42, 44,$
$50, 100$

$x = 10?$, giving $2x = 20$ (the most frequent), $4x = 40$, $5x = 50$, $10x = 100$. Perhaps the basic number is $x = 5$, which would double the above ploidy levels, but this is again unlikely, since 20-ploid plants are little-known (see *Tolumnia*).

Zygopetalum Hook. $2n = 48, 48–50, 96$

The basic number of this genus is not clear from the observations made. It could be $x = 12$, giving $4x = 48$, $8x = 96$, with no diploids being known. It is possible to halve the basic number to $x = 6$ and double the ploidy levels of the chromosome numbers recorded, but once again this would seem unlikely because of the general rarity of 16-ploid plants. The records could also be interpreted in terms of $x = 8$, giving $6x = 48$, $12x = 96$. Clearly, further investigations are needed on this genus to determine its precise chromosomal constitution.

PETER BRANDHAM

VIII
MOLECULAR SYSTEMATICS, PARSIMONY, AND ORCHID CLASSIFICATION

Floral morphology has historically been the mainstay of orchid taxonomy; in particular the characteristics of anther organization have formed the basis of the higher categories from the earliest part of the nineteenth century. The use of the terms Diandrae for orchids with two anthers (subfamily Cypripedioideae) and Monandrae (orchids with a single anther, e.g. subfamilies Epidendroideae, Vanilloideae, and Orchidoideae) goes back to Robert Brown (1813), but even before him Olof Swartz (1800*b*) clearly distinguished between these two categories without formally naming both. The point in mentioning this here is simply that although floral traits have a long history of use in orchid classification, it is only recently that these characteristics have been evaluated with independent data analysed in a phylogenetic (e.g. cladistic) context. Recently, higher-level orchid classification has been evaluated by phylogenetic studies of morphological characters (still mostly, but not entirely, floral; Freudenstein and Rasmussen (1999) and DNA sequences (Chase *et al.* 1994; Neyland and Urbatsch 1995; Kores *et al.* 1997; Cameron *et al.* 1999). These studies are highly similar in the patterns they reveal, but they do not have high levels of internal support as estimated by resampling procedures such as the bootstrap (Felsenstein 1985) or the jackknife (Farris *et al.* 1996), both of which are sometimes called 'confidence estimates' (see below). Truly robust estimates of the patterns of orchid phylogeny are still some distance in the future, but not very far.

Whereas an accurate estimate of orchid relationships will lead to changes in the tribes and subtribes we recognize, it is generic delimitation that is the main focus of *Genera Orchidacearum*, and so we will not examine the problems at these higher levels here. We will endeavour to use natural and revised higher-level categories, and we have already seen changes at these levels based on DNA sequence information. For example, we will use Vanilloideae on the basis of the results in Cameron *et al.* (1999)

which demonstrate that this group (*Vanilla*, *Pogonia*, and their relatives, subtribes Vanillinae and Pogoniinae *sensu* Dressler 1993) is clearly distinct from all other monandrous orchids and quite distantly related. Likewise, we have abandoned Spiranthoideae (*sensu* Dressler 1993) because several recent molecular studies (Chase *et al.* 1994; Kores *et al.* 1997; Cameron *et al.* 1999) have demonstrated that they are deeply embedded within Orchidoideae as the close relatives of Diurideae and Chloraeinae, both classified as Orchidoideae by Dressler (1993).

Our emphasis on DNA information in *Genera Orchidacearum* is not based on the belief that morphological information is worthless, but rather that it is extremely time-consuming to collect and difficult to interpret accurately. We need to study the morphology, anatomy, and chemistry of the orchids extensively so that we can understand better the evolutionary factors that have made them the largest family of flowering plants. However, this work is going to take many years, if for no other reason than that the family is so large and diverse. Some day we will know as much about all orchid groups as we now know about the slipper orchids, but we are a long way from such a situation. Molecular data by comparison are quick to collect, and so it is not a completely fanciful expectation that we can evaluate the naturalness of the great majority of orchid genera by DNA analyses (ideally by sequencing both plastid and nuclear regions; mitochondrial sequences evolve too slowly to be useful at generic levels; Freudenstein and Chase 1998).

In addition to being difficult to collect, morphological characteristics are more difficult to assess accurately than DNA characters. Once sequences are aligned, the homology of change is easily assessed, and even alignment is not always problematic because we often use gene sequences (i.e. *matK*, *rbcL*, *ndhF*), which are reasonably length-conserved. Homology of certain morphological features is often impossible to determine without having access to

DNA information. For example, subtribe Oncidiinae are nearly all characterized by the presence of a *tabula infrastigmatica*, a fleshy structure on either the base of the column or at the juncture of the lip and column, but a similar structure occurs on the lip in *Lockhartia* and the genera of Ornithocephalinae and Telipogoninae. It is not exactly the same in these other groups of orchids, so it has never been possible to determine if a *tabula infrastigmatica* is present. These three are otherwise reasonably different from Oncidiinae if one considers their habit (e.g. *Lockhartia*) or pollinia number (four in Ornithocephalinae and Telipogoninae versus two in *Lockhartia* and Oncidiinae). Recent DNA studies (Williams, Whitten, and Chase, unpublished) embed all these groups in Oncidiinae, thus enhancing the hypothesis that these somewhat similar although not identical structures are in fact homologous and can serve to identify all the members of an expanded concept of Oncidiinae. Such ambiguously interpreted features occur in many groups of orchids at all taxonomic levels.

In addition to the problem of simply determining the homology of some orchid structures, generic delimitation has long been focused on a subset of the information available, especially the floral traits associated with pollinator attraction, and this presents even worse prospects for creating an improved generic taxonomic scheme. If we wish our genera to be merely convenient categories in which to organize species, then the use of any and all traits is acceptable, but such systems will lack predictability. If closely related species are placed in different genera at the same time that distantly related species are placed in the same genus, we can never assume anything else about these plants by learning which species belong to which genus. We cannot use such a system for anything other than organizing the species. In modern times most systematists would argue that this lack of predictability is not only unnecessary but also extremely undesirable. Many other scientists use our classifications in their research; when investigating a new chemical compound that has, for example, anti-viral activity, a researcher would like to know if any of the close relatives of this species has a similar and perhaps even more potent form of this compound. If the other species placed in this genus are not in fact the closest relatives of this species, then the researcher cannot depend on the classification to tell them anything useful.

Many previous orchid taxonomic schemes are not only unhelpful but also cast doubts on classification as a branch of science. Construction of generic concepts that seek only to organize species for the purpose of identification is not science and is detrimental. It has led taxonomic botany to the brink of professional extinction. What good are botanists if they cannot or will not produce a predictive system that helps researchers in other branches of science to organize their research better? There is a critique of phylogenetic classification that states that producing a natural classification is also of no use if it is impossible to identify the genera so delimited, and we admit that identification is also an important function of classification. Much of our current taxonomic scheme in orchids are accidents of history: the feature that we use to separate *Laelia* and *Cattleya* is that which Lindley first enumerated, pollinia number; likewise, we still separate *Oncidium*, *Odontoglossum*, *Cochlioda*, and *Miltonia* on the same set of lip/column traits used by Lindley. This is not intended as a criticism of Lindley; he was an admirable taxonomist, and his contribution to botanical science was immense, but we can and must improve upon what Lindley accomplished.

If for the present we reorganize generic limits of orchids to reflect phylogeny based on DNA studies while the much longer studies of other forms of data are being advanced, then we are likely to learn which traits are reliable indicators of relationships and be able to use these for the purposes of identifying and organizing species and genera. It is quite clear that *Orchis s.l.* is not a useful concept because it is composed of three unrelated groups of species (Pridgeon *et al.* 1997; Bateman *et al.* 1997; Cozzolino *et al.* 1998), and from this we can infer that the trivial floral traits that previous authors used to recognize this concept of *Orchis* are unreliable. At present, we know that chemistry, cytology, and crossability are not reflected by being a member of *Orchis s.l.*, so the interim step of splitting the genus and placing the groups of species not related to the type species of *Orchis* into *Anacamptis* and *Neotinea* (the genera to which they are clearly related) makes sense, although it removes the ability to use those floral traits to identify some of the genera of subtribe Orchidinae. Future studies can now be focused on revealing which other characters can be used to identify reliably the natural groups that Bateman *et al.* (1997) have recognized. This change has made identification of the expanded concepts of *Anacamptis* and *Neotinea* more difficult, partly because we are so used to focusing on only a small portion of the variation present in these plants to identify them and now need to expand our perception to include more of what we see as well as those categories of information that the naked eye cannot perceive. *Anacamptis* and *Neotinea* were, of course, easily recognized in the past, partly because they contained only a single species each, so that species characters and generic characters were the same.

At the same time that this change causes us to re-think the traits that we will use to recognize the expanded concepts of *Anacamptis* and *Neotinea*, we must console ourselves that while we have made our work more difficult and challenging, we have provided the rest of society with a more useful classification (how many people outside the botanical community could distinguish the species of *Orchis*, *Dactylorhiza*, and *Traunsteinera* before *Orchis* was split?); their use of these new generic concepts will produce a better understanding of evolutionary patterns and make the studies of other scientists more robust. If a researcher is studying the species formerly known as *Orchis ustulata* (now *Neotinea ustulata*) and comparing it only to other species of *Orchis s.l.*, then he or she would not come to the right conclusions about the distribution of the aspects being studied, and the interpretation of empirical results would be erroneous. Because we can now produce a better and more predictive classification, we should do so and at the same time continue the studies that will eventually help us to determine those characteristics that permit us to identify the genus of each species easily and accurately.

Lastly, generic classification should be based on phylogenetic studies, which ideally should include both molecular data as well as morphology (if the latter has been studied well enough to contribute), so that it is not subject to opinion. Cladistic or phylogenetic classification (the two terms are synonyms) has rules to its application, whereas other types of classification are based on intuition, opinion, and accidents of history. Classification can follow directly upon the production of a robust phylogenetic estimate (i.e. one based on evidence), whereas each new generation of taxonomists applying non-cladistic (sometimes referred to as 'evolutionary') principles has seen fit to reorganize a substantial number of generic concepts. For example, just because someone can split *Oncidium* into *Odontoglossum*, *Miltonia*, and *Cochlioda* does not mean that someone else cannot split them further into *Cyrtochilum*, *Trigonochilum*, *Miltonioides*, and *Anneliesia*. How do we decide which if any of these taxonomies is accurate and serves the needs of research and horticulture better? If instead we conduct phylogenetic studies of morphology, anatomy, and DNA sequences, then we can instead establish a taxonomy that is empirically based and in which opinion, intuition, and historical accident have no role to play.

There will undoubtedly be people who are displeased by having to re-learn the names of their favourite genera and species because of changes required to create natural units (i.e. which are monophyletic, all more closely related to each other than members of other genera), but if they are unhappy about such changes it is up to them to disprove the accuracy of the evidence from these other studies rather than to argue that their classification is superior simply because they like it better. Classifications based on evidence can be reversed if the results can be falsified, but this is not true of those based on opinion alone. The latter is not science, and we reject such an approach completely. Orchid taxonomy has for far too long been formulated by 'experts' who intuit and then offer up their opinion in the form of a classification. This time has come to end this approach. Future work on orchid classification has entered the realm of empirical study.

DNA sequences and phylogenetics

It is one thing to agree that robust (i.e. well-supported) phylogenetic estimates should be the basis of classification at all taxonomic levels and another to believe that DNA sequences are superior to all other forms of data. The problems associated with interpretation of many, perhaps most, morphological data are discussed above, but the weakness of morphological hypotheses also stems from their paucity. It is the general rule that substantially greater numbers of characters than taxa are needed to produce a robust phylogenetic study. Relative to the numbers of orchid species and genera that we wish to assess, this is unlikely ever to occur with morphological and other non-molecular data. However, all plants contain astronomical amounts of DNA characters, so it becomes a simple and technical matter to sequence more base pairs of DNA and generate enough characters to produce a robust phylogenetic tree. Robustness can be assessed in several ways, but the most frequently used are Bremer support (Bremer 1988), which is also known as decay of parsimony (Donoghue *et al.* 1992), the bootstrap (Felsenstein 1985), and the jackknife (Farris *et al.* 1996). The latter two are re-sampling techniques that remove characters (the jackknife) and replace them with duplicates of others randomly selected (the bootstrap) and then re-run the analysis. Presence of a group, for example a genus, in a high percentage of these re-runs is indicative of clear patterns in the data showing that such a grouping of species is valid (with respect to this evidence it is a clear pattern). Bremer support simply asks the question of how many additional changes are required before a group disappears and its component species begin to associate with others in the matrix; the greater the number of steps, the clearer the evidence that the group is natural. Thus, if we find that we lack clear evidence, for example, that *Cochlioda* is embedded within *Oncidium*, then we simply collect another DNA sequence region and re-run the analysis. With the advent of automated sequencing, it is both feasible and reasonably inexpensive to produce large amounts of DNA sequence data at any taxonomic level.

The regions collected in phylogenetic studies should ideally include both nuclear and plastid sequences, although practically speaking this is unnecessary because sequences produced from these two regions have been shown to be highly congruent (Soltis *et al.* 1997, 1998; Chase and Cox 1998). Several orchid studies not yet published (the published abstract of Whitten *et al.* 1996, for example) have obtained the same result. Thus problems of incongruence of different molecular matrices do not have to be addressed, and nearly all of the comparative studies of separate molecular and morphological analyses also indicate basic congruence (Chase *et al.* 1993; Sheahan and Chase 1996; Nandi *et al.* 1998; Rudall *et al.* 1998) and synergism in that the combined matrices perform better (i.e. have higher bootstrap percentages) than the individual results. Thus, whereas collecting theoretically independent data sets is desirable, it has not been demonstrated to be required. It is preferable to have 5000 base pairs of plastid sequence rather then 2000 split equally between plastid and nuclear regions (all, of course, with the same level of variability so that similar things are compared). The larger matrix will provide better support and therefore be more accurate.

DNA sequences are of course not the only form of molecular data. Comparative restriction site mapping of chloroplast genomes has also been used in orchids (Chase and Palmer 1992, 1997; Yukawa *et al.* 1993; Yukawa *et al.* 1996), but compared to many other families of flowering plants (e.g. Asteraceae, the work of Jansen and others; Jansen *et al.* 1992) these studies will never be particularly useful. This is because restriction site studies depend on consistency in size of the restriction fragments as size is the basis for determining homology. Whereas it is clear that a guanine and cytosine residue at base position 1236 of two *rbcL* gene sequences are homologous, it is much more difficult to make this decision about restriction fragments separately on an agarose gel. Both Chase and Palmer (1992) and Yukawa *et al.* (1993) reported such difficulties, and some have abandoned this approach as a reasonable source of phylogenetic data for orchids. Biochemical data suffer from similar problems of interpretation as morphological data, and they are not as information-rich as the DNA sequences that code for them. Furthermore, DNA sequences are easily exchanged and understood between workers in distant laboratories and are transferable electronically (including the raw data themselves as produced by automated sequencers). This ease of transfer and communication is not true of restriction site and biochemical data.

DNA sequence data are the perfect phylogenetic data: their homology is relatively easily assessed, they are abundant (many more variable base pairs of sequence than taxa), and they are easily and accurately communicated. If every genus in a subtribe can be demonstrated to have bootstrap of 90–100% (Whitten *et al.* 1996), then it is difficult to imagine how anyone could argue that species in it should be transferred to another genus; such arguments, frequent if not omnipresent in orchid taxonomic history, can now be avoided, so that we can worry about bigger evolutionary issues.

Cladistics and parsimony

The basic operating principle of cladistics is the application of the tenet of parsimony: accept the simplest hypothesis that is consistent with the evidence. With any data matrix, the best optimal hypothesis is the one in which the length of the tree (each step being a hypothesized transformation in one character) is minimized. Cladists are concerned about finding the shortest tree, but it is generally assumed, even by cladists, that evolution, with respect to one character, is not likely to be parsimonious. To say that we trust the principle of parsimony as a tool in phylogenetics does not mean that we believe evolution is guided by that same principle. When applying parsimony to a suite of characters, we assume that we will find, in fact that we could prove, that some have evolved in a complicated and unparsimonious manner. A series of criticisms have been raised against what is perceived to be the monolith of cladistics, from the fact that evolution must proceed parsimoniously to the belief that speciation must be dichotomous and that the Linnean hierarchical system and cladistic classification are inherently incompatible (Cronquist 1987; Mayr 1988; Brummitt 1997), but none of these criticisms is accurate. Whereas it is true that the output of phylogenetic analysis software is dichotomous, there is no timeframe associated with a cladogram, so it is possible that several consecutive dichotomies took place more or less simultaneously. For example, if we are looking at events that took place in the early Cretaceous, then dichotomies that took place more than one hundred years apart would appear to us today to have been 'simultaneous'.

The cladogram is a framework on which to organize patterns, but it is clearly unable to impart certain kinds of information. Furthermore, just because two species are shown to be sister taxa, it does not imply that their common ancestor disappeared when one of the two appeared. Whether Hennig said this or not does not matter. Wagner's (1980) Groundplan-divergence Method is equally well viewed as founder of the lineage that became cladism, and he explicitly placed some species as progenitors to others. This was carefully qualified to mean that with the data analysed a given species could be viewed as directly ancestral to another, but this assumption is entirely limited to the data at hand, not all possible data that we could eventually collect. Certainly from the perspective of molecular change when a new species evolves, the two lineages that remain, parent and descendent, begin a process of divergence that results in the disappearance of the progenitor when those individuals die. Modern cladistic thinking is not the same as that of either Wagner or Hennig, and rejection of cladistic thinking should not be based simply on what any single cladist, living or dead, is projected to have believed. Darwin's theor-

ies have been significantly reinterpreted by evolutionary biologists in light of modern concepts of genetics, none of which existed in Darwin's lifetime.

Leaving aside most of these controversies (but see Freudenstein 1998 for a lucid explanation of the faults in the argument of Brummitt 1997), we can then turn to parsimony as the best criterion in producing accurate phylogenetic estimates. The term 'phylogeny' should not be interpreted as the output of a parsimony analysis, because no matter how robust the resulting tree is, it is only an estimate of phylogeny and never can be the 'true phylogeny' of that set of organisms. Having produced the cladogram we still do not know the phylogeny because a phylogeny entails knowing more information than can be inferred from the cladogram; the further away in time from the actual events, the less we can determine about them. Parsimony is only one of the commonly used tree-building algorithms; the others are neighbor-joining (NJ; Saitou and Nei 1987), unweighted pair-group method analysis (UPGMA) and maximum likelihood (ML; Felsenstein 1993). Several studies and simulations have investigated how reliable these methods are, and parsimony fares reasonably well (Hillis 1996). Nevertheless, if patterns within a data matrix are clear, all methods, including the one proven to be the most unreliable (UPGMA), find them. The problem of analysis is clearly not a matter of method of analysis but rather one of data quality. If there is sufficient high-quality information, then finding monophyletic groups becomes easy, regardless of the size of the matrix. If there are insufficient or poor-quality data, then the results are not reliable, regardless of the method of analysis. In the end, if the decision about the reliability of a set of relationships is based solely on faith in a particular algorithm's reliability, then a great deal of scepticism is warranted.

The biggest problem in phylogenetics has nothing to do with the principle of parsimony or method of analysis. It is simply the matter of data quality, and to a large extent in molecular systematics quality is synonymous with quantity. Of course, it is possible that some sequence regions are not of such high quality as others, but it is generally true that the larger the matrix, the more robust the result, not the reverse. DNA sequences are not entirely unlike all other forms of data, except that we know that base positions (if they vary at all) may vary many times. Our main objective in DNA studies is to discover accurately how many times each base position has changed, and the more homoplasy (reversals and convergences) we uncover the more accurate our tree. Homoplasy is evidence of relationships, not noise to be discounted or eliminated; it is as worthy of explanation as homology itself (Siddall and Kluge 1997).

Since most methods of analysis detect the same pattern, it would appear that there should not be a favoured one, but plant systematists clearly prefer parsimony over the others. This is due to several factors. The first is that the assumptions of parsimony are well documented, both theoretically and empirically (Hillis 1996), and parsimony finds the 'known' relationships in simulations as well as other methods (Hillis 1996). Secondly, parsimony does not add anything to the raw data themselves (e.g. corrections for multiple hits as in neighbour-joining), so after analysis it is possible to ask questions about patterns of change at specific sites because parsimony does not boil down variation to a percentage; it retains the particulate nature of each substitution at each position. Thirdly, parsimony assumes no model of molecular evolution nor deals in probabilities of change; rather it simply takes the distribution of variation across the matrix and minimizes the number of hypothesized steps (changes) to account for this distribution. Thus it is possible after analysis to ask questions about patterns of change in the sequences, but if one incorporates a model of evolution into the analysis such procedures become circular (as with maximum likelihood). If molecular evolution were completely understood, we could use this information to extract more of the weak signal present in each gene; however, we do not understand these patterns well (Savolainen et al., in press). Hence a maximum likelihood approach to finding the optimal topology can only obscure the actual underlying patterns rather than enhancing our understanding of them. If the simplest method with the fewest assumptions can be demonstrated to be as robust as more complicated methods, then there is no reason to favour these other methods over simple parsimony.

To detect all of the homoplasy (i.e. change), we have two avenues: adding taxa and adding data, both of which are important (Hillis 1996). If we have enough data of sufficient quality and have sampled taxa reasonably well, method of analysis and size of matrix are of little concern. Extremely large matrices (several hundred), both in numbers of taxa and data, now exist, and they are not difficult or time-consuming to analyse with conventional methods (Soltis et al. 1997, 1998; Chase and Cox 1998). It is safe to conclude that sufficient data will be available to address generic relationships of orchids, so we will have at hand robust phylogenetic trees upon which to base revised generic limits. Studies of generic boundaries should make every attempt to include as many species as there is morphological variation in the group under study without regard to worries about matrix size. In fact, large matrices should be encouraged because they have been shown to be more accurate, and analyses of large matrices are now posssible. Then the problem becomes one of how to take a cladogram and derive from it a classification.

If a robust estimate of relationships is available, then turning this into a classification can present some problems. The first rule of cladistic classification is, of course, that all taxa named must be monophyletic. This rule has been vociferously criticized (Sosef 1997; Brummitt 1997), but these objections stem from confusion and lack of understanding of the basic difference between theoretical discussion and the abstraction of a cladogram or any other estimate of relationships. Whereas we can discuss hypothetical ancestors, we cannot ever name them as such. Certainly they existed; their descendants are proof enough of that, but we can never know them.

Cladograms are elaborate hypotheses of relationships and character evolution, but they are not a phylogeny, and certain aspects of evolution will always be entirely speculative. The characteristics of an ancestor and how it should be named are entirely abstract and will continue to change as evidence accumulates. In the end, even after we have studied hundreds of traits and have millions of base pairs of DNA sequence data, discussion of ancestors and what they looked will be no more concrete than today. Although a potentially entertaining exercise, botanists' time would be better spent in more productive activity than defending the concept of paraphyly as necessary to application of Linnean system of nomenclature. There is no practical or theoretical inconsistency in the application of monophyly to naming of taxonomic categories (Freudenstein 1998, and many others).

Species can be paraphyletic to other species, which is sufficient to allow evolution to take place, but we cannot deduce this from a cladogram. The naming of supra-specific categories is a separate matter which has nothing to do with evolution. Naming is entirely an abstraction and can be made to fit whatever criteria we wish. If those who wish to use paraphyletic taxa can develop a system of rules for naming similar to those of Backlund and Bremer (1998) for the application of monophyly, then we might be willing to consider this an option. At present there are no rules for the circumstances in which naming a grade is appropriate, and we suspect that those who wish to recognize paraphyletic taxa are simply indulging in the maintenance of categories to which they have some personal attachment. The great majority of paraphyletic taxa are based on single trivial characters that have become known as the 'defining' or 'key' characters for that taxon, in spite of the fact that nearly all other characters are shared throughout the group. In practice, use of paraphyletic taxa puts us back in the intellectual Dark Ages. If instead we insist on monophyly, then classification becomes scientific and can be the basis of hypotheses that can either be supported or refuted by the production of evidence and in which personal opinion counts for little.

The decision about which levels in a hierarchical set of relationships should be named involves a number of choices that might be said to involve too much opinion and too little rigour, but this is an overly cynical attitude that contributes nothing. No one will argue that applying monophyly will always be simple, but the alternative is to let whim and historical accident determine names. The former may present us with unpopular choices, but only the latter is scientifically sound. Popularity is not a valid concern of science, but accuracy and stability are. Beyond the application of monophyly, the application of names should follow from a consideration of: (i) stability, (ii) information content, (iii) ease of use, and (iv) patterns of support (Lidén et al. 1997; Backlund and Bremer 1998). The first of these considerations is the reason why others (e.g. Brummitt 1997) have criticized the strict application of monophyly. In fact it is those who defend paraphyly who commit us to chaos, not the reverse, because they advocate a system that depends upon whim (masquerading as 'authority') and accidents of history. It is only through the application of monophyly and the other secondary considerations listed above that we can obtain both stability and predictability, and these are the attributes that nomenclatural practice must produce if it is to be considered scientific by other areas of science. Cladistic classification requires a degree of rigour that some currently practising taxonomists resent; this is equivalent

to saying that 'it was not done that way when I learned to do taxonomy, so why is it necessary now?' In short, cladistic classification introduces rigour, evidence, and stability into classification. This is sufficient justification for its application to orchid taxonomy. Our intent in *Genera Orchidacearum* is to standardize and stabilize orchid generic taxonomy as much as possible, and we will apply the tenets of cladistic taxonomy. This will result in some difficult decisions, but in the end everyone will benefit from this program.

Generic delimitation within Orchidaceae has been an erratic process. Some groups have been well studied (e.g. Catasetinae, Cypripedioideae, Diseae, Orchidinae, and Stanhopeinae), whereas others have been saddled by a taxonomic scheme developed when few species of large and diverse groups were available for comparison. Hence, we would predict that the classification of some groups of orchids would need little alteration, whereas others require massive reorganization. Oncidiinae (Maxillarieae; Epidendroideae) and Laeliinae (Epidendreae; Epidendroideae) are prime examples of the latter. The taxonomic scheme as laid down by Lindley used trivial characters that have been largely unmodified to date. Every recent author working in these groups has expressed the same opinion: that the current system of generic limits is unworkable (Chase 1986), but the solution to these problems is not clear because too little work has been completed. Analysis of DNA sequences provides a quick and reliable solution to the lack of data provided by other disciplines. What is quite clear is that many groups of orchids will require little or no generic modification as a result of DNA studies to produce a cladistically acceptable taxonomy.

For example, a study of Cypripedioideae (Cox et al. 1997a) provided complete validation of the current generic scheme (Albert and Chase 1992), as well as the infrageneric classification of McCook (1989) for *Phragmipedium* and Cribb (1987) for *Paphiopedilum*. It refuted the highly artificial system proposed by Braem (1988). No generic alterations were necessary, so producing monophyletic groups and preserving stability were easily combined. Likewise, DNA studies of Stanhopeinae (Whitten et al. 1996) and Catasetinae (Pridgeon and Chase 1998) required little alteration of currently recognized genera to meet the requirements of cladistic classification, but both these subtribes have been well studied for many years, and we would have anticipated little modification. Diseae are another well-studied group (Linder and Kurzweil 1994), and all of the generic realignments proposed by these authors have been supported by analysis of ITS rDNA sequences (Douzery *et al.*, in press). In other cases, conflicts between floral traits on one side and vegetative, cytological, and chemical characters on the other have been present in the literature for many years, but there was an impasse about how to proceed. Making DNA sequences available has led to slight alteration because they identified the non-floral traits as the reliable ones; hence Pridgeon *et al.* (1997) and Bateman *et al.* (1997) split *Orchis* into the three DNA-based groups reflected by the vegetative, cytological, and chemical characters. In this case, two of the clades identified were related to pre-existing genera, so a number of species were each transferred into *Anacamptis* and *Neotinea* rather than having to describe new genera. A few other relatively minor changes were also required: subsuming *Aceras* into *Orchis s.s.*, *Nigritella* into *Gymnadenia*, and *Coeloglossum* into *Dactylorhiza*.

These studies stand in stark contrast to those now underway on Oncidiinae, Pleurothallidinae, and Habenariinae, in which substantial reorganization of generic limits will be required. In the case of Pleurothallidinae, *Pleurothallis* is polyphyletic, and some new genera will probably be required as well as the subsuming of some well-known genera into others, but it is too early to include more about these results. Likewise, in Oncidiinae several genera will be subsumed into *Oncidium* (most of *Odontoglossum*, for example), whereas several groups of *Oncidium* will have to be split off because they are distantly related to the type species, *O. altissimum* (see Chase and Palmer 1992, for results of restriction site analysis that are presently being confirmed with DNA sequence data (Williams, Chase, and Whitten, unpublished)).

DNA sequence studies have proven their reliability in many groups of angiosperms, as well as in several well-studied groups of orchids, so we are now poised to carry out the most thorough set of molecular systematic studies of generic relationships that has ever been attempted in any family of angiosperms. These projects will undoubtedly usher in a period of relative taxonomic instability, after which only minor adjustments should be required. The benefits of such modifications should be apparent: these studies will take the debate about generic delimitation out of the realm of opinion and intuition and into that of evidence. No longer will each generation of orchid taxonomists have the option of changing their predecessor's system at whim, but, if changes are to be made, firm evidence demonstrating the reasons will need to be gathered and presented. Such approaches have long been needed, and now we are set to embark upon the modernizing of orchid biology and propel the family back into the forefront of modern evolutionary studies in all fields of scientific inquiry. This is what a stable and predictive nomenclature holds for groups that are of great evolutionary interest, among which the orchids should rank first.

MARK W. CHASE

A

B

C

D

Plate 1. **A.** Bulbophyllums growing epiphytically on a fallen branch, Sabah (Photo: P. Cribb); **B.** *Prosthechea* (= *Encyclia*) *cochleata*, Mexico (Photo: P. Cribb); **C.** *Dendrobium speciosum* growing lithophytically, Australia (Photo: P. Cribb); **D.** *Caladenia flava*, a colony-forming terrestrial species, Australia (Photo: A. M. Pridgeon).

A

B

C

D

Plate 2. **A.** *Disa uniflora* growing along a stream, South Africa (Photo: A. M. Pridgeon); **B.** *Vanilla pompona*, a climbing vine, cultivated, Germany (Photo: P. Cribb); **C.** *Rhizanthella gardneri*, a subterranean species, Australia (Photo: A. George); **D.** Root of *Phalaenopsis*, cultivated, USA (Photo: A. M. Pridgeon).

A

B

C

D

Plate 3. **A.** *Taeniophyllum trachypus*, a leafless species with photosynthetic roots, New Caledonia (Photo: A. M. Pridgeon). **B.** *Bulbophyllum lasiochilum*, showing sympodial growth with leaves and pseudobulbs, cultivated, Indonesia (Photo: J. Comber); **C.** *Trichoglottis smithii*, showing monopodial growth, Sabah (Photo: P. Cribb); **D.** *Corymborkis veratrifolia* with plicate leaves, cultivated, Bogor Botanical Gardens, Indonesia (Photo: P. Cribb).

A

B

C

D

Plate 4. **A.** *Cymbidium rectum*, with conduplicate leaves, Sabah (Photo: D. DuPuy); **B.** *Macodes sanderianum*, one of the so-called 'jewel orchids' (Photo: R. Mitchell), Solomon Islands; **C.** *Coelogyne pandurata*, with a raceme of flowers, cultivated, Royal Botanic Gardens, Kew (Photo: RBG Kew); **D.** *Oncidium splendidum*, with a panicle of flowers, cultivated, Guatemala (Photo: M. W. Chase).

A

B

C

Plate 5. **A.** *Lockhartia oerstedii*, with a cymoid
inflorescence, Mexico (Photo: P. Cribb); **B.** *Glomera
flammula*, with a more or less globose head of flowers,
Irian Jaya (Photo: P. Cribb); **C.** *Bulbophyllum lepidum*,
showing a subumbellate inflorescence, cultivated,
Bogor Botanical Gardens, Indonesia (Photo: P. Cribb);
D. *Paphiopedilum godefroya*e, transection of unilocular
ovary with parietal placentation, cultivated, Royal
Botanic Gardens, Kew (Photo: RBG Kew).

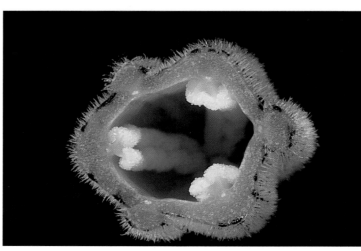

D

A

B

C

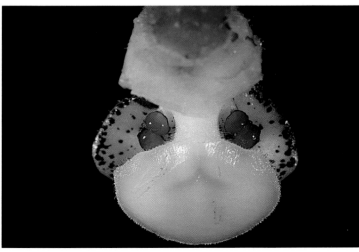

D

Plate 6. **A.** *Cattleya amethystoglossa*, cultivated, USA (Photo: A. M. Pridgeon); **B.** *Cymbidium lowianum*, cultivated, China (Photo: P. Cribb); **C.** *Aerangis clavigera*, with nectar-filled spur, cultivated, USA (Photo: G. Kennedy); **D.** Column of *Paphiopedilum godefroyae*, with two anthers flanking broad shield-shaped staminode (Photo: J. Asher).

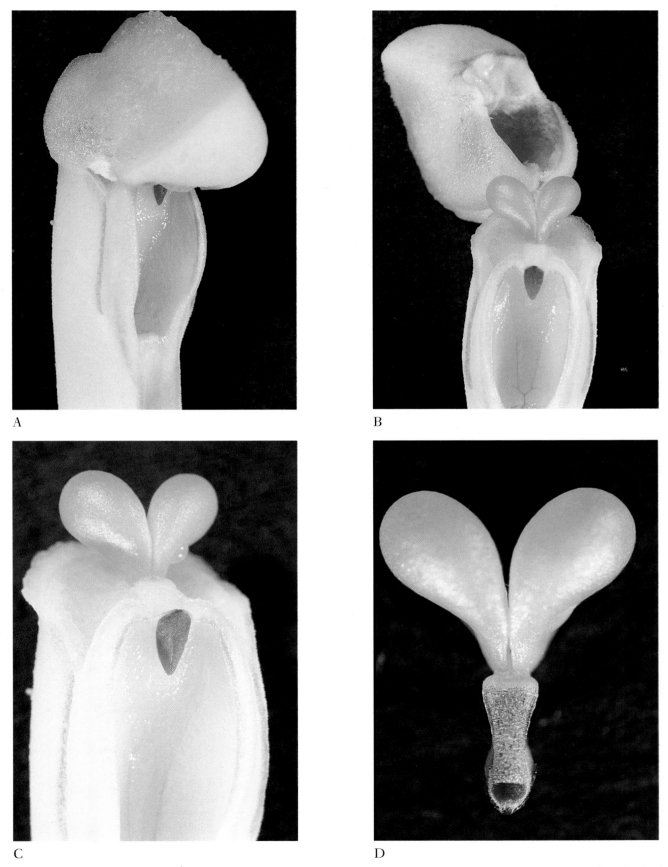

Plate 7. *Aspasia principissa* column (Photos: A. M. Pridgeon). **A.** Anther cap above, intact, with stigmatic cavity below; **B.** Anther cap tilted upward to reveal pollinarium; **C.** Pollinarium intact; **D.** Pollinarium removed, showing two pollinia, 'stipe' (tegula), and brown viscidium.

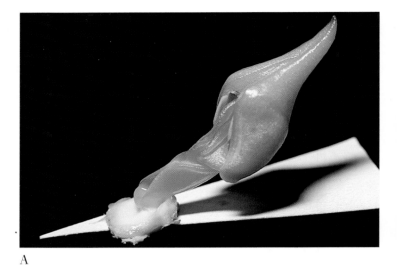

A

B

C

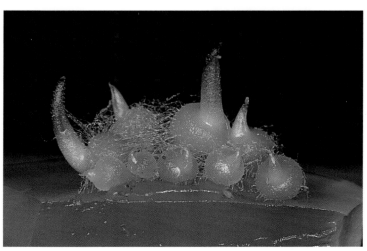

D

Plate 8. **A.** *Catasetum* pollinarium with anther cap intact (Photo: A. M. Pridgeon). **B.** *Dimorphorchis rossii* showing dimorphic flowers, those above scented and differently colored from the unscented flowers below (Photo: RBG Kew); **C.** Immature capsule of *Brassavola nodosa* sectioned to reveal thousands of developing seeds (Photo: A. M. Pridgeon); **D.** Protocorms of *Ophrys apifera* at four weeks (Photo: RBG Kew).

A

B

D

C

Plate 9. **A.** *Neuwiedia veratrifolia*, cultivated, Sabah (Photo: P. Cribb); **B.** *Apostasia wallichii*, Indonesia (Photo: J. Comber); **C.** *Cypripedium irapeanum*, Mexico (Photo: P. Cribb); **D.** *Cypripedium reginae*, USA (Photo: C. A. Luer).

Plate 10. **A.** *Cypripedium flavum*, China; **B.** *Cypripedium calceolus*, Switzerland; **C.** *Cypripedium parviflorum* var. *pubescens*, cultivated, Royal Botanic Gardens, Kew; **D.** *Cypripedium tibeticum*, China (Photos: P. Cribb).

A

B

C

D

Plate 11. **A.** *Cypripedium arietinum*, USA (Photo: P. Cribb); **B.** *Cypripedium formosanum*, cultivated, Royal Botanic Gardens, Kew (Photo: P. Cribb); **C.** *Cypripedium acaule*, USA (Photo: C. A. Luer); **D.** *Cypripedium guttatum*, China (Photo: P. Cribb).

A

B

C

D

Plate 12. **A**. *Cypripedium debile*, cultivated (Photo:
P. H. Davies); **B**. *Cypripedium margaritaceum*, China
(Photo: P. Cribb); **C. D.** *Selenipedium chica*, Panama
(Photos: K. Dressler).

A

B

C

D

Plate 13. **A.** *Paphiopedilum armeniacum*, China
(Photo: P. Cribb); **B.** *Paphiopedilum micranthum*,
cultivated, Royal Botanic Gardens, Kew (Photo:
P. Cribb); **C.** *Paphiopedilum bellatulum*, China (Photo:
P. Cribb); **D.** *Paphiopedilum liemianum*, cultivated,
Indonesia (Photo: P. Cribb).

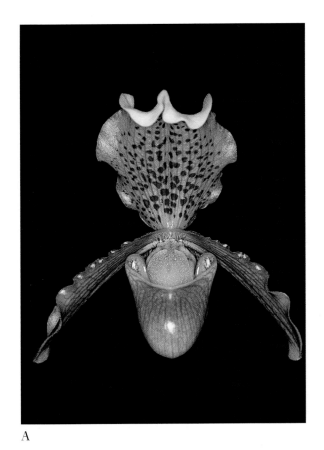

A

C

B

D

Plate 14. **A.** *Paphiopedilum insigne*, cultivated (Photo: M. W. Chase); **B.** *Paphiopedilum rothschildianum*, cultivated, Royal Botanic Gardens, Kew (Photo: RBG Kew); **C.** *Paphiopedilum sanderianum*, cultivated, Royal Botanic Gardens, Kew (Photo: RBG Kew); **D.** *Paphiopedilum lowii*, cultivated, Indonesia (Photo: P. Cribb).

A

C

B

D

Plate 15.   **A.** *Paphiopedilum violascens*, Irian Jaya
(Photos: P. Cribb); **B.** *Phragmipedium besseae*,
cultivated, Eric Young Orchid Foundation, Jersey
(Photo: P. Cribb); **C.** *Phragmipedium lindleyanum*,
cultivated, Royal Botanic Gardens, Kew (Photo: RBG
Kew); **D.** *Phragmipedium longifolium*, cultivated, Royal
Botanic Gardens, Kew (Photo: RBG Kew).

A

B

C

D

Plate 16. **A.** *Phragmipedium pearcei*, cultivated, Royal Botanic Gardens, Kew (Photo: RBG Kew); **B.** *Phragmipedium wallisii*, cultivated, Germany (Photo: P. Cribb); **C.** *Phragmipedium lindenii*, cultivated, Royal Botanic Gardens, Kew (Photo: RBG Kew); **D.** *Mexipedium xerophyticum*, Mexico (Photo: E. Hágsater).

CLASSIFICATION

ORCHIDACEAE

Derivation of name

From the Greek '*orchis*', a testicle, in allusion to the shape of the tubers of European orchids in genera such as *Orchis*, *Ophrys*, and *Serapias*.

Description

Perennial, terrestrial, epiphytic or lithophytic *herbs* or rarely scrambling climbers, sometimes saprophytic. Mycorrhizal fungi in the roots and often elsewhere. Growth either monopodial or more commonly sympodial with rhizomes, tubers, or rootstocks. *Roots* adventitious, often aerial, sometimes assimilatory, with velamen tissue in epiphytic and some terrestrial species. *Stems* usually leafy, the one or more internodes often swollen in sympodial species to form pseudobulbs. *Leaves* glabrous or occasionally hairy; usually entire, rarely palmate or lobed; alternate or occasionally opposite, often distichous, with plicate or convolute vernation; membranaceous to coriaceous, often terete or canaliculate, reduced to scale-like bracts in some species; almost always with a basal sheath, sometimes articulated at the base of the lamina or with a false petiole. *Inflorescences* erect to pendent; spicate, racemose or paniculate; one- to many-flowered; basal, lateral or terminal, the flowers rarely secund, distichous or subumbellate. *Flowers* small to large, often showy; zygomorphic but appearing almost actinomorphic in some cases; bisexual or rarely unisexual and polymorphic; sessile or variously pedicellate, most often twisted through 180°, occasionally not twisted or twisted through 360°. *Sepals* three, usually free but sometimes variously connate, the dorsal sepal often dissimilar to the lateral sepals, the lateral sepals sometimes adnate to the column foot to form a saccate, conical or spur-like mentum. *Petals* three, free or rarely adnate in part to the sepals, the medial petal distinguished from the others as the lip or labellum. *Lip* entire, variously lobed or two- or three-partite, often ornamented with calli, ridges, hair cushions or crests, with or without a basal spur or nectary, margins entire to laciniate. *Column* short to long, formed from stylar and filamentous tissue, with or without a basal foot, occasionally winged or with lobes or arms apically or ventrally; fertile anthers one (rarely two or three), terminal or incumbent, cap-like or dehiscing by longitudinal slits; pollen shed as monads or tetrads, often agglutinated into discrete masses called pollinia; pollinia mealy, waxy or horny, soft to hard, sectile or not, 2, 4, 6 or 8, sessile or attached by caudicles or stipites (tegula or hamulus) to one or two sticky viscidia forming a pollinarium; stigma 3-lobed, the midlobe often modified to form a rostellum, the other lobes either sunken on the ventral surface of the column behind the anther or with two lobes porrect. *Ovary* inferior, unilocular with parietal placentation or rarely trilocular with axile placentation. *Fruit* a capsule, usually opening laterally by three or six slits, rarely baccate; seeds numerous, dust-like, rarely with hard seed coats, lacking endosperm, sometimes markedly winged.

Orchidaceae comprise one of the largest families of flowering plants, with an estimated 800 genera and conservatively almost 20 000 species (Atwood 1986; Dressler 1993). They are distributed throughout all continents except Antarctica but are most numerous in the humid tropics and subtropics (Fig. IX.1). Hundreds of species are grown throughout the world as ornamentals. Two or possibly three species of *Vanilla* Miller are grown commercially to produce the flavouring vanillin. Tubers of several species are used for food or aphrodisiacs; various plant parts of other species are used in folk medicine, particularly in India, China, the Far East, Africa, and Madagascar (Lawler 1984).

PHILLIP J. CRIBB

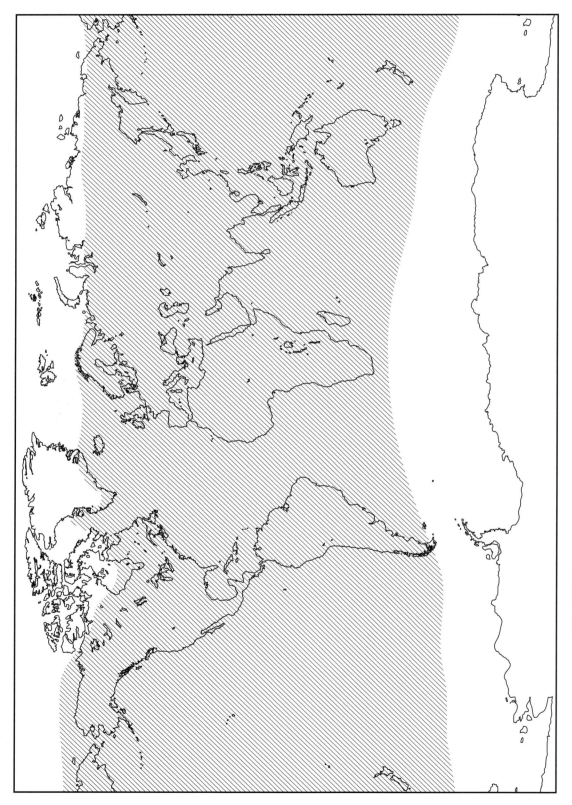

Fig. IX.1. Worldwide distribution of Orchidaceae (redrawn from Dressler 1981).

Artificial key to the subfamilies

1. Pollen mealy or paste-like, not forming coherent pollinia (except some *Phragmipedium* species)
   2. Fertile anthers normally 2 or 3; leaves with sheathing bases
      3. Fertile anthers oblong to ovate; staminodium shield-like;
         lip saccate ....**Cypripedioideae**
      3. Fertile anthers lance-linear; staminodium finger-like; lip similar
         to petals ....**Apostasioideae**
   2. Fertile anther 1, incumbent; stem leaves without sheathing bases ....**Vanilloideae**
1. Pollen coherent, forming definite pollinia
   4. Anther erect or bent back
      5. Leaves convolute but not strongly plicate; roots often
         tuberous or fleshy ....**Orchidoideae**
      5. Leaves usually distinctly plicate; roots rarely
         fleshy .... **Epidendroideae [in part]**
   4. Anther distinctly incumbent .... **Epidendroideae [in part]**

ROBERT L. DRESSLER

APOSTASIOIDEAE

Apostasioideae Garay in *Bot. Mus. Leafl.* **19**(3): 86 (1960). Type: *Apostasia odorata* Blume.

Description

Small to large terrestrial *herbs. Rhizome* scaly, or absent. *Roots* usually aerial and stilt-like, penetrating the base of the lower leaves or rhizome-scales (when present) or both, usually adpressed against the lower part of the stem or rhizome for a short distance, terete, longitudinally grooved or ribbed, subterranean portion branched, sometimes woolly, swollen into nodular storage roots in *Apostasia. Stems* erect or ascending, branched or simple, terete to irregularly angular or ribbed. *Leaves* spirally arranged, convolute, usually plicate, non-articulate, linear, oblong to ovate-oblong, narrowly elliptic or ovate-elliptic, acuminate, glabrous or with scattered hairs, prominently nerved beneath, herbaceous to papyraceous. *Inflorescence* an erect, terminal, peduncled, simple or branched raceme; rachis glabrous or hirsute; floral bracts leafy, persistent, glabrous or hirsute; flowers spirally arranged. *Flowers* only slightly zygomorphic, small to medium-sized, white, yellowish, or deep yellow; pedicel resupinate or non-resupinate; *ovary* cylindrical or ellipsoid, 3-angled to terete, not twisted, 3-celled, with axile placentation. *Sepals* 3, free, midrib forming a thick ridge on the outside terminating in a subulate cusp below the incurved apical margin. *Petals* 2, free, sometimes attenuate towards the base, midrib forming a thick, fleshy, projecting keel on the outside separated from the underlying tissue by a constriction and terminating in a free subulate cusp below the incurved apical margin. *Lip* similar to sepals and petals, although sometimes broader, never modified. *Column* straight to strongly curved, formed by the fusion of the lower portion of the style, the base of the filaments, and the staminode (when present), smooth, glabrous; stamens 2 or 3, median (when present) alternipetalous, laterals epipetalous, filaments mostly apically partly free from the style, anthers with two equal or unequal thecae, each of them bilocular, dorsifixed or sub-basifixed, introrse; staminode (when present) median, alternipetalous; style cylindrical, fleshy; stigma terminal, rounded or pyramidal, sometimes 2- to 3-lobed, apex flattened; pollen grains as monads, powdery, with reticulate or perforate-reticulate sculpturing. *Capsule* thin-walled or fleshy, 3-locular, disintegrating or (in one species) opening loculicidally with three valves; seeds very numerous, ovoid to elliptic, usually on a shorter or longer stalk, apically provided with a minute appendage, which (in two species) is very long or thick, testa dark when ripe, usually alveolate or reticulate, sticky when dry. **(JW)**

Distribution

Two genera distributed in Sri Lanka, north-east India, Nepal, Bhutan, Myanmar (Burma), Thailand, Vietnam, Laos, Cambodia, southern China (including Hainan, Hong Kong), Japan (including Ryukyu Islands), Malaysia, Brunei, Indonesia, the Philippines, New Guinea, Solomon Islands, Vanuatu, and northern Australia (Queensland). **(JW)**

Artificial key to the genera (JW)

1. Anthers 3, free from each other, median one not staminodial. Inflorescences simple, erect. Filaments fused in the lower portion. Ovary and fruit long-ellipsoid to rounded, strongly contracted at apex. Leaves often rather wide ....1. **Neuwiedia**

1. Anthers 2, fused with one or both margins; median one of the inner whorl staminodial, or absent. Inflorescences usually branched, recurved or spreading, not erect. Filaments free from one another. Ovary and fruit narrowly cylindrical, not contracted at apex. Leaves usually narrow..............2. **Apostasia**

Anatomy

Stern *et al.* (1993) investigated the vegetative anatomy of *Apostasia* and *Neuwiedia*, updating the study made by Siebe (1903) and treating the data in a systematic way for the first time.

Leaf

Epidermal cells are rectangular in transection, mostly flat to rounded but domed to papillate near margins and larger veins. Cuticle is thin to very thin.

Stomata in leaves occur on both surfaces but are infrequent on the adaxial surface. Patterns of stomatal organization are basically tetracytic, but that of *Apostasia* differs from that of *Neuwiedia*. In the former the polar subsidiary cells parallel the long axis of the leaf, and in the latter they are perpendicular to it.

Trichomes are absent on both leaf surfaces of both genera.

Chlorenchyma is homogeneous. There is a single row of collateral vascular bundles, and sclerenchymatous bundle caps are strongly developed at both poles. Silica bodies occur in stegmata and are linearly arranged along the bundle fibres.

Stem

Epidermal cells are regular, the flat, outer walls being smooth in *Apostasia* and irregularly sculptured in *Neuwiedia*. Trichomes are absent except in *N. zollingeri* where thin-walled unicellular hairs are raised on multicellular buttresses of thin-walled epidermal cells. Stomata occur sporadically in *Neuwiedia*. Cortical cells are parenchymatous and relatively thick-walled. In *Apostasia* and *Neuwiedia* the cortex is bounded internally by a uniseriate endodermoid layer. In some species the ground tissue consists of a multiseriate layer of thick-walled cells surrounding a central core of thin-walled parenchyma cells; in other species the cortex merges imperceptibly with the pith without a sclerenchymatous layer. Vascular bundles are collateral and scattered throughout the ground tissue; there is both xyloic and phloic sclerenchyma, the latter more pronounced than the former. Stegmata bearing silica bodies are associated with peripheral vascular bundles only.

Root

Roots are of two kinds, terrestrial and aerial, generally similar in structure. Aerial roots arise from the stem, pierce the bases of the lower leaves, and seem to serve as props. Terrestrial roots are generally smaller in diameter than aerial roots and bear minute, stalked tubercles.

Velamen is uniseriate, the outer and radial walls thickened (often sloughed off in aerial roots). Exodermal cells have U-shaped thickenings. Cortical cells are parenchymatous and thin-walled. The endodermis is uniseriate with O-shaped thickenings in *Apostasia* and U-shaped thickenings in *Neuwiedia*. Pericycle is one to three cell layers wide with evenly thickened walls. The vascular cylinder is polyarch and surrounded by a multiseriate layer of very thick-walled sclerenchyma cells. Pith comprises angular, thick-walled parenchyma cells (thin-walled in terrestrial roots of *Neuwiedia*.

The tubercles of terrestrial roots of both genera bear multicellular, irregularly shaped papillae. Epidermal cells of these tubercles are very thin-walled without an obvious exodermis. Cortical cells are thin-walled, many filled with hyphal coils of mycorrhizal fungi. Endodermal and pericyclic cells are thin-walled. The vascular cylinder is tetrarch, embedded in a pith of thick-walled cells. Of special interest are the enlarged cortical cells containing fungal hyphae (which may be multinucleate) as well as the epidermis bearing stomata (Stern and Warcup 1994).

The general anatomy of *Apostasia* and *Neuwiedia* is not unique, most features occurring in other orchids. However, these genera are different from the diandrous and monandrous orchids because they have predominantly simple perforation plates in vessel members of roots, a derived feature that has not developed to any extent in other orchid taxa. Because of this, it would appear that, on an anatomical basis, Apostasioideae originated from a common ancestral stock with other orchids and diverged from them in vessel development and do not represent the primal stock from which other orchids evolved. It would not be possible to derive the other orchid groups with less specialized tracheary elements from the Apostasioideae with their highly evolved simple perforation plates (Judd *et al.* 1993). **(WS)**

Palynology

Pollen grains of both *Apostasia* and *Neuwiedia* are shed as monads and are monosulcate with reticulate sculpturing (Newton and Williams 1978; Schill

1978). The colpus of both genera is operculate (Schill 1978). Wall structure is tectate-columellate with no endexine and a lamellate intine (Schill 1978; Zavada 1990). **(AP)**

Phylogenetics

Dressler (1993) considered the subfamily Apostasioideae to be a sister group of the remaining orchids. Earlier, traditional classifications that included the apostasioids together with the cypripedioids are artificial. The two appear to be only distantly related, the latter being, contrary to popular belief, quite specialized in their own way. Most modern workers treat the apostasioids and cypripedioids as separate groups.

Burns-Balogh and Funk (1986) place *Neuwiedia* in a separate subfamily, despite the fact that in *Apostasia* and *Neuwiedia* similarities in seed, pollen, and anther morphology argue against this. The sculpturing of the pollen wall is similar in both genera and very different from that of the cypripedioids (Newton and Williams 1978).

Neuwiedia displays many unusual features, including a crustose testa (in some species), fleshy fruit (in some species), a three-locular ovary, and an abscission layer between the ovary and the perianth, which some authors suggest are primitive traits for the orchids (but see below). The flowers, however, have the basic orchid symmetry and are resupinate; there is also partial union of the three filaments with each other and with the style.

DNA sequence data place *Apostasia* and *Neuwiedia* together as a sister pair to the rest of Orchidaceae. They are well-supported as members of the family, so there seems little reason to separate them simply based on their abundant autapomorphies (Stern 1993; Stern and Warcup 1994). Although they add significant heterogeneity to the characters that define the orchids, this makes a much more evolutionarily interesting picture; the alternative—separate familial status—would require recognition of at least the slipper orchids as separate as well, and this could lead to the eventual recognition of several more families, for which there are no valid practical reasons. It is therefore preferable to maintain one large, somewhat heterogeneous family.

The features by which the apostasioids differ from other orchids make them a clearly derivative group, and although their androecial organization indicates a degree of primitiveness, they are not otherwise how we would imagine the common ancestor of the orchids. Fleshy fruits are not typical of any families of Asparagales or Liliales except for Asparagaceae and Convallariaceae, which are distantly related, so this is not a primitive trait. Likewise the abscission layer between the perianth and ovary is not found in any of the closest families, leaving only the crustose seed and tricarpellate ovary as potential plesiomorphies, and these are found in several other orchids as well as the outgroups. Whether these are primitive traits will need careful evaluation because one could imagine that they, too, are independently derived in *Neuwiedia*. Overall, these two genera are clearly orchid-like plants with many unusual or unique traits, but other than their androecia they bear little resemblance to anything that we would consider an ancestor of the orchid family. **(JW and MWC)**

1. NEUWIEDIA

Neuwiedia Blume in v.d. Hoeven and de Vriese, *Tijdschr. Natuurl. Gesch. Physiol.* **1**, 140 (1834) and *in Ann. Sc. Nat.* **II**, 2, 93 (1834); Pfitzer in Engler & Prantl, *Nat. Pfl. Fam.* **2**, 6, 80 (1888); Rolfe in *Orchid Rev.* **4**, 328 (1896); Holttum, *Rev. Fl. Malaya* **1**, 62 (1953); de Vogel in *Blumea* **17**, 320–35 (1969). Type species: *Neuwiedia veratrifolia* Blume.

Derivation of name

The generic name is dedicated to Prince Maximilian Alexander Philipp zu Wied-Neuwied (1782–1867).

Description (Plate 9A; Fig. 1.1)

Erect or ascending, glabrous to hairy *herbs*. *Rhizome* absent. Aerial *roots* glabrous to woolly. Stem usually simple, hairy or glabrous distally. *Leaves* initially crowded at the base, then becoming spaced out, the lowermost dying off, the uppermost grading into leafy bracts, blade papyraceous, linear, narrowly elliptic to ovate-oblong, mostly plicate, without a tubular-contracted apex. *Racemes* usually simple, rigid. *Flowers* resupinate, mostly containing white

Fig. 1.1. *Neuwiedia veratrifolia*. A. Habit; B. Flower; C. Sepal, front view; D. Sepal, back view; E. Petal, front view; F. Petal, back view; G. Column; H. Anthers; I. Ovary, transverse section; J. Fruit. Single bar = 1 mm, double bar = 1 cm. Drawn from *Lamb AL155/83* and Kew Spirit Collection no. 46660 by Judi Stone.

crystals in their tissue. Pedicel clearly distinct from the ovary. Torus oblique. Perianth rather nodding, little expanded. *Sepals* slightly convex, fleshy, midrib projecting a little way. Dorsal sepal symmetric. Lateral sepals somewhat asymmetric. *Petals* slightly convex, slightly asymmetric, herbaceous, fragile except for the fleshy midrib, lateral nerves flabellately-branching. *Lip* symmetric, folded, broader than petals. *Column* straight; stamens 3, filaments connate at base, adnate to the style for much of their length, anthers not clasping the style, free from each other, (basi-) dorsifixed, median usually shorter than laterals, loculi of median anther equal, loculi of lateral anthers unequal at base, the loculus facing the median anther being the shorter. *Ovary* ellipsoid, strongly contracted at apex, more or less triangular to rounded with 3 grooves in cross-section. *Capsule* distinctly accrescent, beaked; seeds with a distinct, thin-celled, pale funicle, usually reticulate, dark brown when ripe, with a single, minute, rarely very long hair-like or very broad apical appendage. **(JW)**

Distribution (Fig. 1.2)

The eight (possibly nine) species of *Neuwiedia* are distributed in S. China (Guangdong, Hainan, Hong Kong), Thailand, Vietnam, Malaysia, Singapore, Indonesia, Brunei, the Philippines, and Papua New Guinea, with one species extending east as far as the Solomon Islands and Vanuatu. The richest centre of diversity is found in Borneo, from where five species have been recorded, two of which are endemic. *Neuwiedia inae*, previously thought to be endemic to Borneo, has recently been recorded from Vietnam. A fruiting specimen of a possible ninth species has been collected in the Batu Islands off the west coast of Sumatra (de Vogel 1969), but its status remains uncertain, pending fresh flowering material. *Neuwiedia annamensis* and *N. siamensis* are endemic to Vietnam and Thailand respectively. *Neuwiedia veratrifolia* is the most widespread species, ranging from Peninsular Malaysia in the west and eastward to Vanuatu in the south-west Pacific. **(JW)**

Anatomy

See subfamily description.

Palynology

See subfamily description.

Cytogenetics

$2n = 96$, c. 144 (see Chapter VII).

Phytochemistry

Nothing is known of the phytochemistry of *Neuwiedia*. **(RG and NV)**

Phylogenetics

See subfamily treatment.

Ecology

Species of *Neuwiedia* thrive under conditions of high humidity and deep shade and are typical plants of the forest floor environment. They inhabit differing forest types and can be found from sea level to an elevation of around 1300 m but are most frequent at lower elevations or at up to around 800 m in the hills.

The widespread *N. veratrifolia* is recorded from a wide range of habitats and substrates including limestone, sandstone, shale, and ultramafics. Favoured habitats for the genus as a whole include: lowland and hill dipterocarp forest; coastal mixed dipterocarp forest on clay soil (e.g. *N. borneensis* in Brunei); hill and lower montane *Castanopsis*/dipterocarp ridge forest on sandstone and shale (e.g. *N. veratrifolia* in Brunei); *Agathis borneensis* and peat-swamp forest on waterlogged sandy, acid soils; and secondary regrowth. In Hong Kong *N. zollingeri* var. *singapureana* has been recorded in one locality growing among shrubs near streams at about 1400 m. On Manus Island, the largest of the Admiralty Islands in Papua New Guinea, *N. veratrifolia* can be found at sea level in well-drained *Calophyllum* forest. On Guadalcanal in the Solomon Islands, near the easternmost limit of its range, *N. veratrifolia* forms extensive colonies in deep leaf litter between scrubby *Pandanus* on damp but well-drained calcareous soil.

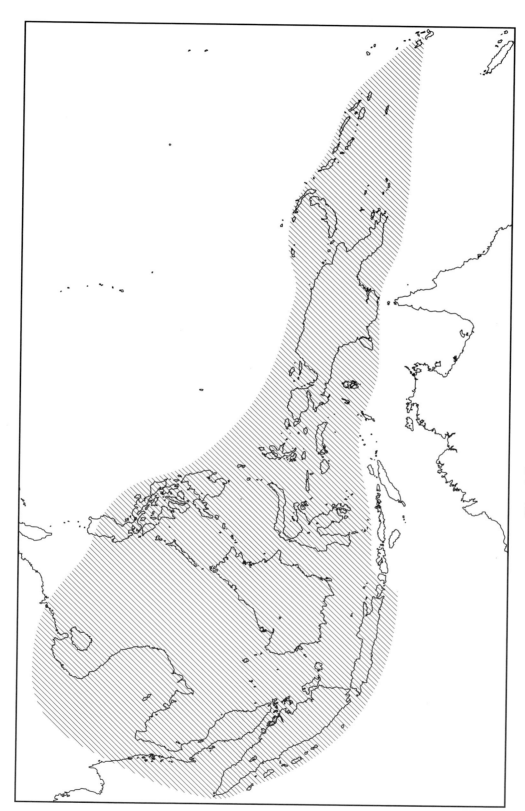

Fig. 1.2. Distribution of *Neuwiedia.*

Three species of *Neuwiedia* have been collected on Mt. Kinabalu in Malaysian Borneo, viz. *N. borneensis*, *N. veratrifolia*, and *N. zollingeri* var. *singapureana*. Here they inhabit mixed dipterocarp, *Agathis borneensis*, Fagaceae and *Gymnostoma* (*Casuarina*) *sumatranum* forest between 600 and 1000 m developed over an ultramafic substrate. *Neuwiedia borneensis* is the least common. **(JW)**

Pollination

No observations on the pollination of *Neuwiedia* are available, although some of the apostasioids are known to be autogamous. Vogel (1981) suggested pollination by female bees vibrating on the flower to release the pollen from the anthers, as in some Solanaceae or the Melastomataceae.

The flowers of *N. borneensis* are reported by Lamb (personal communication) to be scented of chocolate liquor. **(JW)**

Uses

None recorded. **(JW)**

Cultivation

Nothing is known of the cultivation of the apostasioids as few, if any, are represented in collections. Most would presumably thrive under the humid, shady regime required for the successful cultivation of many *Malaxis* or forest floor species of *Calanthe*. **(JW)**

Taxonomic literature

Vogel, E. F. de. (1969). Monograph of the tribe Apostasieae (Orchidaceae). *Blumea* **17**, 2, 313–50.

2. APOSTASIA

Apostasia Blume, *Bijdr*: 423 (1825); R. Brown in Wallich, *Pl. As. Rar.* **1**: 74 (1830); Pfitzer in Engler & Prantl, *Nat. Pfl. Fam.* **2**: 6, 80 (1888); Rolfe in *Orchid Rev.* **4**: 329 (1896); Holttum, *Rev. Fl. Malaya*

1: 64 (1953); de Vogel in *Blumea* **17**: 2, 335–48 (1969). Type species: *Apostasia odorata* Blume.

Niemeyera F.Muell., *Fragm. Phyt. Austr.* **6**: 96 (1867). Type: *Niemeyera stylidioides* F. Muell.

Mesodactylis [Wall., *Pl. As. Rar.* **1**: 74 (1830), *in obs.*] *Index Kew.* **2**: 217 (1895). Type: *Mesodactylis deflexa* Wall., *nom. inval., in syn.*

Adactylus (Endl.) Rolfe in *Orchid Rev.* **4**: 329 (1896). Type species: *Apostasia nuda* R.Br.

Derivation of name

The generic name *Apostasia* is a Greek word meaning separation or divorce and refers to the unique floral structure which some earlier authors have used to separate it from the Orchidaceae.

Description (Plate 9B; Fig. 2.1)

Erect, entirely glabrous *herbs*. *Roots* elongate, producing tubercles. *Rhizome* scaly, scales gradually grading into the cauline leaves, shortly amplexicaul, obliquely spathaceous, parchment-textured, often split at the dried, persistent top, nerves prominent, raised. *Stem* often branched, thin. *Leaves* well-spaced, lowermost smaller, often dying off, blade herbaceous, not plicate, usually narrow, apical margins forming a filiform tubular projection. *Racemes* usually branching and recurved or pendulous. *Flowers* non-resupinate. Pedicel not clearly distinct from the ovary, not twisted, sometimes curved, irregularly ribbed. Torus not oblique. Perianth more or less erect. *Sepals* boat-shaped, slightly fleshy, 3- to 5-nerved. *Petals* narrowed into a short, broad claw, or unclawed, shallowly boat-shaped, (1–)3(–5)-nerved. *Lip* similar to the petals, sometimes slightly more convex and larger. *Column* straight or curved; stamens 2, median one of the inner whorl staminodial, or absent, filaments adnate to the style to a varying degree, anthers clasping the style, loculi distinctly unequal to subequal. *Ovary* narrowly cylindric, obtusely to acutely 3-angled in cross-section, longitudinally grooved, the sepaline midribs continued onto surface as 3 longitudinal ribs. *Capsule* slightly accrescent, disintegrating; seeds on a short funicle, apical appendage small. **(JW)**

Fig. 2.1. *Apostasia wallichii*. A. Habit; B. Flower; C. Sepal, front view; D. Sepal, back view; E. Petal, front view; F. Petal, back view; G. Column, front view; H. Column, back view; I. Column, anthers removed; J. Anther, opened; K. Ovary, transverse section. Single bar = 1 mm, double bar = 1 cm. Drawn from *Lamb & Phillip AL803/87* and Kew Spirit Collection no. 52139 by Judi Stone.

Distribution (Fig. 2.2)

The seven species of *Apostasia* are distributed in south-west China (Yunnan), Sri Lanka, north-east India, Nepal, Bhutan, Myanmar (Burma), southern Japan (Ryukyu Islands), Thailand, Laos, Cambodia, Vietnam, Malaysia, Brunei, Indonesia, and the Philippines, east to northern Australia (Queensland) and Papua New Guinea. Section *Adactylus* is less widespread and does not occur east of Borneo. The richest centre of diversity is again Borneo from where five species have been recorded, one of which, *A. parvula*, is endemic. *Apostasia wallichii* is the most widespread species, occurring from Sri Lanka east to Papua New Guinea. *Apostasia latifolia* is endemic to Peninsular Malaysia and *A. nipponica* to the southern Japanese Ryukyu Islands, at the northernmost limit of the range. **(JW)**

Infrageneric treatment

1. **Apostasia** section **Apostasia**; de Vogel in *Blumea* **17**, 2, 336 (1969).
A. sect. *Mesodactylus* Endl., *Gen. Pl.*, part 3: 221 (1837). Types: *A. odorata* Blume and *A. wallichii* R.Br.

2. **Apostasia** section **Adactylus** Endl., *Gen. Pl.*, part 3: 221 (1837). Type: *A. nuda* R.Br.
Adactylus (Endl.) Rolfe in *Orchid Rev.* **4**, 329 (1896).

Artificial key to the sections of *Apostasia* (JW)

1. Staminode present, a large portion adnate to the style (sometimes replaced by a third stamen). Anthers dorsifixed
..................................section **Apostasia**
1. Staminode absent. Anthers sub-basifixed
..................................section **Adactylus**

Anatomy

See subfamily description.

Palynology

See subfamily description.

Cytogenetics

$2n = 48$ (see Chapter VII).

Phytochemistry

Nothing is known of the phytochemistry of *Apostasia*. **(RG and NV)**

Phylogenetics

See subfamily treatment.

Ecology

Species of *Apostasia* are found in similar habitats to *Neuwiedia*, from around sea level up to an elevation of about 1700 m. Habitats include mixed evergreen forest on granite, and secondary forest in Thailand. In Borneo collections have been made from a wide range of habitats including mixed dipterocarp forest on shale; hill dipterocarp forest; mixed dipterocarp forest with a kerangas element, on sandstone and shales, particularly on steep-sided ridges; dipterocarp, Fagaceae, *Gymnostoma sumatranum*, *Agathis borneensis* forest on ultramafic substrate; mixed forest with *Gymnostoma sumatranum* and an understorey of climbing bamboo on ultramafic substrate; lower montane dipterocarp to oak-laurel transitional forest; kerangas forest developed over podsolic soils; swamp forest; river banks; and secondary forest. In Queensland *Apostasia* has been found in littoral forest at or near sea level, and in Papua New Guinea collections have been made in lower montane *Castanopsis* forest. **(JW)**

Pollination

The flowers of *Apostasia* resemble those known to be adapted to the vibration pollination syndrome, as found in many Solanaceae, for example. One anther has been lost, while the two remaining ones tightly

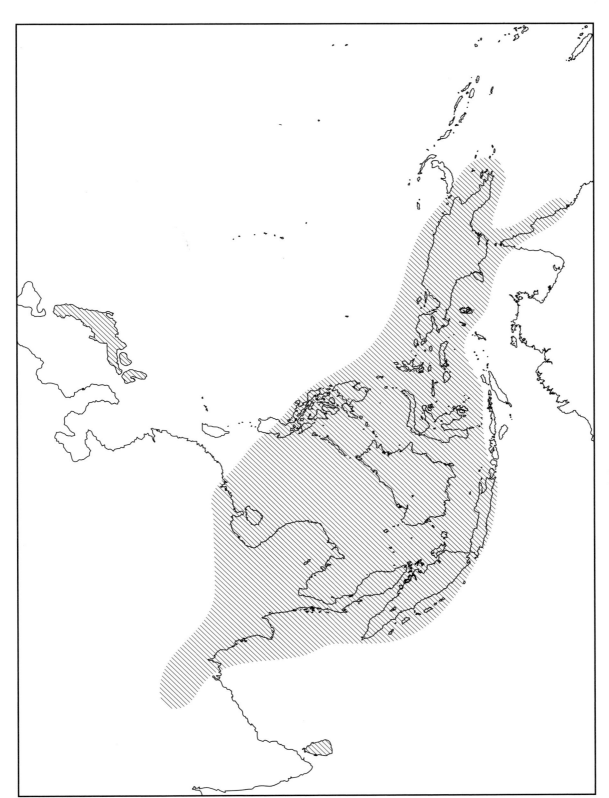

Fig. 2.2. Distribution map of *Apostasia*.

clasp the style forming a tube which is functionally analogous to the tubular anthers of *Solanum*. Bees vibrate on the flower, thereby releasing the pollen from the anthers. (JW)

Uses

De Vogel (1969) recorded that *A. wallichii* is used as an antidiabetic in Peninsular Malaysia. Here, too, the roots of *A. nuda* are boiled and used as a treatment for diarrhoea and sore eyes. It is also a favourite remedy for dog bites. (JW)

Cultivation

Plants in cultivation are rare and should never be removed from the wild. Jones (1988) recommends that plants be potted in a porous soil mix and given ample humidity, air movement, and shade.

Water may be freely supplied throughout the year. Minimum temperatures are 5–10°C.

Taxonomic literature

Dressler, R. L. (1993). *Phylogeny and classification of the orchid family*. Cambridge University Press.

Garay, L. A. (1960). On the origin of the Orchidaceae. *Botanical Museum Leaflets* 19, 57–96.

Seidenfaden, G. (1992). The orchids of Indochina. *Opera Botanica* 114, 1–502.

Seidenfaden, G. and Wood, J. J. (1992). *The orchids of Peninsular Malaysia and Singapore*. Olsen & Olsen, Fredensborg, Denmark.

Vogel, E. F. de. (1969). Monograph of the tribe Apostasieae (Orchidaceae). *Blumea* 17, 313–50.

Wood, J. J., Beaman, R. S., and Beaman, J. H. (1993). *The plants of Mt. Kinabalu. Vol. 2. Orchids*. Royal Botanic Gardens, Kew.

Wood, J. J. and Cribb, P. J. (1994). *A checklist of the orchids of Borneo*. Royal Botanic Gardens, Kew.

CYPRIPEDIOIDEAE

Cypripedioideae Garay in *Bot. Mus. Leafl. Harvard Univ.* **19** (3), 86 (1960). Type: *Cypripedium calceolus* L.

Cypripedia Spreng., *Anleit* **II**, 1, 298 (1817). Type: *Cypripedium calceolus* L.

Cypripedieae Lindl., *Orch. Sceletos*: 1, 18 (1826). Type: *Cypripedium calceolus* L.

Cypripedilinae Pfitzer, *Morph. Stud. Orch. Blüte*: 108 (1886); *Nat. Ord. Orchid.*: 12, 95 (1888); In Engler & Prantl, *Pflanzenf.* II, 6: 765, 82 (1889); in *Bot. Jahrb. Syst.* **19**, 31 (1894) & **25**, 518, 525 (1898); in Engler, *Pflanzenr.* IV, 50. *Orch. Pleonandr.*: 9 (1901). Type: *Cypripedium calceolus* L.

Description (Plates 9C, 9D, 10, 11, 12A, 12B, Figs 3.1, 4.1, 5.1, 6.1, 7.1)

Small to large terrestrial, lithophytic or epiphytic, rhizomatous *herbs*; *roots* elongate, fibrous; rhizome short to elongate. *Shoots* unbranched, very short to long, erect, leafy, clustered or less frequently well-spaced, glabrous, the lower part enclosed by two to several sterile sheaths, three- to many-leaved above. *Leaves* one to many, relatively thin-textured, fleshy or coriaceous, plicate or conduplicate, spreading or suberect, ovate, lanceolate, elliptic, ligulate, or oblong, obtuse, acute or acuminate, often tridenticulate at apex, yellow-green, green, or bluish green, plain, chequered or tessellated with dark and lighter green on upper surface or spotted with blackish maroon on upper surface, lighter green below, sometimes finely spotted or flushed with purple at base or all over on lower surface, glabrous or variously pubescent, ciliate or not on the margins. *Inflorescence* terminal, one- to many-flowered; rachis terete, hairy, glandular or glabrous; bracts conduplicate, linear, elliptic, lanceolate, ovate or oblong, green, sometimes spotted or flushed or striped with purple, glabrous, pubescent or glandular, ciliate or not. *Flowers* usually showy, concolorous or bi-coloured; pedicel obscure to short; ovary unilocular or trilocular, three-ribbed, glabrous or hairy, placentation parietal or axile. *Dorsal sepal* erect to hooded over lip, ovate, lanceolate, obovate or elliptic, obtuse, acute or acuminate, glabrous or pubescent on the outer surface, sometimes pubescent within at base, ciliate or not, unmarked, spotted or streaked on inner surface. *Lateral sepals* usually fused to form a concave synsepal that is more or less similar to the dorsal sepal, rarely free to base, apex entire or bifid, sometimes two-keeled on outer surface. *Petals* free, flat, inflexed or reflexed, spreading or pendent, spiralling or not, elliptic, ovate, lanceolate, linear-lanceolate, linear or oblanceolate, rounded, obtuse, acute or acuminate at apex, unmarked, glabrous or rarely pubescent on outer surface, often pubescent in basal half on inner surface, usually ciliate on margins, sometimes warty on margins, spotted or streaked on inner surface, sometimes veins coloured differently. *Lip* deeply pouched and inflated, slipper-shaped or urn-shaped, with (or more rarely without) more or less pronounced incurved side lobes which are sometimes reduced to auricles, hairy within especially on lower surface, sometimes spotted within, glabrous or hairy on outer surface; front margin incurved or not, sometimes with short marginal horns; side lobes sometimes warty. *Column* short, stalked, porrect; anthers two, bilocular, borne on short obtuse to acute filaments; pollen powdery or viscid; staminode terminal on column, sessile or short stalked, often shield-like, rarely reduced or absent, transversely reniform, oblong, ovate, obcordate or linear, flat, convex or longitudinally conduplicate, entire, bifid or tripartite at apex, glabrous to papillose or finely pubescent, ciliate or not; stigma stalked or subsessile, dependent, tripartite, more or less papillose. *Capsule* erect to pendent, three-ribbed, cylindrical to almost ellipsoidal, sometimes beaked at apex. *Seeds* spherical to ellipsoidal, with a thin testa except in *Selenipedium* in which the testa is hard and dark-coloured. **(PC)**

Distribution (Fig. B.1)

A subfamily comprising five genera. Widespread in the temperate regions of Eurasia and North America, Central America south to Brazil and Bolivia, tropical Asia from India across to Taiwan, south-east Asia from Indo-China across to the Philippines, New Guinea, and the Solomon Islands. (PC)

Artificial key to the genera of Cypripedioideae (PC)

1. Leaves plicate with convolute vernation; flowers not deciduous, persistent on fruit 2

1. Leaves conduplicate with duplicate vernation; flowers deciduous, not persistent on fruit 3

2. Ovary unilocular; seeds fusiform with a thin seed coat ....3. *Cypripedium*

2. Ovary trilocular; seeds subglobose with a crustose coat ....4. *Selenipedium*

3. Sepals valvate in bud; lip margins broadly involute or induplicate; distributed in tropical America 4

3. Sepals imbricate in bud; lip margins not infolded, slightly incurved or recurved; distributed in tropical Asia ....5. *Paphiopedilum*

4. Ovary trilocular; petals elliptic, subcircular or tapering, spreading to pendent ....6. *Phragmipedium*

4. Ovary unilocular; petals oblong, incurved ....7. *Mexipedium*

Cytogenetics

The cytogenetics of slipper orchids has been studied extensively. Comprehensive accounts have been published for *Paphiopedilum* (Atwood 1984; Karasawa 1979, 1986; Karasawa and Aoyama 1988; Karasawa and Tanaka 1981; Karasawa *et al.* 1997) and *Phragmipedium* (Atwood 1984; Karasawa 1980). Less is known about the chromosomes of *Cypripedium*, although counts for a number of species have been published (Karasawa and Aoyama 1986). The chromosomal constitution of *Mexipedium* has only recently been reported (Aoyama and Karasawa 1997; Cox *et al.* 1997b). *Selenipedium* has not been examined. A summary of the available chromosome data for slipper orchids is presented in Table B.1.

The application of available chromosome data to an explicit phylogenetic hypothesis for the slipper orchids by Cox *et al.* (1997a) clearly demonstrated several conspicuous changes from symmetric to asymmetric karyotypes. Centric fission has clearly played a major role in karyotype evolution in some slipper orchid groups (Karasawa 1979; Karasawa and Tanaka 1980). This raises the question of why such marked differences in karyotype symmetry have arisen. Genome fragmentation may have occurred in different ways. Centromeres may have become unstable in some groups, leading to chromosome breakage. Alternatively, division of the nuclear chromatin among more chromosomes may confer a selective advantage to the plants.

The selection pressures acting to bring about this particular mode of chromosome evolution are presently unknown. In a study of the cycad genus *Zamia* L., which exhibits extensive centric fission, the onset of stressful ecological conditions was suggested as the selective pressure that resulted in Robertsonian chromosome change (Caputo *et al.* 1996). In the slipper orchids Cox *et al.* (1997a) observed that many of the species with high chromosome numbers (i.e. high numbers of telocentrics) are nearly all narrow endemics either far removed from the main centres of generic distribution (i.e. *P. druryi*) or on islands such as Borneo, Java, and Sumatra (Cribb 1987). Some species are known only in single locations. The colonization of such islands may require increased levels of genetic variation to cope with the rigorously selective environments to which island floras may be subjected (Stebbins 1971). An alternative but not mutually exclusive explanation is that island habitats may simply permit a greater fixation rate of chromosomal (and other) variation following founder effect and effective inbreeding within small populations occupying novel ecological niches.

Although genome size data are available for approximately one percent of the angiosperm flora, information for Orchidaceae is under-represented (Bennett and Leitch 1995). The 4C genome sizes of 29 slipper orchids are currently known (Cox *et al.* 1998) and range in the conduplicate-leaved genera

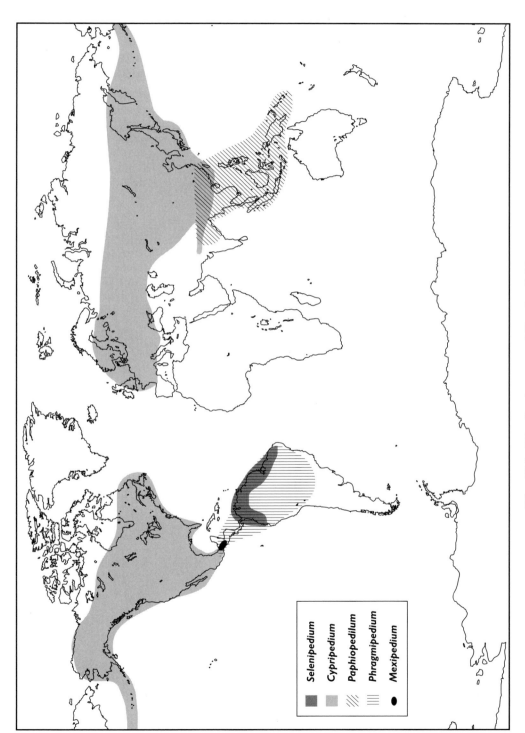

Fig. B.1. Distribution map of Cypripedioideae.

Table B.1. Chromosome number, karyotype, and geographical distribution of slipper orchids (data from: Atwood 1984; Cox *et al.* 1997*a,b*; Cribb 1997; Karasawa 1979, 1980, 1986; Karasawa and Aoyama 1986, 1988; Karasawa *et al.* 1997; Karasawa and Tanaka 1981)

Taxon	Taxonomic placement	Diploid chromosome number (2*n*)	Geographical distribution
Selenipedium	(*sensu* Pfitzer 1889)		
S. chica Rchb.f.		N/A	Panama
Cypripedium	(*sensu* Cribb 1997)		
C. irapeanum Llave & Lex.	sect. *Irapeana*	N/A	Guatemala, Honduras, Mexico
C. californicum A.Gray		N/A	USA (Oregon)
	sect. *Cypripedium*		
C. calceolus L.	subsect. *Cypripedium*	20	Europe, Russian Fed., China
C. candidum Mühl. ex Willd.		20	USA
C. cordigerum D.Don		20	N. India, Nepal, Tibet
C. kentuckiense C.F.Reed		20	USA
C. parviflorum Salisb.		20,22	USA
C. parviflorum var. *pubescens* Willd.		20	USA, Canada
C. segawai Masam.		20	Russian Fed., China
C. himalaicum Rolfe	subsect. *Macrantha*	20	Nepal, Bhutan
C. macranthos Sw.		20	China, Japan, Russian Fed., Korea, Taiwan
C. fasciculatum Kell. ex S.W.Watson	sect. *Enantiopedilum*	20	USA (Colorado)
C. flavum P.F.Hunt & Summerh.	sect. *Obtusipetala*	N/A	China, Tibet
C. reginae Walt.		20	USA (Michigan)
C. passerinum Richardson		20	USA, Canada
C. acaule Aiton	sect. *Acaulia*	N/A	USA, Canada
C. debile Rchb.f.	sect. *Retinervia*	20	China
C. guttatum Sw.	sect. *Bifolia*	20	USA, Russian Fed., China, Japan, India, Korea, Nepal, Bhutan
C. yatabeanum Makino		20	Russian Fed., China
C. formosanum Hayata	sect. *Flabellinervia*	20	Taiwan
C. japonicum Thunb.		20	Japan
C. arietinum R.Br.	sect. *Arietinum*	20	USA
C. plectrochilum Franch.		N/A	Russian Fed., China
C. lichiangense Cribb & S.C.Chen	sect. *Trigonopedia*	N/A	China
C. margaritaceum Franch.		N/A	China
Phragmipedium	(*sensu* McCook 1989)		
P. besseae Dodson & J.Kuhn	sect. *Micropetalum*	24,28	Ecuador, Peru
P. schlimii Linden & Rchb.f.		30 (24)	Colombia
P. caricinum (Lindl. & Paxton) Rolfe	sect. *Phragmipedium*	20,22	Bolivia
P. caudatum (Lindl.) Rolfe		28 (20)	Costa Rica, Guatemala, Panama, Bolivia, Ecuador, Peru
P. exstaminodium (Lindl.) Rolfe		N/A	Mexico
P. lindenii (Lindl.) Dressler & N.H.Williams		28 (20)	Colombia, Ecuador
P. pearcei (Rchb.f.) Rauh & Senghas		20 (4), 21,22	Ecuador, Peru
P. wallisii (Rchb.f.) Garay		28 (20)	Colombia, Ecuador
P. warscewiczianum (Rchb.f.) Garay		28	Panama
P. boissierianum (Rchb.f.) Rolfe	sect. *Lorifolia*	18	Ecuador, Peru
P. czerwiakowianum (Rchb.f.) Rolfe		N/A	Peru
P. longifolium (Rchb.f. & Warsz.) Rolfe		21 (5), 23 (7)	Costa Rica, Panama, Colombia, Ecuador
P. kaieteurum (N.E.Br.) Garay	sect. *Platypetalum*	N/A	Venezuela
P. lindleyanum (Lindl.) Rolfe		22 (10)	Brazil, French Guiana, Guyana, Suriname, Venezuela
P. sargentianum Rolfe		22 (10)	Brazil

Table B.1. Chromosome number, karyotype, and geographical distribution of slipper orchids (data from: Atwood 1984; Cox *et al.* 1997*a,b*; Cribb 1997; Karasawa 1979, 1980, 1986; Karasawa and Aoyama 1986, 1988; Karasawa *et al.* 1997; Karasawa and Tanaka 1981) (continued)

Taxon	Taxonomic placement	Diploid chromosome number (2*n*)	Geographical distribution
Mexipedium	(Albert and Chase 1992)		
M. xerophyticum (Soto Arenas, Salazar & Hágsater) V.A.Albert & M.W.Chase		26	Mexico
Paphiopedilum	(*sensu* Cribb 1987) subg. *Brachypetalum*		
P. bellatulum (Rchb.f.) Stein	sect. *Concoloria*	26	China, Burma, Thailand, Laos
P. concolor (Lindl.) Pfitzer		26	China, Laos, Burma, Thailand, Vietnam
P. godefroyae (God.-Leb.) Stein		26	Thailand
P. niveum (Rchb.f.) Stein		26	Thailand, Malaysia
P. armeniacum S.C.Chen & F.Y.Liu	sect. *Parvisepalum*	26	China
P. delenatii Guill.		26	Vietnam
P. emersonii Koop. & P.J.Cribb		26	China
P. malipoense S.C.Chen & Z.H.Tsi		26	China, Vietnam
P. micranthum T.Tang & F.T.Wang		26	China, Vietnam
	subg. *Paphiopedilum*		
P. adductum Asher	sect. *Coryopedilum*	N/A	Philippines
P. glanduliferum (Blume) Stein 'A'		26	Indonesia, Papua New Guinea
P. glanduliferum (Blume) Stein 'B'		26	Indonesia, Papua New Guinea
P. kolopakingii Fowlie		26	Indonesia
P. philippinense (Rchb.f.) Stein		26	Malaysia, Philippines
P. rothschildianum (Rchb.f.) Stein		26	Malaysia
P. sanderianum (Rchb.f.) Stein		26	Malaysia
P. stonei (Hook.) Stein		26	Malaysia
P. supardii Braem & Loeb		26	Malaysia
P. dianthum T.Tang & F.T.Wang	sect. *Pardalopetalum*	N/A	China
P. haynaldianum (Rchb.f.) Stein		26	Philippines
P. lowii (Lindl.) Stein		26	Indonesia, Malaysia
P. parishii (Rchb.f.) Stein		26	China, Burma, Thailand
P. glaucophyllum J.J.Sm.	sect. *Cochlopetalum*	36 (22), 37	Indonesia
P. primulinum M.W.Wood & P.Taylor		32 (14)	Indonesia
P. victoria-mariae (Rolfe) Rolfe		N/A	Indonesia
P. victoria-regina (Sander) M.W.Wood		32,33,34,35,36	Indonesia
P. barbigerum T.Tang & F.T.Wang	sect. *Paphiopedilum*	26	China
P. charlesworthii (Rolfe) Pfitzer		26	Burma, Thailand
P. druryi (Bedd.) Stein		30 (8)	India
P. exul (Ridl.) Rolfe		26	Thailand
P. fairrieanum (Lindl.) Stein		26	Bhutan, India
P. gratrixianum (Mast.) Guill.		N/A	Laos, Vietnam
P. henryanum Braem		N/A	China, Vietnam
P. hirsutissimum (Lindl. ex Hook.) Stein		26	China, India, Bhutan
P. tigrinum Koop. & N.Haseg.		26	China
P. insigne (Wall. ex Lindl.) Pfitzer		26	India, Nepal
P. villosum (Lindl.) Stein		26	India, Burma, Thailand
P. acmodontum Schoser ex M.W.Wood	sect. *Barbata*	36 (20)	Philippines
P. appletonianum (Gower) Rolfe		38 (24)	China, Cambodia, Laos, Thailand, Vietnam
P. argus (Rchb.f.) Stein		38 (24)	Philippines
P. barbatum (Lindl.) Pfitzer		38 (28)	Malaysia
P. bullenianum (Rchb.f.) Pfitzer		40 (28) 40 (28) 42 (32)	Borneo
P. bougainvilleanum Fowlie		40 (28)	Bougainville Is.
P. callosum (Rchb.f.) Stein		32 (12)	Cambodia, Laos, Thailand, Vietnam

Table B.1. Chromosome number, karyotype, and geographical distribution of slipper orchids (data from: Atwood 1984; Cox *et al.* 1997*a*,*b*; Cribb 1997; Karasawa 1979, 1980, 1986; Karasawa and Aoyama 1986, 1988; Karasawa *et al.* 1997; Karasawa and Tanaka 1981) (continued)

Taxon	Taxonomic placement	Diploid chromosome number (2*n*)	Geographical distribution
P. ciliolare (Rchb.f.) Stein	sect. *Barbata*	32 (12)	Philippines
P. dayanum (Lindl.) Stein	(cont.)	36 (20)	Malaysia
P. fowliei Birk		36 (20)	Philippines
P. hennisianum (M.W.Wood) Fowlie		36 (18)	Philippines
P. hookerae (Rchb.f.) Stein		28 (0)	Indonesia, Malaysia
P. javanicum (Reinw. ex Lindl.) Pfitzer		38 (24)	Indonesia, Malaysia
P. lawrenceanum (Rchb.f.) Pfitzer		36 (20)	Malaysia
P. mastersianum (Rchb.f.) Stein		36 (20)	Indonesia
P. papuanum (Ridl.) Ridl.		N/A	Indonesia, Papua New Guinea
P. purpuratum (Lindl.) Stein		40 (28)	China, Vietnam
P. sangii Braem		28	Indonesia
P. schoseri Braem & H.Mohr		35	Indonesia, Malaysia
P. sukhakulii Schoser & Senghas		40 (28)	Indonesia
P. superbiens (Rchb.f.) Pfitzer		38 (24)	Indonesia
		36 (20)	
P. tonsum (Rchb.f.) Stein		32 (12)	Indonesia
P. urbanianum Fowlie		N/A	Philippines
P. venustum (Wall.) Pfitzer ex Stein		40 (28)	Bhutan, India, Nepal
P. wardii Summerh.		41 (29)	Burma

N/A, Not available.

from 24.4 pg in *Phragmipedium longifolium* to 138.1 pg in *Paphiopedilum wardii*. The three known 4*C* DNA amounts from *Cypripedium* range from 86.2 to 129.5 pg (Table B.2, B.3).

Although large genome size variation exists at the family level, this is not the case at subfamily and genus levels. At the generic level, all three genera examined show less than 2-fold variation in genome size (*Paphiopedilum* 1.9-fold, *Phragmipedium* 1.5-fold, and *Cypripedium* 1.5-fold: see Table B.3), which is low compared with many other genera examined in the angiosperms.

Overall, the data indicate that although genome size varies 5.7-fold in slipper orchids, it does not appear to apply constraints to the other mechanisms of karyotype evolution. This is particularly apparent in the two conduplicate-leaved genera, which differ considerably in genome size but have undergone similar types of Robertsonian change to increase the number of linkage groups in the karyotype. (AC)

Phylogenetics

Numerous classifications for the Cypripedioideae have been proposed over the last two hundred

years (Table B.4). Linnaeus (1753) recognized only one species and several varieties (now elevated to specific rank). By 1840 Lindley recognized 22 species in one genus, *Cypripedium* (Lindley 1830–1840). Reichenbach (1854) described a new genus, *Selenipedium*, on the basis of axile placentation of the ovary. There followed a period in the late nineteenth century when considerable nomenclatural confusion arose about generic names.

The first attempt at a treatment of slipper orchids incorporating ideas of their evolution was made by Rolfe (1896) who recognized the four principal genera. This was reversed by Kränzlin (1897) who reverted to the use of a single genus, *Cypripedium*. Pfitzer (1903) provided the most comprehensive treatment of the slipper orchids and included vegetative as well as floral characters. The classification was based on an evolutionary hypothesis that postulated *Selenipedium* as the ancestral (most 'primitive') genus, a view that still prevails today. Pfitzer's treatment was considered taxonomically inflated by Atwood (1984). Both Atwood (1984) and Brieger (1973) tried to rationalize the host of subgenera and sections erected by Pfitzer. After a phylogenetic analysis of morphological characters, Atwood (1984) supported the suggestion of Rafinesque (1819) that *Cypripedium*

Table B.2. Variation in chromosome number and genome size in the slipper orchids (data from: Cox *et al.* 1997*a*, 1998)

Taxon	Chromosome number (2*n*)	Mean 4C DNA amount (pg)
Paphiopedilum		
Sect. *Barbata*		
P. *appletonianum* (Gower) Rolfe	38(24)	129.7
P. *wardii* Summerh.	41(29)	138.1
P. *callosum* (Rchb.f.) Stein	32(12)	96.2
P. *purpuratum* (Lindl.) Stein	40(28)	108.5
P. *barbatum* (Lindl.) Pfitzer	38(24)	135.0
P. *tonsum* (Rchb.f.) Pfitzer	32(12)	112.6
P. *mastersianum* (Rchb.f.) Stein	36(20)	118.9
P. *sukhukulii* Schoser & Senghas	40(28)	118.9
Sect. *Coryopedilum*		
P. *philippinense* (Rchb.f.) Stein	26	93.0
P. *rothschildianum* (Rchb.f.) Stein	26	90.3
P. *glanduliferum* (Blume) Stein	26	94.9
Sect. *Paphiopedilum*		
P. *insigne* (Wall. ex Lindl.) Pfitzer	26	92.1
P. *druryi* (Bedd.) Stein	30(8)	106.5
P. *gratrixianum* (Mast.) Guill.	26	100.0
P. *villosum* (Lindl.) Stein	26	89.9
Sect. *Cochlopetalum*		
P. *primulinum* M.W.Wood & P.Taylor	32(14)	83.6
P. *victoria-mariae* (Rolfe) Rolfe	36	85.6
Sect. *Brachypetalum*		
P. *concolor* (Lindl.) Pfitzer	26	77.9
P. *godefroyae* (God.-Leb.) Stein	26	71.2
Sect. *Parvisepalum*		
P. *micranthum* T.Tang & F.T.Wang	26	91.0
P. *delenatii* Guill.	26	87.3
Phragmipedium		
Sect. *Phragmipedium*		
P. *caudatum* (Lindl.) Rolfe	28(20)	36.7
Sect. *Platypetalum*		
P. *lindleyanum* (Lindl.) Rolfe	22(10)	32.1
Sect. *Lorifolia*		
P. *pearcei* (Rchb.f.) Rauh & Senghas	20,21,22	25.3
P. *longifolium* (Rchb.f. & Warsc.) Rolfe	21(5)	24.4
Sect. *Micropetalum*		
P. *besseae* Dodson & J.Kuhn	24	28.3
Cypripedium		
C. *parviflorum* Salisb. var. *pubescens* Willd.	20	129.5
C. *formosanum* Hayata	20	113.9
C. *californicum* A.Gray	N/A	86.2

N/A, Not available.

arietinum be removed from *Cypripedium* and placed in its own genus, *Criosanthes* Raf., on the basis that it is only distantly related to other extant species. This suggestion has not been generally accepted.

Generic limits in Cypripedioideae have traditionally been defined using a combination of four characters: leaf type, vernation, locule number, and placentation. Unfortunately, all are individually

Table B.3. Range in chromosome number, karyotype structure (where known), *nombre fondamental*, and range of 4C DNA amount (data from: Karasawa, 1979; Cox *et al.* 1997*a*, 1998)

Genus	Section	Chromosome number (2n)	*Nombre fondamental* if karyotype structure known	Range in 4C DNA amount (pg)
Paphiopedilum	*Barbata*	28–42	52–54	96.2–138.1
	Pardalopetalum	26	52	–
	Coryopedilum	26	52	90.3–94.9
	Paphiopedilum	26 (except *P. druryi*)	52	89.9–106.5
	Cochlopetalum	32–37	50	83.6–85.6
	Brachypetalum	26	52	71.2–77.9
	Parvisepalum	26	52	87.3–91.0
Phragmipedium	*Phragmipedium*	28	36	36.7
	Platypetalum	22	34	32.1
	Lorifolia	18–21	36–37	24.4–25.3
	Micropetalum	24, 30	36	28.3
Cypripedium		20	40	86.2–129.5
Mexipedium		26	46	—

inconsistent. Reliance upon these 'critical' characters has been questioned (Atwood 1984), although no more obvious ones have yet been found (a number of correlated but cryptic characters exist, e.g. perianth aestivation and persistence after flowering). *Paphiopedilum*, *Phragmipedium*, and *Mexipedium* have conduplicate leaves, and *Selenipedium* and *Cypripedium* have plicate leaves. *Paphiopedilum* and *Cypripedium* are unilocular with parietal placentation, whereas *Selenipedium* and *Phragmipedium* are trilocular with axile placentation. The recent description of conduplicate-leaved *Phragmipedium xerophyticum* Soto, Salazar and Hagsater (Soto *et al.* 1990) served only to blur the distinctions among existing genera because it possesses a mixture of *Paphiopedilum*-like and *Phragmipedium*-like features. Its New World distribution (Mexico) is similar to that of *Phragmipedium*, but it possesses unilocular ovaries like the Old World genus *Paphiopedilum*. *Phragmipedium xerophyticum* may be distinguished from *Paphiopedilum* by having branched racemes and valvate sepal aestivation. Because of its mixed suite of characters Albert and Chase (1992) transferred *Phragmipedium xerophyticum* to a new monotypic genus, *Mexipedium*. Although Albert and Pettersson's (1994) combination of all conduplicate-leaved genera under *Paphiopedilum* has not been widely accepted, this union solved many of the apparent inconsistencies in slipper orchid taxonomy.

The boundaries of most genera have been supported by crossing experiments. Extensive horticultural interest in the slipper orchids has resulted in many interspecific and complex hybrids being produced. As yet no verified intergeneric hybrids have been registered, although *Mexipedium xerophyticum* has been crossed with many *Phragmipedium* species (H. Koopowitz, personal communication).

The most recent rigorously phylogenetic treatments of subfamily Cypripedioideae have been presented by Albert (1994) and Cox *et al.* (1997*a*). The former used morphological data and nucleotide sequences of the plastid gene *rbcL* to evaluate generic relationships within the subfamily (Fig. B.2). Albert interpreted his data as supporting the monophyly of currently proposed slipper orchid genera, but the DNA sequences lacked the divergence required to evaluate species relationships.

Systematic inferences drawn from a comprehensive study of rDNA internal transcribed spacer (ITS) nucleotide sequence by Cox *et al.* (1997*a*) are in broad agreement with previous studies of slipper orchid generic relationships (Fig. B.3). Although overall internal support for the topology was not high, confidence was gained from marked congruence with previous taxonomies as well as many morphological, anatomical, and cytological data. Such congruence establishes ITS as a reliable indicator of phylogenetic relationships at higher levels within

Table B.4. Taxonomy of the Cypripedioideae

Lindley (1840)	Pfitzer (1889)	Pfitzer (1894)	Rolfe (1896)	Kränzlin (1897)	Brieger (1973)	Albert and Chase (1992)	Albert and Pettersson (1994)
	Selenipedium	*Selenipedium*	*Selenipedium*	Sect. *Selenipedium*	Tribe *Selenipedieae*	*Selenipedium*	*Selenipedium*
	Cypripedium	*Cypripedium*	*Cypripedium*	Sect. *Calceolaria* *Calceolus* group	Tribe *Cypripedieae*	*Cypripedium*	*Cypripedium*
Arietinum group	*Arietina* group	*Arietinum* group		*Arietina* group			
Foliosa group	*Foliosa* group	*Eucypripedium* group *Trigonopedilum* group		*Macrantha* group *Obtusifolia* group			
Bifolia group	*Diphylla* group			*Bifolia* group			
Acaulia	*Paphiopedilum* (excl. *C. acaule*) *Genuina* group	*Paphiopedilum* *Coelopedilum* group *Eremantha* *Polyantha*	*Paphiopedilum*	Sect. *Barbata* Sect. *Concoloria* Sect. *Insignia*	Tribe *Paphiopedieae*	*Paphiopedilum*	*Paphiopedilum* subg. *Paphiopedilum*
C. caudatum	*C. caudata* group	*Phragmopedilum* group	*Phragmipedium*	Sect. *Lorifolia* Sect. *Caudata*	Tribe *Phragmipedieae*	*Phragmipedium*	*Paphiopedilum* subg.*Phragmipedium*
						Mexipedium	*Paphiopedilum* subg. *Mexipedium*

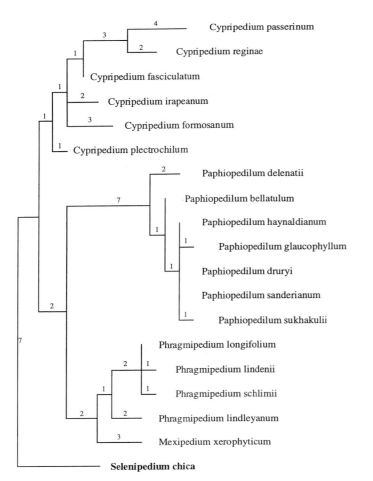

Fig. B.2. Species relationships in the Cypripedioideae deduced from morphological characters and the nucleotide sequence of the plastid gene *rbc*L (Albert 1994). Unpublished data matrix kindly provided by V. A. Albert and M. W. Chase. Figures represent number of inferred nucleotide substitutions along branches (ACCTRAN optimization).

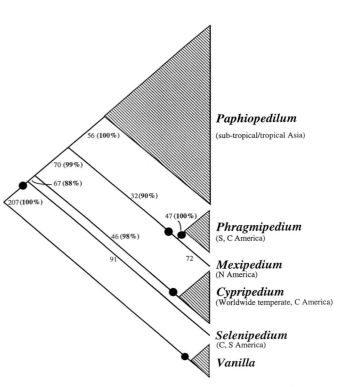

Fig. B.3. Strict consensus of fourteen successively weighted, most-parsimonious trees indicating general cladistic relationships among the five slipper orchid genera, using *Vanilla* as an outgroup (Cox *et al.* 1997*a*). Numbers are inferred nucleotide substitutions along branches (ACCTRAN optimization), and solid circles indicate clades strongly supported (node score >0.63) by the jackknife (Farris *et al.* 1996).

and among genera. It is unlikely that such congruence with the ITS tree should occur by chance, and thus many of the weakly supported groupings appear more reasonable.

The data demonstrate that each genus, as currently circumscribed, is monophyletic (although *Mexipedium* and *Selenipedium* were represented by only a single taxon). Both *rbcL* and ITS sequences data place *Mexipedium* sister to *Phragmipedium*. The plicate-leaved genera, *Cypripedium* and *Selenipedium*, are successive sister groups to the rest of the subfamily respectively, confirming generally held opinions that they display plesiomorphic characters compared to the conduplicate-leaved genera. (AC)

3. CYPRIPEDIUM

Cypripedium L., Sp. Pl. ed.1, 2: 951 (1753); Pfitzer in Engler, *Pflanzenr.* IV, 50 (Heft 12), *Orchid. Pleonan.*: 28 (1903); Keller & Schlechter, *Monogr. Iconogr. Orchid. Eur.* **2**: 11 (1930); S. C. Chen & Xi in *Proc. 12th World Orch. Conf., Tokyo*: 145 (1987); Cribb, *Gen. Cypripedium*: 100 (1997). Type species: *Cypripedium calceolus* L.

Synonymy

Calceolus Mill., Gard. Dict. abr. ed. 4: Calceolus (1754); Adanson, Fam. Pl. 2: 70 (1760). Type: *Calceolus mariae* Mill.

Criosanthes Raf. in *J. Phys. Chim. Hist. Nat.* **89**, 102 (1819); Atwood in *Selbyana* 7: 129–247 (1984). Type: *Cypripedium arietinum* R.Br.

Arietinum Beck, *Bot. North. Midl. States*: 352 (1833). Type: *Arietinum americanum* Beck

Sacodon Raf., *Fl. Tellur.* 4: 46 (1838). Types: *C. macranthos* Sw. & *C. ventricosum* Sw.

Corisanthes Steud., *Nomencl. Bot.* ed. 2, 1: 474 (1840), sphalm. pro *Criosanthes* Raf.

Hypodema Rchb., *Duet. Bot. Herb.-Buch*: 56 (1841). Type: not designated.

Fissipes Small, *Fl. SE United States*: 311 (1903). Type: *Fissipes acaulis* (Aiton) Small.

Derivation of name

From the Greek '*Kypris*', a name for Aphrodite (Venus) who was born on the island of Cyprus, and '*pedilon*', a sandal or slipper, in reference to the shape of the lip. The suffix was incorrectly Latinized as -*pedium* rather than -*pedilum*. The name refers to vernacular names in Europe such as 'Frauenschuh', 'Marienschuh', and 'Lady's Slipper', the lady being variously considered to be Aphrodite, Venus or the Virgin Mary. **(PC)**

Description (Plate 9C, 9D, 10, 11, 12A, 12B; Fig. 3.1)

Terrestrial *herbs*. *Roots* elongate, fibrous, arising from a short to elongate rhizome; *rhizomes* present, short to elongate, creeping. *Shoots* erect, leafy, clustered or well spaced, terete, hairy, glandular or glabrous, the base enclosed by two to four sheathing sterile bracts, one- to several-leaved above; *pseudobulbs* absent. *Leaves* one-several, usually plicate, prostrate, spreading or suberect, lanceolate, ovate, elliptic, oblong or cordate, obtuse to acute or acuminate, green, spotted with blackish maroon in some species, glabrous or hairy, ciliate or not on the margins. *Inflorescence* terminal, one- to many-flowered; *rachis* terete, hairy, glandular or glabrous; bracts usually leaf-like but smaller than the uppermost leaf, rarely linear. *Flowers* usually showy, concolorous or two-coloured; *pedicel* obscure to elongate, elongating after fertilization in some species; *ovary* unilocular, three-ribbed, glabrous,

hairy or glandular. *Dorsal sepal* erect to hooded over lip, ovate, lanceolate or elliptic, obtuse, acute or acuminate, glabrous or pubescent on the outer surface, rarely pubescent within, ciliate or not; *lateral sepals* usually fused to form a concave synsepal that is similar to the dorsal sepal, but free and linear-lanceolate in section *Criosanthes*. *Petals* free, spreading, incurved or clasping the sides of the lip, elliptic, ovate, lanceolate, linear-lanceolate or subpandurate, rounded, obtuse, acute or acuminate, often pubescent in basal half within, usually ciliate. *Lip* deeply pouched and inflated, slipper-shaped or urn-shaped, glabrous or hairy on outer surface; *side lobes* incurved, entire, often much reduced; *midlobe* deeply saccate, incurved or not on front margin, hairy within especially on lower surface. *Column* porrect, short, stalked; *anthers* two, bilocular, borne on short obtuse to acute filaments; *pollen* powdery or viscid; *staminode* terminal on column, sessile or stalked, oblong, ovate, cordate or linear, flat, convex or conduplicate, glabrous to papillose or finely pubescent, ciliate or not; *stigma* stalked, dependent, tripartite, more or less papillose. *Capsule* erect to pendent, three-ribbed, cylindrical to almost ellipsoidal. **(PC)**

Distribution (Fig. 3.2)

A genus of about 47 species widespread in Europe, temperate Asia across to Japan and China and the Himalayas, Sakhalin, the Kurile and Aleutian Islands, North America, and south to Guatemala and Honduras. **(PC)**

Infrageneric treatment (PC)

1. Section **Subtropica** S.C.Chen & K.Y.Lang. Type species: *C. subtropicum* S.C.Chen & K.Y.Lang.
2. Section **Irapeana** Cribb. Type species: *C. irapeanum* La Llave & Lex.
3. Section **Obtusipetala** (Pfitzer) Cribb. Type species: *C. reginae* Walt.
4. Section **Cypripedium**. Type species: *C. calceolus* L.
   Subsection *Cypripedium*.
   Subsection *Macrantha* (Kränzl.) Cribb. Type species: *C. macranthos* Sw.

Fig. 3.1. *Cypripedium macranthos.* A. Habit, ×0.5; B. Flower, ×0.66; C. Dorsal sepal, ×1; D. Petal, ×1; E. Synsepal, ×1; F. Lip, longitudinal section, ×1; G. Column, ×1.5; H. Staminode, ×1.5; J. Ovary, transverse section, ×4. Drawn from Kew Spirit Collection no. 45172 by Eleanor Catherine. Reprinted from Cribb, P. (1997). *The Genus Cypripedium.*

5. Section **Enantiopedilum** Pfitzer. Type species: *C. fasciculatum* Kellogg ex S. Watson.

6. Section **Criosanthes** (Raf.) Pfitzer. Type species: *C. arietinum* R.Br.

7. Section **Flabellinervia** (Pfitzer) Hennessy ex Cribb. Type species: *C. japonicum* Thunb.

8. Section **Acaulia** (Lindl.) Hennessy ex Cribb.

Type species: *C. acaule* Aiton.

9. Section **Bifolia** (Lindl.) S.C.Chen. Type species: *C. guttatum* Sw.

10. Section **Retinervia** (Pfitzer) S.C.Chen. Type species: *C. debile* Rchb.f.

11. Section **Trigonopedilum** (Franch.) Pfitzer. Type species: *C. margaritaceum* Franch.

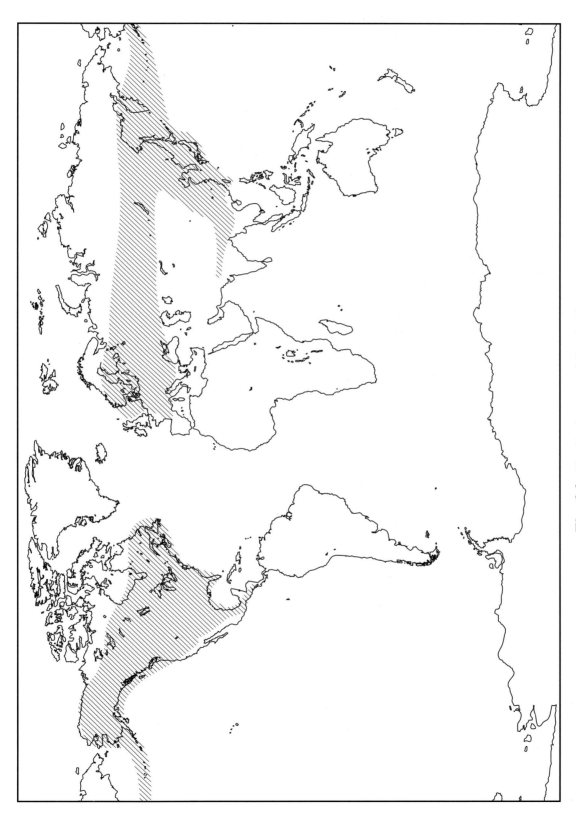

Fig. 3.2. Distribution map of *Cypripedium*.

Anatomy

Leaf

Anticlinal walls of epidermal cells are straight and/or sinuous, either the same on both surfaces or straight adaxially and sinuous abaxially (Möbius 1887; Rosso 1966; Atwood 1984; Pridgeon, unpublished). In transection the cuticle and outer tangential walls are thin, in contrast to those of the conduplicate-leaved genera *Paphiopedilum*, *Mexipedium*, and *Phragmipedium*.

Stomata occur on both leaf surfaces of many species, although in *C. acaule*, *C. margaritaceum*, *C. pubescens*, and *C. arietinum* they are abaxial only (Rosso 1966). Atwood (1984), however, observed adaxial stomata in *C. irapeanum*, *C. californicum*, and *C. reginae*. Pridgeon (unpublished) observed amphistomaty only in *C. candidum* and *C. franchetii*, whereas leaves of *C. acaule*, *C. arietinum*, *C. bardolphianum*, *C. daliense*, *C. debile*, *C. flavum*, *C. henryi*, *C. japonicum*, *C. macranthos*, *C. margaritaceum*, *C. plectrochilum*, and *C. reginae* are hypostomatic. As in *Selenipedium*, guard cells are thin-walled with only weak development of cuticle (Rosso 1966). Subsidiary cells are absent (Rosso 1966; N. H. Williams 1979).

Trichomes, glandular and/or nonglandular, are distributed over the leaf surfaces of leaves of most species (Solereder and Meyer 1930). Notable exceptions are *C. bardolphianum*, *C. debile*, *C. henryi*, *C. margaritaceum*, and *C. plectrochilum* (Pridgeon, unpublished). Nonglandular hairs are most common (Fig. 3.3), but both types are present in *C. acaule*, *C. candidum*, *C. flavum*, and *C. reginae* (Pridgeon,

unpublished) and in *C. pubescens*, *C. guttatum*, and *C. irapeanum* (Rosso 1966). Both types of hairs are uniseriate and comprise 2–7 cells; glandular hairs have in addition a globose or elliptical apical cell.

Chlorenchyma is undifferentiated into palisade and spongy layers (Fig. 3.4, 3.5), although there is a tendency for adaxial layers to be more compactly arranged (Rosso 1966). Vascular bundles are in one row in transverse section, the midvein more prominent, the larger bundles surrounded by bundle sheaths of sclerenchyma fibres. Rosso (1966) notes striking similarities in vasculature between *C. irapeanum* and *Selenipedium*, including suberized endoderm-like cells between fibres and mesophyll, and relatively more extensive bundle sheaths than in other *Cypripedium* species.

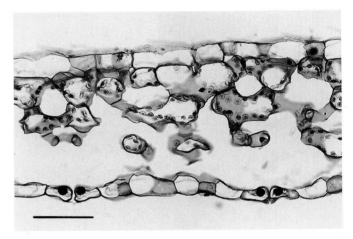

Fig. 3.4. *Cypripedilum debile*. Leaf transection showing undifferentiated mesophyll and two abaxial stomata. Light micrograph. Scale bar = 50 μm. With permission of A. M. Pridgeon.

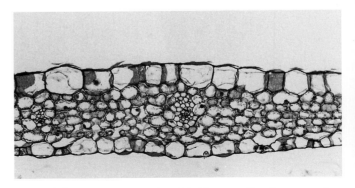

Fig. 3.5. *Cypripedium bardolphianum*. Leaf transection showing vascular bundles, undifferentiated mesophyll, and abaxial stomata. Light micrograph. Scale bar = 200 μm. With permission of A. M. Pridgeon.

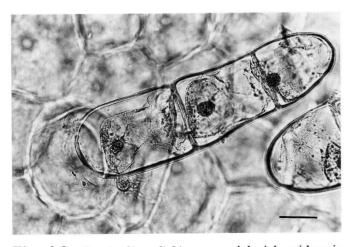

Fig. 3.3. *Cypripedium lichiangense*. Adaxial epidermis of leaf showing multicellular, nonglandular trichome. Light micrograph. Scale bar = 50 μm. With permission of A. M. Pridgeon.

Møller and Rasmussen (1984) reported conical silica bodies in the leaf of *C. irapeanum* and the absence of stegmata in leaves of *C. pubescens* and *C. reginae*.

Stem

Leafy stem

Rosso (1966) provided the only modern account of leafy stem and rhizome anatomy of *Cypripedium*. Epidermal cells are elongated longitudinally, more so than in other cypripedioid genera, and possess straight walls. Stomata are present (as in *Selenipedium*), but subsidiary cells are absent. Trichomes, both nonglandular and glandular, are present, especially in the upper portion of the shoot. Cortical parenchyma abuts both the epidermis and the sclerotic sheath of the vascular cylinder, which bears scattered vascular bundles (Solereder and Meyer 1930; Rosso 1966). *Cypripedium irapeanum* is unique in possessing arm parenchyma in the cortex. A sclerenchyma sheath encloses the vascular cylinder; among the cypripedioid genera *Cypripedium* species have the weakest sheath development. Amphivasal bundles basally give way to collateral bundles above the basal internodes, although amphivasal bundles persist at the nodes (Rosso 1966). Vessels are absent. Larger tracheids have scalariform pitting.

Rhizome

The rhizome of *Cypripedium* is more elongate than that of the conduplicate-leaved genera. The epidermis, often necrotic or sloughed in older plants, lacks hairs and stomates. Cell walls are impregnated with cutin or suberin. Cortical parenchyma is limited, bounded internally by a uniseriate endodermis. However, the endodermis of *C. irapeanum* is associated with several rows of suberized cells (Rosso 1966). The vascular cylinder comprises a uniseriate pericycle and amphivasal vascular bundles embedded in ground parenchyma (Rosso 1966; Atwood 1984). Vessel elements are absent; tracheids have annular, spiral, and scalariform thickenings (Rosso 1966).

Møller and Rasmussen (1984) recorded the absence of stegmata in the leafy stem of *C. irapeanum* and rhizomes of *C. pubescens* and *C. reginae*.

Root

A uniseriate epidermis with persistent root hairs is present in all species except *C. irapeanum*, which has a uniseriate velamen and exodermis like other genera in the subfamily (Rosso 1966). Long cells of the exodermis have U-shaped thickenings (Rosso 1966). Tilosomes are absent. The innermost layer of the cortex, the endodermis, is uniseriate. Endodermal walls are variously thickened depending on the taxon. Roots are 5–13-arch (Meinecke 1894; Rosso 1966), again depending on the species but also at different levels. Vessels with simple perforation plates and metaxylem tracheids, both with scalariform pitting, are present in roots (Rosso 1966). **(AP)**

Palynology

Pollen of *Cypripedium* is in monads, monosulcate, heteropolar, isobilateral, and spheroidal to elliptical (Schill and Pfeiffer 1977). Sculpturing is uniformly psilate except in *C. parviflorum* which is foveolate (Newton and Williams 1978). According to Burns-Balogh and Hesse (1988), the exine of *C. irapeanum* comprises a tectum with fragmented columellae and a lamellated foot-layer in the non-apertural region; in the apertural region the exine is much reduced. However, in other *Cypripedium* species the foot-layer is absent, and columellae are present only as fragmented globules embedded in the intine. The intine in the germination area is thin in all *Cypripedium* species except *C. irapeanum*, in which the intine is much thicker. **(AP)**

Cytogenetics

Much of the cytogenetic information available for *Cypripedium* has been collected by Karasawa and Aoyama (1986). Tanaka and Kamemoto (1974, 1984) have provided the most recent summaries of currently known diploid numbers (Table B.1). The genus is characterized by a near-uniform diploid karyotype of 20 large chromosomes. These may either be all metacentrics in sections *Bifolia* and *Retinervia*, 16–17 metacentrics and 3–4 telocentrics in section *Cypripedium* or 12 metacentrics and 8 telocentrics in section *Flabellinervia*. Earliest investigations by Pace (1907) suggested a base number of 11, but subsequent work by Belling (1924) changed this to a more generally accepted base number of 10. Occasional aneuploid numbers have been reported. **(AC)**

Phytochemistry

Relatively few phytochemical studies of Cypripedioideae have been undertaken, and only a few species belonging to the genera *Cypripedium* and *Phragmipedium* have been studied phytochemically. In the nineteenth century it was known that people who handled certain species of *Cypripedium* could contract severe inflammation of the skin. MacDougal (1894) reviewed a number of these cases, and the same author (1895) carried out experiments on himself and his students using *C. spectabile* (= *C. reginae*) and *C. parviflorum*. Six out of nine persons tested developed symptoms similar to contact dermatitis caused by poison ivy, *Rhus vernix*, whereas three people did not appear to be sensitive to the orchids. MacDougal reported that although these tests were a severe drain on the enthusiasm of his students, they were repeated with *C. parviflorum*, which exhibited the same activity on the skin of sensitive people. The irritant action of the plants appeared to be located in the top glandular cell of glandular hairs which are present in abundance on the plant surface of all three species (leaves, stems, fruits).

Much later, Schmalle and Hausen (1979) isolated a compound with allergenic activity from *C. calceolus* which is likely to be one of the culprits of the allergenic skin reaction described by MacDougal. The compound is a red quinone which, in sensitizing experiments carried out in albino guinea pigs, gave a positive eliciting reaction in sensitized animals. The structure of this allergenic quinone was elucidated and the compound named cypripedin. Thin-layer chromatography (TLC) of an extract of *C. calceolus* showed that the plant contained several more related substances which, together with cypripedin, belong to a group of rare non-terpenoid phenanthraquinones. Closely related compounds, phenolic phenanthrenes, have been found in many other orchid species as 'phytoalexins', antifungal substances which are synthesized in plants when attacked by pathogenic fungi and other microorganisms.

The flower pigments of *Cypripedium calceolus*, *C. cordigerum*, and *C. reginae* were investigated, and the anthocyanin chrysanthemin (= cyanidin 3-glucoside) was found in all three species (see Arditti 1992). *Cypripedium calceolus* and *C. reginae* were also studied for volatile plant constituents, as were *C. acaule*, *C. candidum*, and *C. formosanum* (Holman 1983). The fragrances and odours in *C. calceolus* include 2,6-dimethyloctatriene, benzene ethanol, phenol, anisaldehyde, and methyl anisole in leaves, stems, and roots, whereas *C. reginae* and *C. acaule* contain pyridine in lateral petals, gynostemium, staminode, stem and leaf; dimethyl butyric acid in gynostemium, staminode, and ovary; and methyl anisole in root, stem, and leaf. The volatile plant constituents in *C. candidum* consist of 2,6-dimethyloctatriene, 6-methyl-1-heptanol, and methyl anisole (leaves, stems and roots). Methyl anisole was detected in *C. formosanum*. **(RG and NV)**

Phylogenetics

Several factors complicate an understanding of infrageneric relationships within *Cypripedium*. Wide geographical distribution of the genus and restriction of many species to certain areas hinder a thorough study of all species. Most investigations of interspecific relationships (Table 3.1) have therefore been confined to accessible species such as those by Case (1993, 1994), who used isozyme data to assess the affinities of five North American species: *C. calceolus* and related taxa, *C. candidum*, *C. arietinum*, *C. acaule*, and *C. reginae*.

Despite selection mechanisms constraining the general morphology of the slipper orchid flower, a wide range of structural modifications has arisen in some species of *Cypripedium*. Similarities in these characteristics have been used repeatedly in assessing species relationships. The great diversity in floral morphology makes comparisons among taxa problematic and may prove to be systematically misleading because parallel adaptations to specific pollination mechanisms may have occurred; perceived similarity may be due to convergence on strategies that attract and trap pollinators.

Few phylogenetic analyses have been conducted on members of the plicate-leaved genera of Cypripedioideae. In the most recent phylogenetic studies (Albert 1994; Cox *et al.* 1997a) *Selenipedium* and *Cypripedium* form the first two successively diverging branches of Cypripedioideae (Fig B.2). This agrees with general views that these two genera possess what have been described as plesiomorphic character states, for example plicate leaves, crustose seeds (in the case of *Selenipedium*) and multiple-flowered inflorescences. However, the precise status of these characters in Orchidaceae is uncertain, and more detailed investigations are needed to determine

Table 3.1. Taxonomy of genus *Cypripedium* (modified from Cribb 1997)

Lindley (1840)	Kränzlin (1901)	Pfitzer (1903)	Brieger (1973)	Chen and Xi (1987)	Cribb (1997)
Foliosa group	**Sect. Calceolaria Calceolus group**	**Series Arcuinervia Sect. Eucypripedium Subsect. Obtusipetala**	**Subgen. Cypripedium**	Sect. *Subtropica*	Sect. *Subtropica*
				C. subtropicum	*C. subtropicum*
					C. wardii
				Sect. Cypripedium Subsect. Obtusipetala	**Sect. Irapeana**
	C. californicum	*C. californicum*	*C. californicum*	*C. californicum*	*C. californicum*
C. irapeanum	*C. irapeanum*	*C. irapeanum*	*C. irapeanum*	*C. irapeanum*	*C. irapeanum*
C. molle					*C. molle*
					C. dickinsonianum
		Series Arcuinervia Sect. Eucypripedium Subsect. Acutipetala		**Sect. Cypripedium Subsect. Cypripedium**	**Sect. Cypripedium Subsect. Cypripedium**
C. calceolus	*C. calceolus*	*C. calceolus*	*C. calceolus* (incl. *C microsaccos*)	*C. calceolus*	*C. calceolus*
	C. henryi	*C. henryi*	*C. henryi*	*C. henryi*	*C. henryi*
				C. shanxiense	*C. shanxiense*
					C. segawei
C. cordigerum	*C. cordigerum*	*C. cordigerum*	*C. cordigerum*	*C. cordigerum*	*C. cordigerum*
C. parviflorum (incl. *C. pubescens*)		*C. parviflorum* (incl. *C. pubescens*)	*C. parviflorum* (as *C. pubescens*)	*C. parviflorum* (as *C. pubescens*)	*C. parviflorum* (incl. *C. pubescens*)
	C. candidum	*C. candidum*	*C. candidum*	*C. candidum*	*C. candidum*
C. montanum		*C. montanum*	*C. montanum*	*C. montanum*	*C. montanum*
				Sect. Cypripedium Subsect. Obtusipetala	
	C. fasciolatum	*C. fasciolatum*		*C. fasciolatum*	*C. fasciolatum*
				C. farreri	*C. farreri*
	Sect. Calceolaria Macrantha group			**Sect. Cypripedium Subsect. Cypripedium**	**Sect. Cypripedium Subsect. Macrantha**
		C. himalaicum	*C. himalaicum*	*C. himalaicum*	*C. himalaicum*
				Sect. Cypripedium Subsect. Obtusipetala	
C. macranthos	*C. macranthos*	*C. macranthos* (incl. *C. thunbergii*)	*C. macranthos*	*C. macranthos*	*C. macranthos*
			C. tibeticum	*C. tibeticum*	*C. tibeticum*
	C. corrugatum	*C. corrugatum*	*C. corrugatum*		*C. corrugatum*
					C. smithii
		C. yunnanense	*C. yunnanense*		*C. yunnanense*
					C. ludlowii
					C. franchetii

Table 3.1. Taxonomy of genus *Cypripedium* (modified from Cribb 1997) (continued)

Lindley (1840)	Kränzlin (1901)	Pfitzer (1903)	Brieger (1973)	Chen and Xi (1987)	Cribb (1997)
	Sect. Calceolaria **Calceolus group** *C. fasciculatum* **Sect. Calceolaria** **Obtusiflora group**	**Series Arcuinervia** **Sect. Enantipetalum** *C. fasciculatum* **Series Arcuinervia Sect.** **Eucypripedium Subsect.** **Onbtusipetala**	**Subgen. Fasciculata** *C. fasciculatum*	**Sect. Cypripedium Subsect.** **Enantiopetalum** *C. fasciculatum*	**Sect. Enantiopetalum** *C. fasciculatum* **Sect. Obtusipetala**
C. reginae (as *C. spectabile*) *C. passerinum*	*C. flavum* (as *C. luteum*) *C. reginae* (as *C. spectabile*) *C. passerinum*	*C. flavum* (as *C. luteum*) *C. reginae* *C. passerinum*	*C. flavum* (as *C. luteum*) *C. reginae* *C. passerinum* *C. wardii*	*C. flavum* *C. reginae* *C. passerinum*	*C. flavum* *C. reginae*
Acaulia group		**Series Arcuinervia Sect.** **Eucypripedium Subsect.** **Acutipetalum**	**Subgen. Acaulia**	**Sect. Cypripedium Subsect.** **Obtusipetala**	**Sect. Acaulia**
C. acaule	*C. acaule*	*C. acaule*	*C. acaule*	*C. acaule*	
Bifolia group	**Bifolia group** **Bracteosa subgroup**	**Series Retinervia**	**Subgen. Retinervia**	**Sect. Retinervia**	**Sect. Retinervia**
	C. elegans *C. debile*	*C. debile* **Series Arcuinervia Sect.** **Eucypripedium Subsect.** **Obtusipetalum**	*C. elegans* *C. debile* **Subgen. Guttata**	*C. palangshanense* *C. elegans* *C. debile* **Sect. Bifolia**	*C. palangshanense* *C. elegans* *C. debile* **Sect. Bifolia**
C. guttatum	*C. guttatum*	*C. guttatum* **Series Flabellinervia**	*C. guttatum* *C. yattabeanum* **Subgen. Flabellinervia**	*C. guttatum* **Sect. Cypripedium subsect.** **Flabellinervia**	*C. guttatum* *C. yattabeanum* **Sect. Flabellinervia**
C. japonicum	*C. japonicum*	*C. japonicum*	*C. japonicum*	*C. japonicum*	*C. japonicum* *C. formosanum* **Sect. Arietinum**
Arietinum group	**Arietinum group** *Criosanthes*	**Series Arcuinervia Sect.**	**Subgen. Cypripedium**	**Sect. Criosanthes**	
C. arietinum	*C. arietinum*	*C. arietinum*	*C. arietinum*	*C. arietinum* *C. plectrochilum*	*C. arietinum* *C. plectrochilum*
Bifolia group Ebracteata subgroup *C. micranthum*	**Series Arcuinervia** **Sect. Enantiopedilum** *C. micranthum*	**Subgen. Ebracteata** *C. micranthum* *C. bardolphianum*	**Sect. Trigonopedium** *C. micranthum* *C. bardolphianum*	**Sect. Trigonopedia** *C. micranthum* *C. bardolphianum*	
C. margaritaceum *C. fargesii* (as *C. ebracteatum*)	**Series Arcuinervia Sect.** **Trigonopedilum** *C. margaritaceum* *C. fargesii* (as *C. ebracteatum*)	*C. margaritaceum* *C. fargesii* (as *C. ebracteatum*)	*C. margaritaceum* *C. fargesii*	*C. margaritaceum* *C. fargesii*	
			C. wumengense	*C. wumengense* *C. forrestii* *C. lichiangense*	

their general distribution. Until such characters are better understood it is too speculative to assume that these are 'primitive'; instead they could simply be autapomorphic. Many of these assessments depend on knowing more about outgroup characters, which to date have been poorly examined.

Infrageneric relationships in *Cypripedium* have been little-studied due largely to the scarcity of living material with which to work. However, molecular (Cox 1995; Cox *et al.* 1997*a*), morphological (Cribb 1997), and floral fragrance (Barkman 1993) analyses have recently been performed on representative subsets of the species. Unfortunately, species representation in these studies does not fully overlap, and therefore interpretation of chemical and morphological data in the context of molecular evidence (and vice versa) is not possible. Despite such restrictions, several points have been highlighted by these studies. *Cypripedium irapeanum* and allied species are sister to the rest of the genus (Fig. 3.6). These taxa resemble members of the earliest diverg-

ing slipper orchid genus, *Selenipedium*, in general plant habit, flower lip morphology, possession of two purple spots on an involute lip margin, and multiple-flowered inflorescences. *Cypripedium* has been considered by some (e.g. Rosso 1966) to grade into *Selenipedium* via *C. irapeanum*. *Cypripedium fasciculatum* also appears consistently to diverge near the base of the topology although current data sets do not place it consistently.

The 5S rDNA data of Cox (1995) identify two further major species groups that receive some degree of support from subsequent studies (Fig. 3.7). The first comprises all the members of yellow- and red-flowered, multi-leaved species. This group appears to show separation of the North American yellow-flowered slipper orchids from Eurasian yellow- and red-flowered species.

The second major group contains the remaining *Cypripedium* species. These are a morphologically diverse and geographically widespread group of taxa. Generally there appears to be a trend in reduction of leaf number in the more derived taxa. Species such as the derived *C. margaritaceum* have undergone a marked reduction in vegetative parts, retaining a single leaf and an enlarged floral bract. Currently available phylogenetic data fail to produce any congruent picture of species relationships, and further studies are required to provide an accurate phylogeny for the genus. **(AC)**

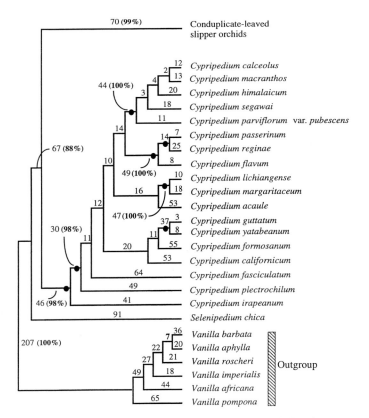

Fig. 3.6. One of the most-parsimonious weighted trees from Cox *et al.* (1997*a*) showing cladistic relationships of the plicate-leaved genera *Cypripedium* and *Selenipedium*. Solid circles indicate clades strongly supported by the jackknife procedure (node score >0.63).

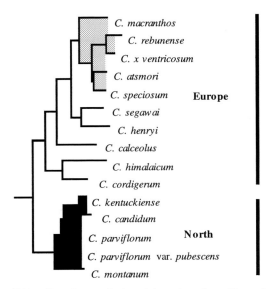

Fig. 3.7. Species relationships in the *C. calceolus/ C. macranthos* clade interpreted by Cox (1995). The *C. macranthos* clade is indicated by stippled shading. The North American yellow-flowered clade is shown with solid shading.

Ecology

Our knowledge of the ecology of most hardy slipper orchids is still poor. This is particularly true of Asiatic and Mexican species, including those which are most narrowly endemic. The current vogues in many countries for habitat restoration and management and for reintroduction of rare species have emphasized the need for more information in order to provide optimal conditions for reintroduction. Detailed ecological observations and analyses are lacking for most species of *Cypripedium*, but the general ecological preferences of a number of species have been recorded. Most species grow in colonies. The colonies are small in some species, such as *C. elegans*, *C. margaritaceum*, and *C. lichiangense*, but can be extensive, covering hundreds of hectares, in species such as *C. macranthos* and *C. tibeticum*. Much depends on the extent of suitable habitat because most species seem to have rather specific edaphic, climatic, and shade requirements.

Seasonality

All temperate slipper orchids are found in climates with a distinct seasonality, growing in the spring, flowering in early to mid-summer and setting seed in the autumn. They survive dormancy by storing food reserves in their fleshy rhizomes, and the dormant shoot lies with its tip at or just below the soil surface. Growth commences as the weather improves in early spring, influenced by increasing warmth and moisture. The majority of species flower in May, June, and early July. However, flowering can be as early as April in species such as *C. kentuckiense* and *C. formosanum* and as late as August in *C. irapeanum* and *C. molle*.

A number of Asian species, such as *C. guttatum*, *C. himalaicum*, and *C. tibeticum*, are found most commonly in montane grassland and meadows. All grow at elevations of up to 4000 m or more, and the grassland is covered by deep snow in the winter and early spring. As the snow melts in May, *C. tibeticum* emerges among the brown grasses and flowers as its young leaves develop.

Habitats

Temperate slipper orchids are found over an extensive latitudinal range, extending from north of the Arctic Circle (about 70° North) in Alaska to 25° North in south-west China and even farther south (about 14° North) in Central America. In the northern part of their distribution slipper orchids are found from sea level into the lower hills, occupying coniferous forests, mixed deciduous woodlands, bogs, fens, grasslands or prairies. Farther south they usually inhabit montane woodlands and grasslands at higher elevation. *Cypripedium himalaicum* has been reported from as high as 4900 m in the Himalayas.

They usually grow in light, moderate or deep shade. *Cypripedium tibeticum* can thrive in full sun but, in contrast, Griesbach and Asher (1983) reported that *C. acaule*, *C. parviflorum* var. *parviflorum*, and *Cypripedium arietinum* grow in deep shade on raised acid bogs and their margins. However, in other woodland species, forest clearance may allow temperate slipper orchids to persist in more open situations, but reproductive success may be much reduced. Few thrive in full sun, but *C. candidum* is a species of the North American prairies. *Cypripedium guttatum* and *C. tibeticum* are also commonly found in high-elevation grasslands in Asia. A number of species grow in bogs, fens or along the sides of streams. *Cypripedium reginae* and *C. parviflorum* var. *parviflorum* are characteristic plants of the calcareous fens of North America. *Cypripedium californicum* grows along turbulent mountain streams in northern California and southern Oregon and must have its rhizomes in damp soil most of the growing season.

Perner (personal communication) examined the distribution of *C. macranthos*, *C. calceolus*, and *C. shanxiense* in Primorsk in Siberia. He also noted that *C. calceolus* and *C. shanxiense* grow in slightly drier sites that *C. macranthos*. *Cypripedium shanxiense* grows in more shaded situations than *C. calceolus*.

Substrate and pH

In Europe, *C. calceolus* is a well-known calcicole growing in alkaline or rarely neutral soils. In England its former localities were on oolitic and magnesian limestone on north-facing slopes in light shade of oak and hazel woodland (Summerhayes 1968). However, Perner (personal communication) reports that he found it growing in slightly acidic conditions (pH 6) in eastern Siberia. In central Europe it grows in conifer forest and on the edges of mixed deciduous woods, often in flat or sloping sites, up to an elevation of 2000 m or more. In places it forms large colonies with individual plants bearing many shoots.

Other plants often in the same habitat can comprise a single shoot, the propensity to produce clumps related to the branching of the rhizome.

Many other Eurasian species are apparently calcicoles, the extensive limestone regions of eastern Asia being host to many *Cypripedium* species. Species such as *C. flavum*, *C. margaritaceum*, *C. lichiangense*, *C. forrestii*, *C. henryi* and *C. debile* grow in mixed deciduous woods of poplar, birch, pine, fir, and larch. The substrate is hard limestone and the soils thin, comprising leaf litter and humus derived from the leaves of the forest trees with its understory dominated by rhododendrons. Such soils may not be alkaline; the author has measured a pH of around 5.5. The fact that the habitats often sustain a rich ericaceous flora confirms this. More work is needed to establish the effects of substrate on the ecology of Asiatic species. Perner (personal communication) measured the pH of soils where *C. macranthos* and *C. shanxiense* grow in eastern Siberia and found the soils ranged from neutral to slightly acidic (pH 6).

The ecological preferences of the North American species have been catalogued by a number of authors (e.g. Morris and Eames 1929; Correll 1950; Luer 1975; Sheviak 1983; Griesbach and Asher 1983; Case 1987). Several species grow in acidic conditions, but Case makes the point that some of these grow in other habitats as well, for example *C. acaule* is recorded from acidic, mineral-poor *Larix-Picea*-dominated bogs and other marshy habitats but also in pine-barrens on deep dry sands; *C. parviflorum* var. *pubescens* can grow in neutral or alkaline bogs; acidic, sterile, damp sandy soil with little vegetation; and in neutral to alkaline soils in deciduous woods and beach flats. Sheviak (1983) considered that soil chemistry is the most critical factor in the establishment of orchids because of its effect on the mycorrhizal fungi. The solubility of nutrients is particularly affected by pH, and levels of critical nutrients may allow a species to grow in both acidic and alkaline habitats but not in neutral ones where elements such as phosphorus, manganese, and boron are present in higher concentrations. Griesbach and Asher (1983) noted that *C. reginae* grows equally well in slightly acid (pH 6.3) or slightly alkaline soils (pH 7.8).

Rainfall and moisture

All temperate slipper orchids need adequate moisture during the growing season, but their tolerance of moisture in the substrate can vary substantially. *Cypripedium acaule* is recorded from both very wet habitats such as *Sphagnum* bogs and dry deep sands. Sharp drainage is a feature of most substrates that slipper orchids occupy. Adequate water supply during the growing season is essential, but too much moisture during dormancy may lead to shoot rot, as many growers have found to their cost.

Succession

The effects of vegetational succession can be dramatic on temperate slipper orchid populations. Case (1987) reported that a thriving colony of *C. reginae* growing in a balsam fir-cedar bog declined dramatically when deer grazing was reduced. The deer heavily grazed the cedar, creating open lightly shaded or sunny glades in the forest where, in 1946, over 2000 plants flourished. By 1957 the reduction of the deer population by hunting had led to the regeneration of cedar and the shading of the glades. By 1960 most of the plants had disappeared or become spindly, almost like seedlings. Clearance of cedar by hand allowing more light into the forest resulted in the reappearance of flowering plants by 1962.

It is not unusual to find areas where a few old clumps *of C. calceolus*, *C. parviflorum* var. *pubescens* or *C. acaule* survive in old woodland but with little or no sign of young seedlings. Conversely, in neighbouring areas where older trees have been cleared, seedlings may be common but large plants absent. Habitat management has been successful in restoring slipper orchid habitats to optimal level. In Europe there are several sites of *C. calceolus* in Germany and Switzerland that are carefully managed to maintain orchid populations at high levels. Clearance of heavily shaded areas improves light levels and provides bare soil where seed germination and seedling growth is encouraged. In North America prairie management regimes have increased the populations of the rare *C. candidum* in several middle American states. **(PC)**

Pollination

Darwin (1862), in his hugely influential account of orchid pollination, mistakenly suggested that pollination of *Cypripedium calceolus* was effected by an insect sitting on the outside of the slipper-shaped lip and reaching the pollen by inserting its mouth-parts through the basal orifices of the lip.

Asa Gray (1862) was the first to study the pollination of *Cypripedium* in nature. He examined the pollination of three American species, noting that bees entered the lip of these species through the large opening on its upper surface, but he did not observe effective pollination. Müller (1868, 1873) observed bee pollination of *Cypripedium calceolus* in the wild and confirmed Delpino's (1867) theory that pollination occurred when the bees crawled out of the lip through the small openings on either side of the column. Darwin (1877) also confirmed this by placing a small *Andrena* bee in the lip of a flower of *C. calceolus*. It failed to escape the way it had entered because of the infolded lip margins and eventually crawled out through one of the openings at the base of the column and lip. On examination it was found to be smeared with the glutinous pollen. Gray had noted that the stigmatic surfaces of most species were covered by 'minute, rigid, sharp-pointed papillae, all directed forwards, which are excellently adapted to brush off the pollen from the insect's head or back'. He also noted that *C. acaule* has a slightly concave stigma which is viscid and also drier, rather granular pollen.

Nilsson's (1979) observations on the pollination of *Cypripedium calceolus* on the Swedish island of Öland have been the most detailed. He studied seasonal and diurnal activity, approach, alighting, entering, and escape of different insects visiting the orchid's flowers over a four-year period. A number of visiting insects, notably bees and hover flies, entered the lip. They were attracted by flower colour, the floral fragrance which is rich in acetates that appear to mimic pheromone secretions of some bees, and also by the spotting patterns of the staminode and lip which are false nectar guides. He suggested that the fragrance chemicals might upset the landing control pheromones of bees alighting on the lip, thereby increasing the likelihood of them slipping into the pouched lip.

The presence or absence of food in the lip has been much debated. Darwin (1862) was intrigued by the flower's attraction and noted that the hairs within the lip secreted little drops of a slightly viscid fluid. He noted that this liquid formed a brittle crust on the summits of the hairs and suggested that if this were sweet or nutritious it would explain the attraction of the flowers for bees. The occurrence of 'nectar' in the lip has been widely and erroneously reported (Müller 1869; Summerhayes 1951; Proctor and Yeo 1973) and has been interpreted as food for pollinators. Nilsson refutes this because the hairs contain oil not nectar. According to Ziegenspeck (1936) and Daumann (1968) the hairs within the lip are not eaten or chewed; however, Stoutamire (1967) reports chewed hairs. Overall, it seems probable that pollinators are attracted by deceit to the trap flowers of *Cypripedium*.

Once inside the lip the insects seldom escape by the route through which they entered. They exit, if at all, out of the base of the lip through the basal orifices. The expanded stigma acts as the essential support to allow pollinators to bend down the lip and thence pass under the anthers and out through the basal orifices. Only bees of the right size, neither too large nor too small, can pass out of the lip this way to effect pollination. Nilsson found that the most frequent and regular pollen vectors were female bees of *Andrena haemorrhoa* (F.) but that other species can also occasionally pollinate flowers.

The lips of many species have areas of unpigmented tissue, so called 'windows', in the lateral part of the posterior region of the lip. These have been suggested as inducing phototaxis in pollinators to lead them out the correct way (Webster 1886; Troll 1951; Faegri and van der Pijl 1971). However, the evidence for this is inconclusive (Nilsson 1979).

Although the pollination of *Cypripedium calceolus* in Europe has attracted most attention, that of other species has also been studied or observed, particularly in North America. Stoutamire (1967) and Catling (1985) have provided summaries of these observations. The former commented that the sizes of the lip mouth and basal orifices and of the escape route under the stigma and anthers determined the likely pollinators.

Bee pollination has been suggested as typical of the pollination syndrome of *Cypripedium* by van der Pijl and Dodson (1966). Certainly, observations of most North American species have confirmed that bees are the most frequent pollinators: *C. acaule* is pollinated by *Bombus vagans* (Stoutamire 1967); *C. reginae* by *Megachile melanophaea* and *M. centuncularis* (Guignard 1886); *C. arietinum* by *Dialictus* sp. (Stoutamire 1967); *C. parviflorum* var. *pubescens* by *Ceratina calcarata* males (Stoutamire 1967); *C. candidum* by *Andrena placida* and other short-tongued bees in the Andrenidae and Halictidae (Stoutamire 1967; Catling and Knerer 1980).

Stoutamire (1967) suggested that the widespread *C. debile* from China, Japan, and Taiwan, which has flowers with a mushroom scent, might be fly-pollinated. The flowers of the Chinese *C. margaritaceum*, *C. lichiangense*, and their allies have a coloration and fragrance that indicates fly pollination, and flies have been seen approaching the flowers of both species.

Catling (1985) reported self-pollination in *C. passerinum*. This has also been suggested for the Mexican species *C. dickinsonianum* by Hágsater (1984).

Natural hybridization has been reported on several occasions, e.g. between *C. calceolus* and *C. macranthos* (Stoutamire 1967; Slyusarenko 1981); *C. parviflorum* and *C. candidum* (Klier *et al.* 1991); and *C. parviflorum* var. *pubescens* and *C. montanum*. **(PC)**

Uses

Cypripedium species are becoming popular subjects for growers of hardy orchids. Plants have been gathered for years from the wild to satisfy this demand, but the ease with which plants can now be grown from seed as asymbiotic germination techniques improve should reduce the impact of this trade on wild populations. While the majority of cypripediums will perhaps always remain a subject for the specialist, some of the more easily grown species and the newly available hybrids may well be grown more widely in the future.

The horticultural interest in slipper orchids has tended to overshadow other aspects of their use. However, the herbal use of *Cypripedium* in North America has been well documented (Rafinesque 1828; Bown 1995). The orchid most frequently cited is *C. parviflorum* var. *pubescens*, the rhizomes being 'collected in the autumn, or early in the spring, carefully dried and reduced to a powder' and administered as 'a teaspoon of powder, diluted in sugar water, or any other convenient form' (Rafinesque 1828). Nowadays, a tincture (1 part orchid rhizome to 5 parts of an ethanol/water mixture) is extracted from the dried rhizomes. The tincture can be stored indefinitely and used to treat several disorders such as insomnia, anxiety, headache, neuralgia, emotional tension, palpitations, tremors, irritable bowel syndrome, delirium, and convulsions due to fever. Generally, it has a sedative, nervine, anodyne or antispasmodic effect. The Cherokee Indians used it to cure worms in children. It has also been reported

as being used by native North Americans in childbirth and as a sedative in serious illness having the relaxing capabilities of opium without the side effects. The ingredients of the tincture are glycosides, volatile oil, tannin, and resin. Care has to be taken because overdosing can cause hallucinations, and harvesting the plant can cause contact dermatitis.

In North America other taxa such as *C. parviflorum* var. *parviflorum*, *C. acaule*, *C. reginae*, and *C. candidum*, with decreasing efficacy, have also been reported as being collected for this use. Modern herbalists tend to emphasize that plants must not be collected from the wild and that generally other plants with similar properties can be successfully substituted for *Cypripedium* (Bown 1995).

Cypripediums are also used in north-east China (Zhu 1989). *Cypripedium guttatum* has many uses. The stems, rhizomes, and flowers can be used medicinally. A soup of the stems has the effect of dilating the blood vessels, stimulating appetite, and curing gastritis. A tincture of the flowers is a strong sedative, effective in various neurological and mental disorders, and especially effective for epilepsy. It is also effective against childhood convulsions, caused by high fever, headaches, and pain in the upper abdomen. It has uses as a diuretic and diaphoretic. There are rumours that it has also been effectively used against cancer.

The rhizomes, stems, and leaves of *C. macranthos* are also used medicinally in China as a diuretic, detumescent, for removing blood clots, improving circulation, and for relieving rheumatic pains. It has been used for curing dropsy, oedema of the legs, leucorrhea, gonorrhea, dysentery, rheumatic pains of the back and legs, fractures, and internal lesions caused by overexertion. The recommended dosage is 10–15 g boiled in water or as a tincture. A powder made of the dried flowers can be used to stop bleeding. **(PC)**

Cultivation

Here we discuss only the fundamentals of cultivating cypripediums. For detailed information on cultivation under lights, compost, watering, fertilizing, pests and diseases, and cultural recommendations for species and hybrids, please refer to Perner (1997).

General requirements for cultivation

Cypripediums are adapted to the temperate and cold climates of the northern hemisphere. Their delicate leaves cannot stand high temperatures and low humidity for long. Most species prefer open shade with some sun during part of the day. Usually sunnier sites give more stems and more flowers, but these tend to be smaller and more compact than in shadier situations. Full sky light (north side of a building with nothing blocking the sky) is very good, with direct sunlight only in the early morning and evening. The soil should stay cool and the humidity high (not below 50% for long periods) with good air movement to prevent rotting. Too strong a wind can bend the stems over. To avoid this and to support the ambient microclimate, companion plants such as ferns are helpful. However, these plants must not overcrowd the cypripediums. Sites must be well drained, but watered as necessary for the species. Bog and fen species do not need to be wet, and overly wet conditions are dangerous.

When in full dormancy only very few species are unable to withstand severe frosts in winter. They require a cold, constant winter. The more that dormancy is disturbed by warm spells initiating growth, the more difficult it is to overwinter cypripediums. A mediterranean climate with mild wet winters and dry hot summers is disadvantageous for cultivating them. The grower has to correct such conditions with sophisticated technical equipment in an artificial environment. However, in the climate zones colder than the Mediterranean the grower will find adequate conditions if the summers are not too dry and hot. In the United States *C. parviflorum* var. *pubescens* can be grown as far south as northern Louisiana, and *C. irapeanum* grows as far south as Central America. When warm spells occur in winter, plants adapted to a constant continental winter (i.e. species from the Himalaya and its ridges, from Siberia and boreal North America) should be protected from the warmth by a cover that is to be removed immediately the frost returns. If such warm spells are frequent in a given area, it would be easier to grow these plants in pots that are plunged in a bed or open frame during summer, and stored in a cold room (between 0° and 4°C) or in a closed frame (protected from sunshine and rain), in winter. Species from temperate north-east America and coastal eastern Asia can usually withstand these warm spells and do not need a special treatment in winter, at least if they are established.

In spring, late frosts may harm the developing growth. The more the growth has developed the more harmful is frost. Flower buds and flowers are usually killed by even the slightest frost. At this period, the grower should watch the weather forecast and protect the plants overnight if a frosty night is announced.

In summer the average daytime temperature should not be far above 20°C and the humidity not too far below 50% for a long period. In particular around noon the plants should be protected from direct sun. Good air movement is necessary to prevent rotting.

The beginning of dormancy in autumn differs within the genus but can also be disrupted by low moisture in combination with high temperature in the latter part of the season, producing the same results as early frost. The plants should be encouraged to stay green as long as possible.

For most species it is not necessary to give any special protection in winter if the plants are well established. A slight mulching with pine needles, fresh fallen beech leaf litter or other materials is sufficient. If the mulch is too deep it can become too compact or it may occasionally attract pests such as mice. In late winter and early spring an airy cover of spruce twigs can be beneficial with some species such as *C. macranthos*, *C. henryi* or *C. formosanum*. Some growers prefer to clean the bed of the dead growths as a precaution against pests and diseases. It should be done after the stems are completely dead. If the soil is frozen the old stems can pulled out; when soil is not frozen it is necessary to cut the dead and dried stems. The dead material should not be composted but placed with the domestic rubbish for disposal. Other growers usually leave the stems of healthy plants until spring to use them as a natural protection that also helps to keep some mulch in place. However, with the emergence of the new growths the old stems should be removed.

Cultivation under glass

Species or clones that are adapted to alpine or arctic climates, i.e. long constant winters without warm spells, short springs and cool damp summers are suitable subjects for a greenhouse that provides cool conditions. Here the growing conditions can be controlled better than in the garden. Any other *Cypripedium* that can be grown in the garden can also be grown in these conditions. However, in suitable climates there is no better place to grow most cypripediums than in the open air. This can be the garden, an open frame or a shadehouse that ideally unites the advantages of greenhouse and garden.

An alpine house is necessary to grow plants successfully under glass. A location providing shade around noon and good shading of the house itself is important to keep temperatures low in summer. Ideal temperatures are 18–20°C (max. 25°C) by day in summer with a humidity of 60–80%. In the night the temperature should fall about 10°C. Good air movement is important for cultivation in the glasshouse, so sufficient fans are essential. In winter the temperature should be around 0°C, i.e. it should not rise above 5°C and not fall under –2°C.

Direct sunlight not only heats up the atmosphere in the glasshouse but also the pots and substrates. A general requirement is to keep the compost cool. Besides sufficiently shading the glasshouse, plunging the pots in raised sand beds can be helpful. However, pathogenic fungi are likely to occur in glasshouse cultivation, and they can spread rapidly from pot to pot if these are plunged.

A shadehouse is probably the safest way to grow cypripediums. The plants are protected from most climatic hazards. In summer the walls consist of wire netting or screen wire, allowing a full exchange of air, while pests are excluded. In winter the walls can be closed for better temperature control. The roof should be made of polycarbonate or glass, with sufficient shading from late spring until autumn. Ideally plants are grown in beds at ground level or in suitable pots and containers placed on benches in summer, then plunged in sand beds in winter to protect roots and rhizomes from too much frost. The shadehouse is often used by commercial growers.

A reliable compromise for the amateur is a cold frame. In many respects it is similar to shadehouse. Protection from digging animals, hailstorms, late frosts, etc. is easily given in a frame, and it provides overwintering even for delicate species. A frame can cover a growing bed for plants or a sand bed into which to plunge pots and containers. Usually there is room, even in the smallest backyard, for a frame. The right spot for one is the north wall of a house or any other place that provides relatively cool and humid conditions. The frame should only be closed during frost. With higher temperatures the cover should be lifted but not removed, providing protection from sun and rain and avoiding stagnant air. In summer the cover should be removed completely (except for hailstorms, etc.) and, if necessary, only shading should be attached.

Garden cultivation

In suitable climates the best method of growing cypripediums is in the open garden. All the conditions necessary to promote robustness and health of the plants, such as wind, temperature changes, and ultraviolet light, are provided in abundance in the open. Certainly the shadehouse is better if it should become too chilly, but nothing matches the aesthetics of lady-slippers in flower under the open sky. If in late spring a frost is announced for the night, the grower can simply place a bucket over the exposed plants in the evening and remove it the next morning to save the flowers and leaves.

For garden cultivation there is also the choice of growing cypripediums in either pots or in a bed. Pots are recommended if the garden soil is unsuitable and the grower does not want to prepare a special bed, or if the hardiness of a given *Cypripedium* is suspect in the local climate. In late autumn pots can easily be removed from the garden and put in a sheltered place (e.g. an unheated room in the basement). If digging rodents are a problem, growing cypripediums in plunged pots can help to protect the rhizomes and roots.

The most vigorous and prolific specimens are usually produced in specially prepared beds where the plants can grow undisturbed for many years, provided the compost is still open and not exhausted. A few robust species like C. *parviflorum* var. *pubescens* and most clones of var. *parviflorum* can grow in any decent garden soil, as long as it is loose and open, with much air and not overwatered. However, for a given species such treatment is satisfactory only for regions with a suitable climate. There is good reason to expect that in the coming years more selected clones of species and hybrids, suitable to thrive in the border or woodland bed with other perennials, will become available. However, a specially prepared bed is recommended for most cypripediums. It should be situated in a north, north-east or east-facing situation, protected by shrubs, ferns or open trees, but not overcrowded. A micro-climate that provides cool soil, fresh air, humidity, and open shade with some sun in the morning or evening is suitable for most species. The bed should be dug out to about 40 cm, then filled with a 10 cm layer of drainage material, e.g. pebbles or coarse gravel. If the ground is free-draining, it is enough to dig only to a depth of 30 cm and not provide extra drainage. If the surrounding soil is open, light, and not very different in its chemical composition to the growing compost, it is not necessary to insulate the walls of the bed with plastic or other materials. Otherwise this should be done. After filling the bed with prepared compost to about 2 or 3 cm above the level of the surrounding soil, it is ready for planting. After planting the surface should be mulched with pine needles to prevent the compost from becoming compressed by rain or by watering. It also helps to keep away blackbirds or cats. A cover with living moss (a common moss found under shrubs or on tree stumps, e.g. *Brachythecium* or *Eurhynchium*, but not *Sphagnum*) is a good alternative to mulch. It acts as an excellent indicator of the correct humidity, but it attracts blackbirds which will destroy an unprotected moss cover within minutes.

Good companions for cypripediums in open beds are ferns. One of the best displays for hardy slipper orchids is a rockery planted with ferns such as *Dryopteris filix-mas*, *Athyrium filix-femina*, and other non-invasive ferns. *Matteucia struthiopteris* should be avoided because the need to remove its strong runners can injure nearby cypripediums, yet if not removed they overgrow everything. Groups of *C. macranthos*, *C. calceolus*, *C. ventricosum* etc., in full flower in front of the young green of recently unfolded fronds, give an impressive display. Enough room has to be given between the fern stems and the cypripediums because the adult ferns can reach enormous dimensions when fully developed in summer.

Some moisture-tolerant species such as *C. reginae* and *C. parviflorum* var. *parviflorum* can also be grown in artificial bogs. To some degree they can tolerate high moisture but they don't need it in cultivation. In North America the so-called Holman bog is sometimes used. This is constructed like a small garden pond but filled with substrate. In the centre a mound of compost is created to offer different levels of moisture, from wet at the base to mesic at the top. This hummock should be the growing site of the cypripediums. Such an artificial bog does not provide the optimal conditions for growing them in the garden, but it helps to keep the compost moist when the grower is absent for a longer period in summer.

Vegetative propagation

An easy way of propagating cypripediums is to divide a clump, if it has a reasonable number of growths. The more growths, rootstock-mass, and roots each division has the better. But a single growth can also be cut from the clump. Such a division needs at least two or three scars showing former growths. This ensures that enough food reserves and roots are provided for the separated growth. With species having compact-growing rhizomes it means the new division has at least 3 cm of rootstock remaining before the bud. For creeping rhizomes (e.g. *C. formosanum*) pieces of 15–25 cm in length must be cut. Generally it is wise to let a specimen grow into a clump rich in stems. Such a specimen plant makes a better show and is usually much stronger and develops more flowers.

When dividing an old rhizome in the autumn, the back part can be treated like the backbulb in tropical orchids. If it is cut (make sure that the division is not too small), reserve buds will often start to grow. This can be promoted by putting the back part in living sphagnum moss, wrapped in a plastic bag, for a few months.

Divisions should be made with a clean sharp knife. The cut can be powdered with charcoal to dry it and to protect it from infections. Some growers recommend that the cut be allowed to dry in the air before replanting. If this is done, it is necessary to protect the roots from the harm of drying out. Other growers immediately plant the divisions in the new compost without special treatment for the cuts to prevent too much disturbance.

With tropical orchids the best time to divide a plant is before the onset of the new year's growth. The same is true for cypripediums. Immediately after flowering the new growth hardens up, the food reserves have been depleted from the roots to a maximum extent, and the new food reserves are moving down again from the leaves. The new roots are starting to grow at this time. Hence, the division will establish itself quickly. The divisions have to be replanted immediately to ensure water supply and prevent wilting of the leaves. If the remaining root system is too small in comparison with the leaf system, cutting off the top half of the stem (with half the leaves) can be beneficial.

Division in autumn is the standard technique with commercial growers because these divisions can be shipped. However, division in autumn does not permit establishment of the plant until the following summer, so there is a long period during which problems may develop.

Micropropagation

Propagation of cypripediums from seed is routine today. However, it is not very easy and needs some hands-on experience and technical equipment. In this chapter only some general information can be given. Details are provided by Steele (1995) and Rasmussen (1995).

As in tropical orchids the seeds are sown out on a sterile medium without symbiotic fungi. The deflasking would be easier with symbiotically raised seedlings but, at present, no reliable symbionts of *Cypripedium* species are available; and seedlings often have different symbionts from mature plants.

Two different methods of harvesting and sowing the seeds are available. One is to take immature seeds in green capsules (green pod method), sterilize the capsule, and take out the sterile seeds. They have not developed any inhibition to germination (necessary to survive the winter in nature) if harvested at the right time and will germinate quickly. Depending upon the species, condition of the specimen and climatic influence the right time lies between 5 and 12 weeks after fertilization. At this time the embryo is still attached to the placenta and comprises between 9–12 cells. Werner Frosch in central Germany recommends the following times to harvest green pods: *C. reginae* and *C. calceolus*: 6 weeks; *C. parviflorum* var. *parviflorum* and var. *pubescens* and *C. kentuckiense*; 8 weeks; *C. macranthos*; 10 weeks; and *C. formosanum*; 12 weeks. A change in the local average temperature of about 2°C can mean a change of an entire week to harvest the capsules.

The other method is to use mature seeds. If the seeds are not sterile, they have to be sterilized with a bleaching of a weak sodium hypochlorite solution. Mature seeds are usually dormant, and this has to be overcome, either with hormone treatment or by storing the seeds near freezing for at least two, but preferably four, months. In some cases only a short treatment (e.g. with sulphuric acid of 2–4% for a few minutes) helps to destroy the inhibitors, located in the inner integument of the seed capsule. The seeds are sown on a jelly-like medium stabilized with agar. The jelly constitution has to be very weak and not as strong as for tropical orchids. The components of a typical medium are sugar, miner-als, vitamins, and other complex organic compounds. The literature offers several formulas (see Rasmussen 1995; Arditti and Ernst 1993).

Planting out seedlings and growing them to maturity is nearly as problematic as producing the seedlings in flasks. First of all, *Cypripedium* seedlings do not produce leaves unless they have had a cold-treatment, the so-called vernalization (like the adult plants, which need this vernalization every winter). This is carried out in the refrigerator with temperatures of about 4°C for at least three months. After this the seedlings can be planted out in pots or other containers. They can be grown in the same compost as the mature plants. However, the seedlings are very susceptible to droughts and rot, and the right conditions are not always achieved in a compost suitable for mature plants.

If optimal conditions are given a *Cypripedium* can be grown from seed to flower within three years. Usually it takes four to five years. If conditions are good, this is probably the same time needed for a *Cypripedium* in the wild to reach flowering size. **(HP)**

Taxonomic literature

Aiton, W. (1789). *Hortus Kewensis*. G. Nichol, London.

Albert, V. A. (1994). Cladistic relationships of the slipper orchids (Cypripedioideae: Orchidaceae) from congruent morphological and molecular data. *Lindleyana* 9, 115–32.

Albert V. A. and Chase, M. W. (1992). *Mexipedium*: a new genus of slipper orchid (Cypripedioideae: Orchidaceae). *Lindleyana* 7, 172–6.

Albert, V. A. and Pettersson, B. (1994). Expansion of the genus *Paphiopedilum* Pfitzer to include all conduplicate-leaved slipper orchids (Cypripedioideae: Orchidaceae). *Lindleyana* 9, 133–40.

Atwood, J. T. (1984). The relationships of the slipper orchids (subfamily Cypripedioideae, Orchidaceae). *Selbyana* 7, 129–247.

Baumann, H., Kunkele, S., and Lorenz, R. (1989). Die nomenklatorischen Typen der von Linnaeus veröffentlichen Namen europäischer Orchideen. *Mitt. Arbeitskreis Heimische Orchid. Baden-Württemberg* 21: 452.

Bentham, G. and Hooker, J. D. (1883). Cypripedieae. *Genera Plantarum* 3, 487–8.

Brieger, F. G. (1973). *Cypripedium*. In R. Schlechter, *Die Orchideen* (3rd edn), 3–4, 185–98.

Brown, R. (1813). *Cypripedium*. In Aiton, *Hortus Kewensis* (2nd edn), 5, 220–2.

Chase, M. W., Cameron, K. M., Hills, H. G., and Jarrell, D. (1994). DNA sequences and phylogenetics of the Orchidaceae and other monocots. In *Proceedings of the 14th World Orchid Conference* (ed. A. Pridgeon), pp. 61–73. HMSO, Edinburgh.

Chen, S. C. and Xi, Y. Z. (1987). Chinese cypripediums, with a discussion on the classification of the genus. In *Proceedings of the 12th World Orchid Conference* (ed. K. Saito and R. Tanaka), pp. 141–6. Twelfth World Orchid Conference, Tokyo.

Cox, A. V. (1995). *The utility of 5S r-DNA in phylogenetic reconstructions: development of the polymerase chain reaction in plant systematics*. Ph.D. thesis. University of Reading, Reading, UK.

Cribb, P. J. (1997). *The genus* Cypripedium. Timber Press, Portland, Oregon.

Dressler, R. L. (1981). *The orchids: natural history and classification*. Harvard University Press, Cambridge, Massachusetts.

Dressler, R. L. (1983). Classification of the Orchidaceae and their probable origin. *Telopea* 2, 413–24.

Dressler, R. L. (1993). *Phylogeny and classification of the orchid family*. Cambridge University Press.

Kränzlin, F. (1901). *Cypripedium* L. Orchidacearum Genera et Species 1, 11–86.

Lindley, J. (1840). *Cypripedium*. In *Genera and Species of Orchidaceous Plants*, pp. 525–32. Ridgways, London.

Linnaeus, C. (1753). *Species plantarum*. L. Salvius, Stockholm.

Pfitzer, E. (1889). Diandrae Cypripedilinae. In *Die natürlichen Pflanzenfamilien* (ed. A. Engler and K. Prantl) II, 6, 82.

Pfitzer, E. (1894). Beitrage zur Systematik der Orchidaceen. *Botanisches Jahrbücher für Systematik, Pflanzengeschichte und Pflanzengeographie* 19, 1–42.

Pfitzer, E. (1903). *Cypripedium*. In *Das Pflanzenreich* (ed. A. Engler), IV, 50 (Heft 12) *Orchidaceae Pleonandrae*, 28–42.

Rafinesque, C. S. (1819). *Criosanthes*. *Journal de physique, de chimie, d'histoire naturelle et des arts* 89, 102.

Rafinesque, C. S. (1838). *Sacodon*. In *Flora Telluriana* 4: 46. Philadelphia.

Rasmussen, F. N. (1985). Orchids. In *The families of the monocotyledons: structure, evolution, and taxonomy* (ed. R. M. T. Dahlgren, H. T. Clifford, and P. F. Yeo), pp. 247–74. Springer-Verlag, Berlin.

Rolfe, R. A. (1896). The *Cypripedium* group. *Orchid Review* 4, 327–34, 363–7.

Schlechter, R. (1926). Das System der Orchidaceen. *Notizblatt des Botanischen Gartens und Museums zu Berlin-Dahlem* 9, 563–91.

Swartz, O. (1800b). *Cypripedium* L. Kongl. *Vetenskaps Academiens Nya Handlingar* 21, 250–1.

Swartz, O. (1805). *Genera et Species Orchidearum. Neues Journal für die Botanik* 1, 1–108.

Vermeulen, P. (1966). The system of the Orchidales. *Acta Botanica Neerlandica* 15, 224–53.

4. SELENIPEDIUM

Selenipedium Rchb.f. in *Bonplandia*, ii: 116 (1854) & in *Xenia Orch.* 1: 3 (1854); Rolfe in *Orch. Rev.* **4**: 331(1896). Lectotype: *Selenipedium chica* Rchb. f.

Solenipedium Beer, *Prakt. Stud. Orch.* 310 (1854), sphalm. pro *Selenipedium* Rchb. f.

Selenipedium section *Foliosae* Benth. & Hook. f., *Gen. Pl.* iii: 635 (1883).

Selenipedilum Pfitzer in Engl. & Prantl., *Natürl. Pflanzenfam.* 2(6): 82, 83(1888) and in Engl., *Pflanzenr.* IV., 50 (Heft 12): 27 (1903); Sprague and Summerh. in *Bull. Misc. Inf. Kew* 1927: 308.

Derivation of name

From the Greek 'selene', a moon or crescent, and 'pedilon', a sandal or slipper, in allusion to the incurved margins of the lip which are crescent-shaped if viewed from above.

Description (Plate 12C, 12D; Fig. 4.1)

Tall terrestrial *herbs*. *Roots* elongate, fibrous, arising from a short to elongate rhizome; *rhizomes* present, short, creeping, rather fleshy. *Stems* erect, 2–5 m tall, often branching, rather woody, leafy, clustered or well spaced, terete, hairy, enclosed by sheathing sterile bracts below and leaf bases above, several-leaved; *pseudobulbs* absent. *Leaves* spreading or suberect, several–many, plicate, many-veined, elliptic-lanceolate

Fig. 4.1. *Selenipedium chica*. A. Habit; B. Roots; C. Dorsal sepal. D. Synsepal; E. Petal; F. Lip, longitudinal section; G. Column, side view; H. Column, front view; I. Ovary, transverse section. Single bar = 1 mm, double bar = 1 cm. Drawn from *Maas and Westra 3962*, *Forest Department 2728*, and Kew Spirit Collection no. 37409 (*Philcox 6120*) by Judi Stone.

to linear-lanceolate, acute or acuminate, green, glabrous or sparsely hairy, ciliate or not on the margins, contracted at the base to form a cylindrical sheath. *Inflorescence* terminal, racemose or branching from base, many-flowered; *rachis* terete, densely hairy; bracts usually leaf-like but much smaller than the uppermost leaf, lanceolate, hairy. *Flowers* with a persistent perianth, small to medium-sized, resupinate, concolorous or bicoloured; *pedicel* obscure to short, hairy; *ovary* trilocular, with parietal placentation, three-ribbed, hairy, crowned with a marcescent perianth. *Sepals* valvate. *Dorsal sepal* erect to hooded over lip, ovate or elliptic, obtuse or acute, pubescent on the outer surface, ciliate or not; *lateral sepals* usually fused to form a concave synsepal that is similar to the dorsal sepal, but free at the tips, pubescent on outer surface, ciliate or not. *Petals* free, spreading or incurved, lanceolate or linear-lanceolate, blunt, acute or acuminate, often pubescent in basal half within, not ciliate. *Lip* deeply pouched and inflated, slipper-shaped, glabrous on outer surface; *side lobes* obscure, incurved, entire; *midlobe* deeply saccate, calceiform or urceolate, incurved on front margin, hairy within especially on lower surface. *Column* porrect, short, stalked; *anthers* two, bilocular, borne on short obtuse filaments; *pollen* powdery or viscid; *staminode* terminal on column, more or less sessile or stalked, ovate or trullate, somewhat conduplicate, glabrous; *stigma* very short stalked, dependent, trisulcate, papillose or not, wider than the staminode. *Capsule* spreading to dependent, three-ribbed, cylindrical to fusiform, smelling of vanilla; seeds subglobose, crustose. **(PC)**

Distribution (Fig. 4.2)

A small genus of about five or possibly six species in the tropical Americas from Panama and Trinidad, south to Venezuela, the Guyanas, Brazil, Colombia, and Ecuador.

Anatomy

Leaf

Anticlinal walls of epidermal cells are straight (Rosso 1966), although Atwood (1984) suggests that they are sinuous. Epidermal cell volume is the smallest among the cypripedioid genera (Rosso

Fig. 4.2. Distribution map of *Selenipedium*.

1966), but cell volume of *S. palmifolium* is larger than that of *S. chica* (Atwood 1984).

Stomata occur on both leaf surfaces of leaves and bracts (Rosso 1966). Guard cells are on average the smallest of all the slipper orchid genera (Rosso 1966). Subsidiary cells are present (N. H Williams 1979), with two laterals and two polars (Rosso 1966), although the latter hesitated to describe them as such.

Trichomes, glandular and nonglandular, are distributed over both leaf surfaces of leaves of *S. palmifolium* (Rosso 1966).

Chlorenchyma is undifferentiated into palisade and spongy (Rosso 1966). Vascular bundles are in one row in transverse section, the midvein more prominent, the larger bundles surrounded by bundle sheaths of sclerenchyma fibres. Rosso (1966) notes striking similarities in vasculature between *Cypripedium irapeanum* and *Selenipedium*, including suberized endoderm-like cells between fibres and mesophyll.

There are no reports of stegmata in any organs of *Selenipedium*.

Stem

Leafy stem

Rosso (1966) provided the only modern account of leafy stem and rhizome anatomy of *Selenipedium*. Epidermal cells possess straight walls. Stomata are present (as in *Cypripedium*), but subsidiary cells are absent. Trichomes, both nonglandular and glandular, are present and surrounded by a thick cuticle. A very small zone of cortical parenchyma abuts both the epidermis and the relatively large sclerotic sheath of the vascular cylinder, which bears scattered vascular bundles (Solereder and Meyer 1930; Rosso 1966). A sclerenchyma sheath with very thick-walled fibres encloses the vascular. As in *Cypripedium*, amphivasal bundles basally give way to collateral bundles above the basal internodes, although amphivasal bundles persist at the nodes (Rosso 1966). Vessels are absent. Larger tracheids have scalariform pitting.

Rhizome

The rhizome of *Selenipedium* is particularly fleshy and bears large starch reserves. The epidermis, often necrotic or sloughed in older plants, lacks hairs and stomates. Cell walls are impregnated with cutin or suberin. Cortical parenchyma is limited, bounded internally by a uniseriate endodermis. The vascular cylinder comprises a multiseriate pericycle of sclerotic fibres and collateral vascular bundles scattered near the periphery and having sclerenchyma caps at both xylem and phloem poles (Rosso 1966). Vessel elements are absent; tracheids have annular, spiral, and scalariform thickenings (Rosso 1966).

Root

A uniseriate velamen lacking root hairs and subtended by a uniseriate exodermis is present in *S. palmifolium* and *S. chica* (Rosso 1966). Long cells of the exodermis have U-shaped thickenings. Tilosomes are absent. The innermost layer of the limited cortex, the endodermis, is uniseriate with U-shaped thickenings. There is a multiseriate layer of thickened sclerenchyma cells immediately internal to the endodermis. Roots are 24-arch in *S. palmifolium* (Rosso 1966). Vessels with simple perforation plates and metaxylem tracheids, both with scalariform pitting, are present in roots (Rosso 1966). As in the conduplicate genera, there is a well-developed pith in *Selenipedium*. (AP)

Palynology

Pollen grains of *Selenipedium* are ellipsoidal with a pronounced sulcus at one pole. Sculpturing is uniformly psilate (Newton and Williams 1978). The exine of *Selenipedium* comprises a large tectum, coherent columellae, and a compact-foot layer in the non-apertural region and a lamellated foot-layer, a much-reduced tectum, and columellae in the apertural region (Burns-Balogh and Hesse 1988). Like that of *Cypripedium irapeanum*, the intine of *Selenipedium* is relatively thick in the germination area, a feature which sets them apart from all other cypripedioids. (AP)

Cytogenetics

No cytological studies of *Selenipedium* have been published to date. (AC)

Phytochemistry

Nothing is known of the phytochemistry of *Selenipedium*. (RG and NV)

Phylogenetics

See Phylogenetics under Cypripedioideae above.

Ecology

Of all the slipper orchids the least known is the genus *Selenipedium*. There are probably several causative factors for the dearth of knowledge. None of the species is particularly attractive and so are not sought by collectors. Out of flower they appear very much like the palms, bamboos, and tall grasses with which they grow, seeming to hide within the subtle camouflage. The species exhibit even more patchy distributions than their counterparts and so are difficult to locate. As with so many orchids, they are usually found when looking for something else.

Selenipedium contains perhaps six species which range from central Panama to Peru, central Brazil, and Trinidad but not the Caribbean Islands. Reports from Costa Rica have not been substantiated, yet a gap in range of plicate-leaved slipper orchids from central Panama to Guatemala seems odd. It is curious that none is known from southern tropical or temperate South America, a fact that may reflect a rather recent immigration through Central America. The subtle differences in species suggest that they may all have descended from a single recent ancestor. Isolation on mountain slopes should not be a major obstacle to dispersal as these are lowland plants with the exception of *S. steyermarkii* that ranges to 1800 m (Foldats 1969; Carnevali and Ramírez de Carnevali 1993). Perhaps the relatively large crustose seeds seriously impede dispersal, despite the fragrance of vanilla recorded in some species that has been suggested to entice collection by insects.

As with *Cypripedium*, all species of *Selenipedium* are terrestrial but are tropical, and all may prefer acid soils as none is documented in association with limestone or serpentine. Most attempts at transplantation have failed, suggesting a high reliance on mycorrhizae, but the single *S. aequinoctiale* has been successfully transplanted (C. H. Dodson, personal communication).

At least one of the species (*S. chica*) seems to prefer some kind of disturbance as do many *Cypripedium* species. In the late 1970s Atwood (unpublished) visited a known station of *Selenipedium chica* on Cerro Campana, Panama, discovered by Robert Dressler. There were five clumps growing clearly in a secondary forest with palms and bamboos of similar size. Despite several extensive searches, no other plants were seen. Either these plants had persisted as relicts or had become established by seed from some other source. Even more bewildering was the equal spacing of the plants, as though they had been planted. Dressler (1989a) related that *S. chica* was once used as a flavouring (*vainilla chica*) by the indigenous peoples. Any history of their attempts to cultivate it seems to have been lost. **(JA)**

Pollination

The floral biology of these orchids is probably similar judging by the uniform morphology and colour of the flowers. The obligate sequential flowering habit allows for both production of fruits and flowers at the same time and over an extended period. Flowers are short-lived and probably attract bees (Dressler 1993). In support of this, Dodson (personal communication) has seen anthophorine bees pollinating *S. aequinoctiale*. **(JA)**

Taxonomic notes

The genus *Selenipedium* was established in 1854 by H. G. Reichenbach to accommodate the tropical American slipper orchids with trilocular ovaries. He recognized two sections: one with tall leafy stems and pleated leaves, the other stemless with conduplicate leaves. In 1882 Pfitzer removed the conduplicate leaved species and united them with the Old World species in his genus *Paphiopedilum* because of their conduplicate leaves and deciduous perianth. Rolfe (1896) reviewed the evidence and redefined *Selenipedium* for those species with tall stems, pleated leaves, a persistent perianth, a trilocular ovary with parietal placentation, valvate sepals, and subglobose seeds. He removed all of the conduplicate-leaved tropical American species to *Phragmipedium*. Rolfe accepted three species in *Selenipedium*, further species having been described by Foldats (1969) in *Flora de Venezuela* and Garay (1978) in *Flora of Ecuador*. **(PC)**

Uses

Fruits have been used for flavouring having much the same fragrance as vanilla (Dressler 1989a). **(PC)**

Cultivation

Very little is known about the cultural requirements of this genus, and very few if any plants are currently grown. If one were to obtain material to grow, we suggest that a standard terrestrial mix such as that which has proved successful with either *Paphiopedilum* or *Phragmipedium* be attempted. It is probably easier to establish seedlings than divisions of adult jungle plants. **(HK)**

Taxonomic literature

Albert, V. A. (1994). Cladistic relationships of the slipper orchids (Cypripedioideae: Orchidaceae) from congruent morphological and molecular data. *Lindleyana* **9**, 115–32.

Albert, V. A. and Chase, M. W. (1992). *Mexipedium*: a new genus of slipper orchid (Cypripedioideae: Orchidaceae). *Lindleyana* **7**, 172–6.

Albert, V. A. and Pettersson, B. (1994). Expansion of genus *Paphiopedilum* to include all conduplicate-leaved slipper orchids (Cypripedioideae: Orchidaceae). *Lindleyana* **9**, 133–40.

Atwood, J. T. (1984). The relationships of the slipper orchids (subfamily Cypripedioideae, Orchidaceae). *Selbyana* **7**, 129–247.

Brieger, F. G. (1973). *Selenipedium*. In *Die Orchideen* (3rd edn), (ed. R. Schlechter), 3–4.

Garay, L. A. (1978). *Selenipedium*. In *Flora of Ecuador* (ed. G. Harling and B. Sparre) **9**, 12–13. University of Göteborg, Stockholm.

Kränzlin, F. (1901). *Cypripedium* L. *Orchidacearum Genera et Species* **1**, 11–86.

Lindley, J. (1840). *Cypripedium*. In *Genera and Species of Orchidaceous Plants*, pp. 525–532. Ridgways, London.

Rolfe, R. A. (1896). The *Cypripedium* group. *Orchid Review* **4**, 327–34, 363–7.

5. PAPHIOPEDILUM

Paphiopedilum Pfitzer, *Morphol. Stud. Orchideenblüten*: 11 (1886); in Engler & Prantl, *Pflanzenfam.* II, 6: 82 (1889); & in Engler, *Pflanzenr.* IV, 50, *Orch. Pleonandr.*: 54 (1903). Type species: *Paphiopedilum insigne* (Wall. ex Lindl.) Pfitzer.

Cordula Raf., *Fl. Tell.* **4**: 46 (1838), *nom. rej.* Type species. *Cordula insignis* (Wall. ex Lindl.) Raf.

Stimegas Raf., *loc. cit.* 45 (1838), *nom. rej.* Type species. *Stimegas venustus* (Wall. ex Sims) Raf.

Derivation of name

From Paphos, the birthplace of Aphrodite on the island of Cyprus in the Mediterranean, and the Greek '*pedilon*', a sandal or slipper. See comments on *Cypripedium*.

Description (Plate 13, 14, 15A; Fig. 5.1)

Small to large terrestrial, lithophytic or epiphytic *herbs*; *roots* elongate, fibrous; *rhizome* short to elongate. *Shoots* short, erect, leafy, clustered or less frequently well spaced, glabrous, the base enclosed by two to four sheathing sterile bracts, three- to several-leaved above. *Leaves* one to several, coriaceous, conduplicate, spreading or suberect, ligulate, elliptic, or oblong, obtuse to acute, often tridenticu-

late at apex, green, bluish green or chequered or tessellated with dark and lighter green on upper surface, lighter green below, sometimes finely spotted or flushed with purple at base or all over, glabrous, ciliate or not on the margins. *Inflorescence* terminal, one- to many-flowered; rachis terete, hairy, glandular or glabrous; bracts conduplicate, elliptic, lanceolate, ovate or oblong, green, sometimes spotted or flushed or striped with purple, ciliate or not. *Flowers* usually showy, concolorous or bicoloured; pedicel obscure to short; ovary unilocular, three-ribbed, glabrous or hairy. *Dorsal sepal* erect to hooded over lip, ovate, lanceolate, obovate or elliptic, obtuse, acute or acuminate, glabrous or pubescent on the outer surface, sometimes pubescent within at base, ciliate or not. *Lateral sepals* usually fused to form a concave synsepal that is more or less similar to the dorsal sepal, sometimes keeled on outer surface. *Petals* free, spreading or pendent, flat, reflexed or spiralling, elliptic, ovate, lanceolate, linear-lanceolate, linear or oblanceolate, rounded, obtuse, acute or acuminate at apex, often pubescent in basal half within, usually ciliate. *Lip* deeply pouched and inflated, slipper-shaped or urn-shaped, with more or less pronounced incurved side lobes, hairy within especially on lower surface, glabrous or hairy on outer surface; front margin incurved or not. *Column* short, stalked, porrect; anthers two, bilocular, borne on short obtuse to acute filaments; pollen powdery or viscid; staminode terminal on column, sessile or shortly stalked, transversely reniform, oblong, ovate, obcordate or linear, flat, convex or longitudinally conduplicate, glabrous to papillose or finely pubescent, ciliate or not; stigma stalked, dependent, tripartite, more or less papillose. *Capsule* erect to pendent, three-ribbed, cylindrical to almost ellipsoidal. **(PC)**

Distribution (Fig. 5.2)

The range of *Paphiopedilum* extends from India eastward across southern China to the Philippines and throughout south-east Asia and the Malay Archipelago to New Guinea and the Solomon Islands. Of the seven Indian species, six are confined to the north-east, along the foothills of the Himalaya from eastern Nepal to the Naga Hills, and also in the Khasia Hills. One species, *P. druryi*, is found only in southern India, in Kerala State, over 2000 km from the nearest *Paphiopedilum* locality to the north.

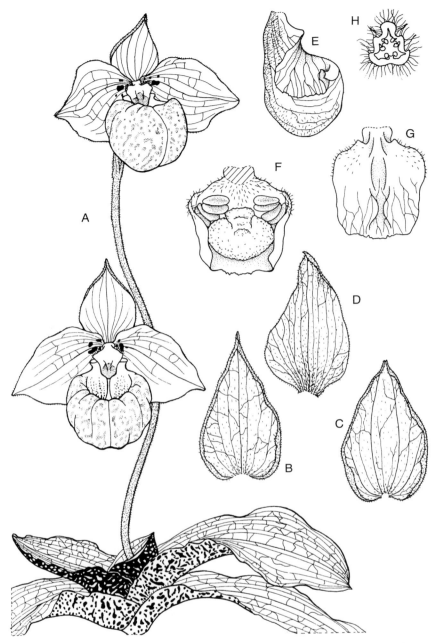

Fig. 5.1. *Paphiopedilum malipoense*. A. Habit and flower, ×2/3; B. Dorsal sepal, ×1; C. Synsepal, ×1; D. Petal, ×1; E. Lip, longitudinal section, ×1; F. Column, ventral view, ×2; G. Staminode, ×2; H. Ovary, transverse section, ×4. Drawn from cultivated material by Valerie Price. Reprinted from Cribb, P. (1987). *The Genus Paphiopedilum.*

Eighteen species have been reported from China. Of these, *P. micranthum*, *P. malipoense*, *P. dianthum*, *P. emersonii*, *P. barbigerum*, and *P. hirsutissimum* var. *esquirolei* are the most northerly distributed in Guizhou Province with *P. purpuratum* the most easterly in Hong Kong and adjacent Guangdong.

The genus is well represented in south-east Asia, notably in Thailand and Vietnam which can boast 12 species but with few in adjacent countries. Numbers increase again in the larger islands of the Malay Archipelago. Borneo, like China, has produced several novelties recently; twelve species, of which seven are endemic, have now been recorded there. Sumatra with ten species and six endemics follows closely behind, whereas Java has only three species and one endemic. *Paphiopedilum javanicum* extends the range of the genus to Bali and to Flores, but no other species is found in the Lesser Sunda islands.

The Philippines are another rich centre of diversity of the genus with eight species, seven of which

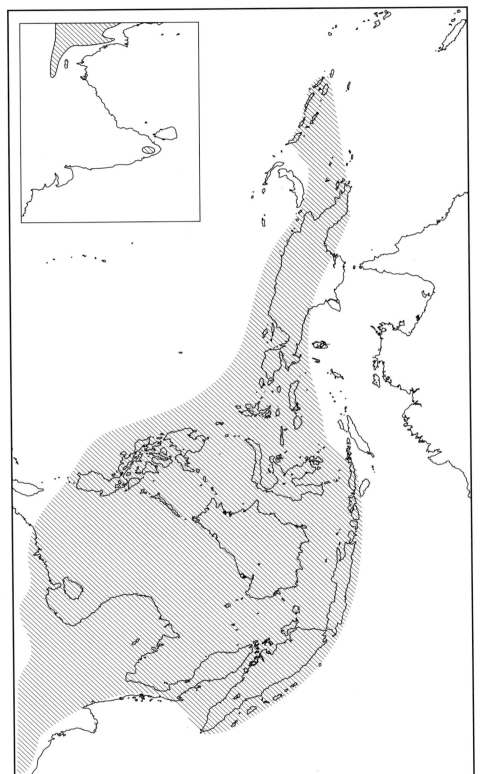

Fig. 5.2. Distribution map of *Paphiopedilum*.

are endemic. To the south and east the number of species drops rapidly. Four species have been recorded from Sulawesi and the Moluccas, one from each being endemic. In New Guinea four species, all endemic, are found with two reported from the Solomons. Both of the latter are closely related to the New Guinea species *P. violascens*. *Paphiopedilum bougainvilleanum* is apparently confined to the island of Bougainville, the northernmost of the Solomon Islands, but *P. wentworthianum* has been reported from there and also from Guadalcanal.

Paphiopedilum is certainly a genus of tropical Asiatic origin, and its extension eastward across Wallace's Line into Sulawesi, the Moluccas, New Guinea, and the Solomon Islands is probably the result of long-distance dispersal of the light seeds from the west and possibly from the north. The immediate affinities of the New Guinean species *P. glanduliferum* appear to lie with Bornean species *P. rothschildianum* and that of *P. violascens*, *P. mastersianum*, and *P. papuanum* with *P. javanicum* rather than with Philippine species.

The affinities of the Philippine species again seem to lie with those of the Malay Archipelago. *Paphiopedilum haynaldianum*, for example, is a narrow endemic closely allied to the widespread *P. lowii*.

A better understanding of the distribution of the genus can be obtained by considering each section separately. Subgenera *Brachypetalum* and *Parvisepalum* and section *Paphiopedilum* are confined to mainland Asia. The first two are found in southern China, Indo-China, Burma, Thailand, and northern Malaya, with subgenus *Parvisepalum* concentrated in south-west China and adjacent north Vietnam and subgenus *Brachypetalum* in south-west China, north Burma, and Indo-China. The last ranges from India to southern China, Thailand, and Indo-China with the greatest concentration of species in north-east India.

Section *Pardalopetalum* with four species is widespread in south-east Asia, the Malay Archipelago, as far east as Sulawesi, and the Philippines. The greatest diversity in section *Coryopedilum* lies in Borneo where six species are found, with three species in the Philippines and two in New Guinea.

Section *Cochlopetalum* has a restricted range in the islands of Sumatra and Java where five closely related taxa have evolved. Karasawa and Saito (1982) have suggested that the loss of a chromosome prior to chromosome fission has been a major factor in the evolution of this section. It seems likely that

this occurred subsequent to the separation of the islands from mainland Asia.

The tessellated-leaved species of section *Barbata* are widespread from eastern Nepal, where *P. venustum* grows, to Hong Kong and the Philippines and south to the Malay Archipelago, New Guinea, and the Solomons. Two species each are found in Thailand, peninsular Malaysia, New Guinea, and the Solomons, one each in Java, Sulawesi, and the Moluccas. The greatest concentration lies in Borneo with five and the Philippines and Sumatra with four species.

Few species in the genus are widespread, but *P. parishii*, *P. callosum*, *P. villosum*, *P. hirsutissimum*, *P. concolor*, *P. purpuratum*, and *P. appletonianum* are exceptions in mainland Asia, whereas *P. lowii*, *P. philippinense*, *P. bullenianum*, and *P. javanicum* range widely in the adjacent islands. The majority of species seem to be narrow endemics, and a few such as *P. druryi*, *P. rothschildianum*, *P. stonei*, *P. dayanum*, and *P. bougainvilleanum* are known from only one or a few localities. **(PC)**

Infrageneric treatment (PC)

1. *Paphiopedilum* subgenus *Parvisepalum* Karasawa & Saito in *Bull. Hiroshima Bot. Gard.* **5**, 31 (1982). Type: *P. delenatii* Guill.
Paphiopedilum section *Parvisepalum* (Karasawa & Saito) Cribb, *The Genus* Paphiopedilum, **86** (1987).

2. *P.* subgenus *Brachypetalum* (Hallier) Pfitzer in Engler, Pflanzenr. *Orch. Pleon.* **55** (1903); Karasawa & Saito in *Bull. Hiroshima Bot. Gard.* **5**, 32 (1982). Type: *P. concolor* (Bateman) Pfitzer.
Paphiopedilum Eremantha Tessellata Pfitzer in *Bot. Jahrb. Syst.* **19**, 40 (1894) *pro parte*.
Cypripedium section *Concoloria* Kränzl., *Orchid. Gen. Spec.* **1**, 13, 73 (1897). Type species: *Paphiopedilum concolor* (Bateman) Pfitzer.
Paphiopedilum Aphanoneura Brachypetalum Hallier in *Ann. Jard. Bot. Buitenzorg* **14**, 34 (1897).
P. section *Concoloria* (Kränzl.) V.A.Albert & B.Petters. in *Lindleyana* **9**, 137 (1994) .

3. *P.* subgenus *Paphiopedilum*; Atwood in *Selbyana* **7**, 244 (1984).

3.1. *P.* section *Coryopedilum* Pfitzer in Engler, Pflanzenr. *Orch. Pleon.* **59** (1903); Atwood in

Selbyana **7**, 245 (1984). Lectotype: *P. glanduliferum* (Bl.) Stein.

P. Polyantha Pfitzer in *Bot. Jahrb. Syst.* **19**, 14 (1894) *pro parte.*

P. section *Gonatopedilum* Pfitzer in Engler, Pflanzenr. *Orch. Pleon.* **58** (1903). Type: *P. rothschildianum* (Rchb.f.) Stein.

P. section *Prenipedilum* Pfitzer in Engler, Pflanzenr. *Orch. Pleon.* **62** (1903). Type: *P. stonei* (Hook.) Stein.

P. section *Mastigopetalum* Karasawa & Saito in *Bull. Hiroshima Bot. Gard.* **5**, 36 (1982). *nom. superfl.* Type: *P. stonei* (Hook.) Stein.

3.2. *P.* section *Pardalopetalum* Hallier & Pfitzer in Engler, Pflanzenr. Orch. Pleon. 66 (1903); Atwood in *Selbyana* 7, 245 (1984). Lectotype: *P. lowii* (Lindl.) Stein.

P. Polyantha Pfitzer in Engler, *Bot. Jahrb.* **19**, 41 (1894) *pro parte.*

P. section *Mystropetalum* Pfitzer in Engler, Pflanzenr. *Orch. Pleon.* **65** (1903); Karasawa & Saito in *Bull. Hiroshima Bot. Gard.* **5**, 38 (1982). Type: *P. parishii* (Rchb.f.) Stein.

3.3. *P.* section *Cochlopetalum* Hallier ex Pfitzer in Engler, Pflanzenr. *Orch. Pleon.* **68** (1903); Atwood in *Selbyana* 7, 245 (1984).

P. subgenus *Cochlopetalum* (Hallier ex Pfitzer) Karasawa & Saito in *Bull. Hiroshima Bot. Gard.* **5**, 55 (1982). Type: *P. victoria-regina* (Sander) M.W.Wood.

3.4. *P.* section *Paphiopedilum*; Karasawa & Saito in *Bull. Hiroshima Bot. Gard.* **5**, 40 (1982); Atwood in *Selbyana* 7, 244 (1984). Type: *P. insigne* (Wall. ex Lindl.) Pfitzer.

P. section *Stictopetalum* Hallier ex Pfitzer in Engler, Pflanzenr. *Orch. Pleon.* **69** (1903); Karasawa & Saito in *Bull. Hiroshima Bot. Gard.* **5**, 41 (1982). Type: *P. hirsutissimum* (Hook.) Stein.

P. section *Neuropetalum* Hallier ex Pfitzer in Engler, Pflanzenr. *Orch. Pleon.* **71** (1903), *nom. illeg.* Type: *P. insigne* (Wall. ex Lindl.) Pfitzer.

P. section *Thiopetalum* Hallier ex Pfitzer in Engler, Pflanzenr. *Orch. Pleon.* **75** (1903); Karasawa & Saito in *Bull. Hiroshima Bot. Gard.* **5**, 41 (1982). Type: *P. druryi* (Bedd.) Stein.

P. section *Cymatopetalum* Hallier ex Pfitzer in Engler, Pflanzenr. *Orch. Pleon.* **76** (1903). Type: *P. spicerianum* (Rchb.f.) Pfitzer.

P. section *Ceratopetalum* Hallier ex Pfitzer in Engler, Pflanzenr. *Orch. Pleon.* **77** (1903); Karasawa & Saito in *Bull. Hiroshima Bot. Gard.* **5**, 42 (1982). Type: *P. fairrieanum* (Lindl.) Stein.

P. subgenus *Paphiopedilum* Karasawa & Saito in *Bull. Hiroshima Bot. Gard.* **5**, 39 (1982).

3.5. *P.* section *Barbata* Kränzl., *Orchid.* **1**: 53 (1897). Type: *P. barbatum* (Lindl.) Pfitzer.

P. section *Spathopetalum* Pfitzer in Engler, *Pflanzenr. Orch. Pleon.* **78** (1903).

P. section *Blepharopetalum* Pfitzer, *loc. cit.* 82 (1903).

P. section *Phacopetalum* Pfitzer, *loc. cit.* 87 (1903).

P. subgen. *Sigmatopetalum* Hallier ex Karasawa & Saito, in *Bull. Hiroshima Bot. Gard.* **5**, 42 (1982). Type *P. venustum* (Wall.) Pfitzer.

P. section *Punctatum* Karasawa & Saito, *loc. cit.* 49 (1982). Type: *P. tonsum* (Rchb.f.) Stein.

P. section *Planipetalum* Karasawa & Saito, *loc. cit.* 54 (1982). Type: *P. sukhakulii* Schoser & Senghas.

Anatomy

Leaf

The most important treatments of leaf anatomy of subfamily Cypripedioideae and *Paphiopedilum* in particular are those by Möbius (1887), Solereder and Meyer (1930), and Rosso (1966). Möbius (1887) examined six species of *Paphiopedilum* as well as two species now included in *Phragmipedium*. He noted that in all the leaves were uniform in structure, with adaxial epidermal cells elongated anticlinally like palisade mesophyll, and suggested that they probably function in water storage. Adaxial epidermal cells are four times the length and twice the width of abaxial epidermal cells. Although the cuticle is thin, the uncutinized portions of the cell walls are thickened. Stomata are always abaxial and lack subsidiary cells (Solereder and Meyer 1930; Atwood 1984), although Rosso (1966) cited the presence of 'accessory cells' in some species. Atwood and Williams (1979) suggested several epidermal and cuticular features such as macropapillae and micropapillae that might be used to identify sterile material of *Paphiopedilum* to species. Trichomes are absent from leaves but often present on bracts. Chlorenchyma is generally undifferentiated into palisade and spongy layers, but Atwood (1984) reported poorly developed palisade layers in *Paphiopedilum lowii* and *P. insigne*. Vascular bundles are in one row in transverse section, the midvein more prominent, surrounded by bundle sheaths of suberized fibres. In the few species of *Paphiopedilum* investigated for the presence of silica cells (Solereder and Meyer 1930; Møller and Rasmussen 1984), stegmata are absent.

Stem

The axis of the plant comprises the leafy stem and rhizome. The leafy stem is bounded by a uniseriate epidermis lacking stomata (Rosso 1966). Stomata and trichomes are absent in *Paphiopedilum* but present in the plicate-leaved genera *Selenipedium* and *Cypripedium* (Rosso 1966; Atwood 1984). Cortical parenchyma abuts both the epidermis and the sclerotic sheath of the vascular cylinder, which bears scattered vascular bundles (Solereder and Meyer 1930; Rosso 1966). The rhizome has a uniseriate epidermis, often necrotic or sloughed, the cell walls impregnated with cutin or suberin. Cortical parenchyma is limited, bounded internally by a multiseriate endodermis with O-shaped thickenings (Rosso 1966). The vascular cylinder comprises a uniseriate pericycle and amphivasal vascular bundles embedded in ground parenchyma (Atwood 1984). Vessel elements are absent; tracheids have annular, spiral, and scalariform thickenings (Rosso 1966).

Root

The root is enclosed in a velamen of up to 12 layers, although 4–7 layers are most common (Meinecke 1894; Rosso 1966; Pridgeon 1987). Unicellular root hairs are persistent on the outermost layer (Rosso 1966). Velamen is of the *Calanthe*-type according to the terminology of Porembski and Barthlott (1988). Long cells of the exodermis generally have U-shaped thickenings, although O-shaped thickenings have been reported for *P. insigne* and *P. niveum* (Rosso 1966; Pridgeon 1987). Tilosomes are absent except for *P. fairrieanum*, which has the lamellate type (Pridgeon *et al.* 1983). The innermost layer of the cortex, the endodermis, is uniseriate (but biseriate locally in *P. venustum*; Rosso 1966). Both Atwood (1984) and Rosso (1966) reported a multiseriate peri-

cycle of sclerotic parenchyma cells. All endodermal walls may be thickened (O-shaped) or only the radial and inner tangential walls (U-shaped) depending on the taxon. Roots are 8–13-arch (Meinecke 1894; Rosso 1966). Vessels are absent (Rosso 1966). (**AP**)

Palynology

Schill and Pfeiffer (1977) described the pollen of several *Paphiopedilum* taxa as laevigate or smooth, but Newton and Williams (1978), sampling more widely, described and illustrated four sculpturing types in *Paphiopedilum*: psilate, foveolate, fossulate, and scabrate. Burns-Balogh and Hesse (1988) showed with TEM that the smooth appearance of the grains is explained by the fact that the exine comprises only sporopollenin 'lumps' on the intine. Well-developed columellae and a foot-layer are absent. (**AP**)

Cytogenetics

The most common diploid chromosome number in *Paphiopedilum* is 26 metacentric chromosomes as displayed in subgenera *Parvisepalum* and *Brachypetalum*, the first two successive sister groups, and also in sections *Pardalopetalum*, *Coryopedilum*, and *Paphiopedilum* (excluding *P. druryi* and *P. spicerianum* which both have 30 chromosomes; Table B.1). Species in sections *Cochlopetalum* and *Barbata* depart from this stable karyotypic pattern with diploid numbers usually ranging from $2n = 32$ to 36 in section *Cochlopetalum*, and $2n = 28$ to 42 in section *Barbata*. In cases for which accurate karyotype data are known, telocentrics may be arranged so as to form 26 metacentric chromosomes. Studies of some species in section *Cochlopetalum* reveal karyotypes that equate to 25 metacentric chromosomes. This is thought to be due to the loss of a pair of telocentric chromosomes (Karasawa 1979). It would therefore appear that 26 is the ancestral number in *Paphiopedilum*.

There are two exceptions to the generally consistent pattern of chromosome numbers in slipper orchid genera (apart from those mentioned above for some species in sections *Cochlopetalum* and *Barbata*). The first example of chromosomal change is displayed by *Paphiopedilum hookerae*. Whereas nearly all other members of section *Barbata* exhibit greatly elevated chromosome numbers, *P. hookerae* possesses 28 chromosomes, none of which is telo-centric. Its karyotype does, however, include two sub-telocentric chromosomes. These may have been derived from telocentrics with the centromeres being repositioned through the gain of terminal chromatin or by a pericentric inversion.

A counter-example is provided by *Paphiopedilum druryi* which has 30 chromosomes (including 8 telo-centrics), whereas the other species comprising *Paphiopedilum* section *Paphiopedilum* have 26 meta- or sub-metacentric chromosomes. *Paphiopedilum druryi* is the only member of the section that appears to have undergone centric fission. This may be cor-related with its narrow endemicity; it is the only slipper orchid to occur in southern India, restricted to the Travencore Hills, near Quilon. **(AC)**

Phytochemistry

Little is known of the phytochemistry of *Paphio-pedilum*. Saponins (which are very rare in Orchida-ceae) and traces of alkaloids occur in *P. javanicum* (Boorsma 1902). Two allergenic quinones have been reported for *P. haynaldianum* (Hausen 1979, 1980). **(RG and NV)**

Phylogenetics

The horticultural popularity of *Paphiopedilum* has ensured the intense interest of taxonomists (Table 5.1). After more than a hundred years of nomenclatural instability the subgeneric class-ification of *Paphiopedilum* is now attaining some degree of stability. It currently comprises three sub-genera and seven sections, as shown in Fig. 5.3.

Subgenus *Brachypetalum* was erected to circum-scribe a group of 'primitive' or 'basal' taxa within *Paphiopedilum*. Originally comprising only the species of section *Brachypetalum*, the subgenus was divided and expanded to include section *Parvisepalum* follow-ing their relatively recent discovery. Although most species in section *Brachypetalum* (*Concoloria sensu* Albert and Pettersson 1994) have been known for a hundred years, those in section *Parvisepalum* have only been known since 1950 and most since 1980. The two sections were grouped because they share flowers with similar involute lip margins (P. Cribb, personal communication), although detailed examina-tion of floral structures does not yield evidence of

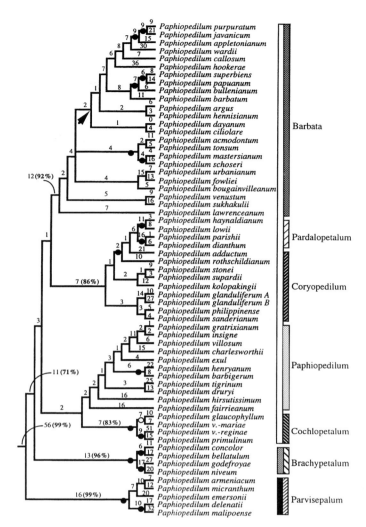

Fig. 5.3. One of the most parsimonious ITS DNA trees showing subgeneric relationships in *Paphiopedilum* (Cox *et al.* 1997a). The open bar indicates subgenus *Paphiopedilum*, and the solid black and grey bars sub-genera *Brachypetalum* and *Parvisepalum* as defined here. Sectional limits as defined here are indicated by various shaded bars. Solid circles indicate clades strongly sup-ported by the jackknife procedure (node score >0.63); open circles indicate those that are weakly supported (node score >0.5 but <0.63). Arrow indicates node that collapses in the strict consensus of all most-parsimonious weighted trees. ITS sequences from two accessions of *P. glanduliferum* were determined, labelled 'A' and 'B', respectively.

other similarities. This type of labellum morphology would appear to be plesiomorphic for the Cypripe-dioideae based on DNA evidence. Cribb (1987) sug-gested that the resemblance in lip morphology may be the result of parallel pollinator pressure rather than being the truly plesiomorphic condition; this is not a parsimonious explanation from the perspective

Table 5.1. Infrageneric treatments of *Paphiopedilum* (revised from Cribb 1987)

Pfitzer (1894)	Hallier (1896)	Pfitzer (1903)	Brieger (1973)
a. *Eremantha Tessellata* (in part)	*Aphanoneura Brachypetalum*	*Brachypetalum*	*Brachypetalum*
b. *Polyantha*	*Chromatoneura Viridia Polyantha*	*Anotopedilum*	*Polyantha*
	XI *Streptopetalum* (in part)	sect. *Coryopedium*	sect. *Streptopetalum*
	XII *Mastigopetalum*	sect. *Gonatopetalum*	sect. *Mastigopetalum*
		sect. *Prenipedilum*	
		Otopedilum	
	XI *Streptopetalum* (in part)	sect. *Mystropetalum*	sect. *Polyantha*
	X *Pardalopetalum*	sect. *Pardalopetalum*	
	XIII *Cochlopetalum*	sect. *Cochlopetalum*	sect. *Cochlopetalum*
a. *Eremantha Viridia*	*Chromatoneura Viridia Eremantha*		*Paphiopedilum*
	VIII *Stictopetalum*	sect. *Stictopetalum*	sect. *Stictopetalum*
	IX *Neuropetalum*	sect. *Neuropetalum*	sect. *Paphiopedilum*
	V *Thiopetalum*	sect. *Thiopetalum*	
	VII *Cymatopetalum*	sect. *Cymatopetalum*	
	V *Ceratopetalum*	sect. *Ceratopetalum*	
a. *Eremantha Tessellata* (in part)	*Chromatoneura Tessellata*		*Barbata*
	II *Sigmatopetalum*	sect. *Spathopetalum*	sect. *Sigmatopetalum*
	IV *Drepanopetalum*	sect. *Blepharopetalum*	sect. *Blepharopetalum*
		sect. *Phacopetalum*	sect. *Barbata*

of available data. These indicate that subgenus *Brachypetalum* is not monophyletic; species of a monophyletic subgenus *Parvisepalum* are together sister to a clade containing the remaining *Paphiopedilum* species.

Subgenus *Parvisepalum* was considered by Chen and Tsi (1984) to be a 'link' group between *Paphiopedilum* and *Cypripedium*, based on observations that these species have flowers that bear a strong resemblance to those of both *Cypripedium* and *Selenipedium*. Subgenus *Parvisepalum* may legitimately be considered a 'link' because it occupies a position that is sister to the rest of the genus which has also retained *Cypripedium*-like flower characteristics. However, to view the group as 'primitive' simply on the basis of gross floral morphology is highly questionable; subgenus *Parvisepalum* has undergone considerable vegetative and molecular divergence from both *Cypripedium* and *Selenipedium* (Fig. 5.3).

Currently available data support circumscriptions of other sections of subgenus *Paphiopedilum* as described by Cribb (shown in Table 5.1) with two exceptions; *Pardalopetalum* and *Coryopedilum* (Fig. 5.3). The data presented by Cox *et al.* (1997*a*) indicated that the taxa of these two sections form a single clade with species in section *Coryopedilum* forming a series paraphyletic to those of section *Pardalopetalum*. Both *P. rothschildianum* and *P. adductum* are successive sister taxa to section *Coryopedilum* (*sensu* Cribb 1987).

As suggested by Atwood (1984) section *Barbata* is the most derived section in *Paphiopedilum* and includes mainly species that are narrow endemics often found only on single Malaysian islands. Section *Barbata* has probably undergone a recent, rapid radiation which may account for the short branch lengths in this clade.

Braem (1988) presented a taxonomic treatment of *Paphiopedilum* that erected numerous subsections, some containing only one or two species. This system contributes nothing to systematic or phylogenetic understanding of the group. Few of the proposed groups he identified are supported by available independent data, thus refuting his (and any other) narrow approach based either on a few trivial characters or intuition, neither of which is now acceptable in rigorous scientific treatments. (**AC**)

Ecology

Species of *Paphiopedilum* are found from India east across south-east Asia to the Philippines in the west

...arasawa and Saito (1982)	Atwood (1984)	Braem (1988)	Cribb (1987)	Cox *et al.* (1997a)
...rachypetalum	*Brachypetalum*	*Brachypetalum*	*Brachypetalum*	*Parvisepalum*
...arvisepalum		*Parvisepalum*	sect. *Brachypetalum*	
			sect. *Parvisepalum*	
...olyantha	*Paphiopedilum*	*Polyantha*	*Paphiopedilum*	*Paphiopedilum*
sect. *Mastigopetalum*	sect. *Coryopedilum*	sect. *Mastigopetalum*	sect. *Coryopedilum*	sect. *Pardalopetalum*
sect. *Mystropetalum*	sect. *Pardalopetalum*	sect. *Mystropetalum*	sect. *Pardalopetalum*	
sect. *Polyantha*				
...ochlopetalum	sect. *Cochlopetalum*	*Cochlopetalum*	sect. *Cochlopetalum*	sect. *Cochlopetalum*
...aphiopedilum	sect. *Paphiopedilum*	*Paphiopedilum*	sect. *Paphiopedilum*	sect. *Paphiopedilum*
sect. *Stictopetalum*		sect. *Stictopetalum*		
sect. *Paphiopedilum*		sect. *Paphiopedilum*		
sect. *Thiopetalum*		sect. *Thiopetalum*		
sect. *Ceratopetalum*		sect. *Ceratopetalum*		
...igmatipetalum	sect. *Barbata*	*Sigmatopetalum*	sect. *Barbata*	sect. *Barbata*
sect. *Spathopetalum*		sect. *Spathopetalum*		
sect. *Sigmatopetalum*		sect. *Blepharopetalum*		
sect. *Blepharopetalum*				
sect. *Punctatum*		sect. *Punctatum*		
sect. *Planipetalum*		sect. *Planipetalum*		
sect. *Barbata*		sect. *Barbata*		
				sect. *Brachypetalum*

and from southern China in the north, south to the Malay Archipelago, New Guinea and the Solomon Islands. Several species are found at sea level, for example *P. philippinense* and *P. glanduliferum*, but the majority are found in hill country, sub-montane and montane conditions. In the Himalayas, *P. fairrieanum* grows at elevations up to 2200 m, and *P. villosum* up to 2000 m; likewise in western Yunnan in China, both *P. tiginum* and *P. armeniacum* have also been reported from 2000 m. In Borneo *P. hookerae* var. *volonteanum* has been found at 2300 m elevation on the slopes of Mt Kinabalu.

Five species of *Paphiopedilum* have been reported as growing epiphytically. *Paphiopedilum parishii*, *P. lowii*, and *P. villosum* are usually found growing on trees, whereas *P. hirsutissimum* and *P. glanduliferum* are facultative epiphytes. The remaining species are either terrestrial or lithophytic.

Most of the species are to be found growing in small colonies with their roots in leaf litter, often spreading some distance from the base of the plant. In some species of subgenus *Brachypetalum* (*P. concolor*, *P. bellatulum*, *P. niveum*) and in subgenus *Paphiopedilum* sects. *Corypedilum* (*P. stonei*, *P. sanderianum*, *P. philippinense*), *Pardalopetalum* (*P. dianthum*) and *Paphiopedilum* (*P. hirsutissimum* var. *esquirolei*,

P. barbigerum), for example, the plants will grow with their roots attached to a rocky substrate and are true lithophytes.

The *Brachypetalum* species are well-known calcicoles, all being found growing on limestone but usually in detritus in the cracks and shallow hollows of the eroded rock (Fowlie 1989a). *Paphiopedilum concolor* has also been reported as growing in lime-rich sand in Thailand (Fowlie 1977). It grows at 300–1200 m in light to deep shade on north- or north-east facing limestone boulders or cliffs or at their base in shallow soil and leaf litter (Fowlie 1989a; Tsi *et al.* 1999). The plants grow in soil in cracks in the rock, in moss beds, or in soil in the shade of shrubs and grasses on the upper lip of rocks. Soil pH is usually on the alkaline side of neutral, ranging from 7.26–8.3. It flowers in late April and May in the wild.

Fowlie provided some ecological information on a number of Chinese *Parvisepalum* species, notably *P. micranthum* (1989b), *P. emersonii* (1990a), and *P. malipoense* (1991a) in Guangxi Province, which all grow on limestone hills. In Guangxi and Guizhou *P. malipoense* grows in evergreen forest on the north- or north-east-facing slopes of karst limestone hills at 850–1140 m elevation (Fowlie 1991a;

Tsi *et al.* 1999). It tends to grow in soil and leaf litter just below the summit ridge and at the foot of steep limestone boulders. Soil pHs of around 7.5 were recorded. It associates with other orchids such as *Cymbidium*, *Paphiopedilum concolor*, *P. micranthum*, *Vanilla annamensis*, *Phalaenopsis wilsonii*, *Vanda concolor*, and *Cheirostylis chinensis*.

In Guangxi and south-west Guizhou *P. micranthum* is found toward the summit ridges of north- or more or less north-facing slopes of karst limestone hills at between 1025–1400 m elevation (Fowlie 1989*b*, Tsi *et al.* 1999). Its grows in deep shade on the top edges of rocks in thin soil, in crevices in thin soil, or on the steeply sloping faces of the rocks with thin soils, mosses and leaf litter. The range of pH is more or less neutral from 7.0 to 7.5. It flowers from mid-March to early May in the wild, following a dry cool winter period.

Fowlie (1990*a*) examined a population of *P. emersonii* in north-west Guangxi grows on wet north-east-facing limestone cliffs. It grew in moss-covered 'clay, sand, and calcareous soil'. Flowering occurs in late April and May.

Paphiopedilum armeniacum grows on steep, rocky, north-facing slopes between 1380–2000 m elevation, in oak woodland or scrub in soil or in crevices on limestone boulders, always in the shade (Tsi *et al.* 1999). It flowers in the wild from late March until early May, following a dry winter and spring.

Averyanov *et al.* (1996*a,b*) reported on the habitat in Vietnam of *P. malipoense*, and *P. delenatii*. *Paphiopedilum malipoense* is found on karst limestone in north Vietnam close to the Chinese border and also in the south-west of northern Vietnam, growing on eroded limestone at 850–1200 m in open forest. It experiences a dry winter and wet summer. All the species of subgenus *Parvisepalum*, except for *P. delenatii*, are also found growing on limestone or calcareous substrates. *Paphiopedilum delenatii* grows on granite cliffs and slopes at between 800–1200 m elevation in the Bi Dup Mountains in southern Vietnam.

In section *Coryopedilum*, *P. gigantifolium*, *P. stonei*, *P. supardii*, and *P. sanderianum* always grow on limestone cliffs; *P. glanduliferum* usually does. Atwood (1985) studied the pollination biology of *P. rothschildianum* in Sabah, North Borneo. His study site had a 'large' population of over a thousand mature plants growing on shelves on a steep ultrabasic cliff under the shade of *Gymnostoma* trees. Rogers *et al.* (1998) conducted a population study of the litho-phytic *P. sanderianum* in Sarawak. None of three populations had more than 100 mature plants, although larger populations of up to 500 mature plants have been reported. In Highland New Guinea, *P. wilheminiae* grows in calcareous clay on steep banks among grasses. Other orchids, such as *Spathoglottis*, *Phaius*, and terrestrial species of *Bulbophyllum*, are frequent in the same habitat (Cribb 1997).

The species of section *Cochlopetalum* are also found in Sumatra and Java on limestone or calcareous rocks. In Java *P. glaucophyllum*, for example, grows with its roots attached firmly to the rock on vertical cliff faces of conglomerate with a high lime content.

Those species of section *Paphiopedilum*, such as *P. fairrieanum* and *P. spicerianum*, for which good habitat information is available, are also found on limestone soils and rocks. Fowlie discussed the ecology of *P. hirsutissimum* var. *esquirolei* (1990*b*) in Guangxi and *P. barbigerum* from south-west Guizhou (1991*b*).

Paphiopedilum hirsutissimum var. *esquirolei* is a lithophyte, growing on sheer rock-faces and cliffs where its stout roots can gain purchase in the cracks and crevices and on the narrow ledges (Fowlie 1990*b*; Tsi *et al.* 1999). The roots form a platform that gathers leaf litter which eventually forms a shallow soil around and on top of the roots where they enter the crevices of the rock. Its narrow, elongate green leaves overhang the cliff. This orchid is a true chasmophyte, most of its habitats being very steep or sheer cliffs and rocks in gorges or on karst limestone hills. Most face more or less to the north, but in the depth of gorges where it is well shaded they are less particular about aspect. They tolerate a pH range of 7.50 to 7.86. Some colonies are extensive with over a hundred large multi-growthed plants. Flowering from April until early June can be prolific in large plants that are not in deep shade, and fruit set is often high.

Paphiopedilum barbigerum grows in similar habitats on sheer limestone cliffs with its roots in narrow cracks and crevices (Fowlie 1991*b*; personal observation). It is found from 300–1200 m elevation, growing in light to deep shade of evergreen forest. The closely related *P. helenae*, recently described by Averyanov *et al.* (1996*a*), is one of the rarest Vietnamese species the habitat of which, *Keteleeria davidiana* forest on north-facing cliffs of karst limestone hills, is greatly endangered by logging and burning.

Sarojini Manon *et al.* (1995) examined seven populations of the rare *P. druryi* in Kerala State in southern India. It is relatively xerophytic compared with other Indian species. It grows on south-eastern and south-western facing slopes between 1400–1550 m elevation in thin sandy loam (pH 4.6–5.5) in montane grassland with scattered shrubs and small trees and on weathered rocks. Its known habitats range from 20 to 1500 square metres in extent, the largest having 3459 plants and the smallest only eight. It tolerates full sunshine during March and April when in flower, the vegetation shading it as the season progresses. Fire is a hazard for those populations that grow in close proximity to *Ochlandra* thickets.

Bänziger (1996) studied the pollination biology of two populations of *P. villosum* in northern Thailand, one on a southern slope at 1540–1790 m elevation, the other on north-east facing slopes between 1270–1570 m. It is one of the few epiphytic slipper orchids, although it can occasionally be found as a lithophyte. His study population comprised 98 mature clumps in seven distinct clusters in 34 trees, the main hosts being *Eugenia angkae*, *Podocarpus neriifolius*, and *Nyssa javanica*. A mean annual rainfall of 1786 mm was recorded over a four-year period, the majority falling in the period between May and November. During the orchid's main flowering season (December to April) little rain fell, but dew formation was regular.

In section *Barbata*, the only record of a species found on limestone is that of *P. hookerae* in Kalimantan and Sarawak. However, in Sabah, *P. hookerae* is found on a variety of substrates, growing in leaf litter, including serpentine on Mt Kinabalu at about 2300 m elevation. *Paphiopedilum rothschildianum* and *P. dayanum*, both Kinabalu endemics, are confined to ultrabasic substrates (serpentine), but both grow with their roots in leaf litter rather than attached to the rock.

The majority of the species of section *Barbata* are found growing in leaf litter on a variety of substrates from granite (e.g. *P. javanicum* var. *virens* on Kinabalu) to volcanic rocks (e.g. *P. superbiens* in Sumatra and *P. wentworthianum* in the Solomons). *Paphiopedilum callosum* has been reported growing on sand in Thailand.

In Hong Kong *P. purpuratum* grows on north and north-west facing slopes exposed to a cold, dry north wind from China in the winter months when the temperature may drop to freezing (Lee 1975). It is found in small colonies on steep rocky slopes in light shade beneath a low cover of scrub at about 500 m elevation. Graham (personal communication) reported it growing on the wooded slopes near streams flowing over large expanses of rock. The plants were growing in shade on moss-covered banks to within a foot (30 cm) of the stream where they were liable to occasional flooding. Elsewhere, they were growing in deep leaf litter among the roots of bamboo and between rocks, always with good shade and moisture. Chau and Siu (personal communication) have found it in several places between 120–750 m elevation, mostly on north and north-west facing slopes but also occasionally on south-west and south-east facing ones along streams. Population size is low, ranging mostly from 20 to 30 plants.

The genus is found over an elevational range from sea-level to about 2300 m, but only in *P. hookerae*, *P. bullenianum*, and *P. lowii* has a species been found over a wide elevational range, the first of these with the most extensive range. *Paphiopedilum hookerae* var. *volonteanum*, in Sabah, grows on ridges in lower montane forest at 900 m and again on Mt Kinabalu on a landslide at *c.* 2300 m elevation. On Mt Kinabalu *P. lowii*, usually a lowland species, has been recorded from *c.* 1600 m.

The majority of *Paphiopedilum* species grow in lower montane evergreen or seasonally deciduous forest, often on the forest floor in shade. A few species such as *P. exul* and *P. philippinense* have been recorded as growing in exposed sunny places. More commonly, though, species such as *P. niveum*, *P. philippinense* and *P. rothschildianum*, which will grow in more open positions, are seldom exposed to direct sunlight for any length of time during the day. On Mt Kinabalu in Sabah, *P. rothschildianum* grows where dappled sunlight hits the plants only in the afternoon. In contrast some of the tessellated-leaved species grow in deep shade. In Java and Borneo, *P. javanicum* and its var. *virens* grow on the forest floor in places where no direct sunlight penetrates.

In many areas where paphiopedilums grow, the rainfall and consequent humidity are high, but rainfall is usually seasonal, and plants in such areas often have to survive considerable dry periods. The thick coriaceous leaves are ideally adapted to survive periodic droughts, and plants recover rapidly when the rains return. Most growers who have received plants through the post will support the view that paphiopedilums can survive well in such adverse conditions. **(PC)**

Pollination

Van der Pijl and Dodson (1966) suggested that *Paphiopedilum* and *Phragmipedium* have trap flowers that are, at least in part, fly-pollinated, whereas *Cypripedium* and possibly *Selenipedium* flowers are bee-pollinated. More recently Bänziger (1996) provided a detailed survey of theories and studies of slipper orchid pollination discussed below.

Van der Pijl and Dodson (1966) and Atwood (1985) cited the lurid floral coloration and the 'curious furry warts' and spots on the floral segments of some species as typical of adaptations in flowers that attract flies. The long tail-like petals of *P. sanderianum*, for example, might serve the same function.

Cross-pollination by insects is certainly prevalent in *Paphiopedilum*. Self-pollination such as that seen in *Phragmipedium lindenii* in which the lip is petaloid is absent in the genus. Delpino (1873) mentioned an 'odore spermatico-urinoso' in *P. villosum* and *P. purpuratum* and suggested basic myophily for the group, corroborated by his discovery of bluebottle flies trapped in the lip of *P. barbatum* in a greenhouse in Italy. Schlechter (1927) made similar observations, and Ziegenspeck (1928) found the same in *P. insigne*. Atwood (1985) has observed syrphid flies in the lip of *P. hennisianum* in cultivation in Florida.

The only detailed studies in the field of pollination are of *P. rothschildianum* and other Bornean species on Mount Kinabalu by Atwood (1985) and of *P. villosum* in north Thailand by Bänziger (1996). The syrphid fly, *Dideopsis aegrota*, was found to be the pollinator of *P. rothschildianum*. The flowers, emitting a 'peppery or spicy fragrance', are considered to mimic brood sites to attract gravid female flies to the staminode where they lay their eggs. Atwood found as many as 76 eggs on a single staminode. After alighting on the staminode, which has a glabrous front, the fly falls into the lip and can only escape by crawling up the hairs, which lead it beneath the stigma and anthers in turn, and emerging eventually from one of two exits on either side of the base of the column. Pollination is effected when the fly repeats the manoeuvre in a second flower. The fly apparently gains no benefit from its experience as the eggs either do not hatch or else the larvae die on hatching from lack of food. Likewise, the fly receives no food benefit from its visit because the flower lacks nectar. Atwood suggested that the glandular hairs on the staminode mimic a colony of aphids, the normal brood site of the fly.

Atwood (1985) also studied colonies of *P. hookerae* var. *volonteanum* and *P. javanicum* var. *virens* which grow on Mount Kinabalu as well but was unable to determine the pollinators. From eighteen observed flowers of the latter he noted that three had been pollinated in a six-week period after anthesis. The high seed-production in successful pollinations no doubt makes up for this low success rate. A single syrphid egg was found on the staminode of *P. hookerae* var. *volonteanum*, but as it was growing near *P. rothschildianum* little could be concluded from the observation.

Cribb (1998*b*) reported collecting a dead syrphid fly trapped between the base of the lip and column of *P. javanicum* var. *virens* on Kinabalu. The fact that it had been trapped and died suggests that a smaller species might be the normal pollinator. He also found syrphid eggs on the staminode of *P. philippinense* in cultivation in Vanuatu in the south-west Pacific.

Atwood's (1985) study gave no clue as to the function of the long petals of *P. rothschildianum*, and his work suggested that the floral fragrance emanates from the centre of the flower rather than from the glandular petal tips as noted in *Phragmipedium* by Vogel (1962).

Bänziger (1996) also provided strong circumstantial evidence that syrphid flies pollinate *P. villosum*. He studied several colonies of this epiphytic species in the evergreen hill forests of north Thailand. *Paphiopedilum villosum* flowers over a period of four and a half months, and individual flowers can last several weeks before dying. Pollen capture was observed in 15 cases, mainly involving female hoverflies of *Episyrphus alternans*, *Syrphus fulvifacies*, and *Betasyrphys serarius*. He surmised that the hover-flies are lured mainly by food deception. Long-distance attraction is probably by the urine-like scent of the flowers; close-range attraction is by the glittering staminode which may mimic droplets of honeydew or moisture. The wart in the centre of the flowers appears to provide an attractive potential perch for the hover-flies. However, it is slippery and they fall into the slipper-shaped lip, the incurved margins of which prevent wing action so that the insect cannot fly out of the lip's orifice. Escape is along a ladder of hairs up to the stigma and anthers in turn and then out by the base of the column. The pollen, which is very sticky and remains viable on the flower for about eight weeks, adheres to the back of the hover-fly. If the hover-fly visits another flower and is trapped, the pollen can be transferred to the stigma of that flower as the insect squeezes beneath the stigma, the flexibility of the lip pressing the insect against the papillose stigma surface.

The flowers of *P. delenatii*, *P. micranthum*, *P. armeniacum*, and their allies strongly resemble those of some species of *Cypripedium* and *Phragmipedium*. These species of *Paphiopedilum* may be bee-rather than fly-pollinated. In *P. micranthum*, for example, the staminode resembles that of *Cypripedium henryi* not only in shape and attitude but also in coloration. The latter is a close relative of *C. calceolus*, which has been the subject of several studies (e.g. Nilsson 1979) of bee pollination. **(PC)**

Uses

Paphiopedilums are one of the most popular groups of orchids in cultivation. They are grown as either pot plants or for their cut flowers, both trades being substantial. They are prized for their long-lasting, substantial, and unusual flowers. Leaf shape and colour are variable: species and hybrids of subgenera *Parvisepalum* and *Brachypetalum* and of section *Barbata* have tessellated leaves, sometimes marked with purple beneath; the rest have green leaves that are paler below. Plants with single flowers or with several-flowered inflorescences are available among both the species and hybrids. Flowers range in size greatly, the smaller species having flowers of 5 cm or less across, whereas some hybrids have enormous, almost circular flowers up to 15 cm in diameter or more. The extraordinary and highly sought-after *P. rothschildianum* has a petal spread of up to 32 cm. The range of flower colour is immense, from white through yellows, greens, browns, purples to reds, or combinations of any of these. Some of the species and complex hybrids have spotted sepals and/or petals and glossy floral segments. **(PC)**

Cultivation

The many species in this genus fall into several discrete groups depending on their cultural requirements (see below). There are, however, a few general rules that should be followed. All species need to have both adequate water and adequate air at the roots. All species appreciate continuous air movement around the leaves and flowers, and most succeed best under low light levels. Diurnal and seasonal temperature fluctuations are needed both for optimal growth and flower initiation in many species.

The essentials of a potting medium is that it should be both freely draining and at the same time retain sufficient moisture for the plant's needs. Currently the most popular potting mixes contain a mixture of organic and inert components. The inert constituents are usually volcanic lava rock, granular rock wool, Styrofoam particles and/or perlite. These are mixed with fir or pine bark, hapuu (tree-fern fibre), sphagnum peat or coconut husk fibre. Variable proportions of these components are used. The mix is also often supplemented with powdered dolomite or other calcium-containing compounds such as crushed oyster shell or marble chips. Many growers also add granulated activated charcoal in their mixes.

Dimensions of the component particles depend on the size and type of plants to be potted. For seedlings and small species, particles should range between 6–9 mm in diameter. For optimum drainage and root aeration, components should be more or less of uniform size. Large plants such as the strap-leaved species and the bigger complex hybrids can be potted into a mix with particles up to 20 mm in diameter.

One mix that has proved successful in California contains five parts washed lava rock pieces, three parts washed Douglas fir bark, and two parts granulated rockwool. To ten gallons of mix approximately half a cup of powdered dolomite is added. Plants also grow in sterilized bark without other components and in the past have been grown in chips of cork and broken walnut or macadamia nut shells. If nut shells are used, the pieces must be small enough so that they do not have cavities that retain water. In some countries, e.g. Japan, paphiopedilums are grown in pure sterilized sphagnum moss or fibres made from cedar bark.

The wide variety of mixes used indicates that the plants tend to be easy to grow and tolerate a wide variety of conditions at their roots. The orchids themselves will indicate if conditions are unsuitable. Thin-textured and flaccid leaves indicate a lack of roots. Leaf-tip dieback indicates chemical problems with the water or potting soil. Pale yellow leaves indicate too much light, and very large dark green leaves indicate too little.

Nearly all species respond positively to repotting with the production of new leaves, and in part this may be a response to increased air spaces around the roots. It is recommended that all plants be repotted at least once a year. The extent of the root system and not the size of the leaves should determine the size of the pot to be used. One common mistake is to overpot. With their sparse root systems many slipper orchids are unable to absorb enough water to drain a pot adequately, and soggy mixes usually lead to rapid decomposition and anaerobic conditions around the roots. In turn the roots die. When in doubt always use a smaller-sized pot. Local conditions determine the type of pot to be used. In areas that have high levels of ambient humidity clay pots are preferred to plastic, but the latter are better for drier climates. Most species can be repotted at any time of the year even when inflorescences are in sheath, although one would normally not disturb the plants at that time of the cycle as it might affect flower quality. However, plants respond to asexual division differently, and plants of some species suffer when divided.

Most tropical slipper orchid species succeed in the temperature range between 13°C (55°F) and 32°C (90°F), although they will tolerate an additional 10°F on either side of the recommended range. The plants appreciate a 10–20°F fluctuation between day and night temperatures. In particular this change should be administered in the autumn if it cannot be produced year-round. It is recommended that temperature be regulated both day and night under those conditions. In tropical low-elevation climates some collections are grown in air-conditioned greenhouses where day temperatures can be brought down to 85°F. However, if the air-conditioning is shut off at night the drop in evening temperature may not be sufficient to induce flower bud formation. Some species such as those belonging to the *Parvisepalum* group can tolerate short exposures to subfreezing temperatures if they are kept dry.

All tropical slipper orchids seem to appreciate considerable air movement, and this should be a permanent feature of the growing conditions for these plants, particularly where and when humidity is high. Small supplementary fans are quite useful in this regard. Slight movements of flower stems indicate adequate air flow.

Many species of *Paphiopedilum* grow as understory plants in forests and are found in quite deep shade, whereas other species are found in exposed sunny situations. Nevertheless, all species can succeed under medium-shade conditions. Light intensities of approximately 1000 ft-candles seem to be optimum. Generally during winter months plants should have somewhat higher light intensities, for example 1300 ft-candles, to compensate for the shorter daylength. During the longer days of the summer months light can be reduced to 900 ft-candles. It is sometimes suggested that the strap-leaved multifloral species such as *Paphiopedilum stonei* and *P. philippinense* can be flowered more successfully at higher light intensities, but a level of approximately 1000 ft-candles is sufficient. In the United States, many hobby growers grow and flower their plants well using artificial lights.

Water quality is important for growing high-quality plants of the various species, and all seem to respond well to pure water with low salt content. Plants also respond to weak concentrations of standard well-balanced fertilizers during periods of active growth. Plants appear to be able to absorb both water and fertilizer through their leaves. Fertilizing the plant by watering the pot may not be as effective as foliar feeding. The leaves of the plant should be drenched with half-strength fertilizer. Rates of feeding vary depending on growing conditions and season. With fertilizers it is better to err on the side of moderation. A few species such as those in the *Brachypetalum* group are intolerant of high salt concentrations or very low pH, indicated by leaf-tip dieback. The best culture is achieved when water contains less than 100 ppm of ions. Many commercial nurseries and amateurs use reverse-osmosis water or water that has been passed through ion-resin beds to achieve those levels of purity.

When plants begin to produce an inflorescence, care should be taken to ensure that there is no standing water, with or without supplemental feeding, around the rachis. Excess moisture can be blown off the sheath. Watering should occur early in the morning to give sufficient time for evaporation or absorption of water droplets. Once the rachis starts to elongate, the pot should not be moved. Developing flowers are particularly sensitive to changes in light intensity, so it is important to maintain the same orientation to light sources. Moving the pot can cause twists in the rachis.

Paphiopedilums are relatively disease-free, and there are almost no documented occurrences of viral infections in them. However, there are several types of fungal and bacterial rots that can occur. Most of these diseases are contagious and can be transmitted to adjacent plants when handled or by being splashed during watering. The best controls are to remove and destroy any infected plants immediately. Valuable infected plants should be isolated from healthy plants before attempts are made to save them. All infected parts should be cut off beyond the point of infection, and special care should be taken not to touch infected parts and then touch healthy tissue. Plants should be kept dry after surgery. There are few phytochemicals available to act as a prophylactic against these infections. Fresh powdered cinnamon has proved effective for many rot problems in *Paphiopedilum* and also *Phragmipedium*. The cinnamon is dusted liberally on cut surfaces and in the crown of the plant. Powdered cinnamon applied to the base of a fan of leaves seems to have a stimulating effect on growth of the treated plant and can be tried on plants that are not growing and appear unresponsive to repositioning or repotting. Proper hygiene in the greenhouse will go a long way toward minimizing outbreaks of these problems.

Most *Paphiopedilum* species are now artificially propagated from seed. Seedlings should be removed from the flask when they are still actively growing. Flasks with overgrown seedlings will grow more slowly and will be more difficult to establish out of flask. Seedlings should be washed and excess water shaken off. There appears to be a cooperative effect that produces better growth of seedlings in community pots than obtained when they are planted singly. Seedlings can be positioned in the community pot actually touching each other. A good seedling potting mix is one of small washed bark (seedling grade) with about 10% fine perlite. During potting the mix should be barely damp. Plants are not watered on potting but covered with a clear container or plastic bag for about one to two weeks. The container can then be removed and the plants watered the following day. Seedlings can remain in these community pots or trays for 6–12 months before being moved to small individual pots. The time needed to reach flowering size will depend on the species involved and actual growing conditions. Under optimum conditions a minimum of three years is required for seedlings of the *Brachypetalum* group and some of the smaller species of section *Paphiopedilum* to flower. Other species, particularly those from the *Coryopedilum* group, may require more than ten years.

Although these cultural conditions apply in general to most tropical slipper orchids, there are groups of related species that have some cultural peculiarities and require special attention. It is good practice to cluster plants with similar growing requirements together in appropriate locations in the greenhouse to ensure that all plants receive the correct culture. Some groups with special requirements are discussed below.

Subgenus *Brachypetalum*

These species are often considered to be among the most difficult to keep going in a collection. Plants have thick succulent leaves and perform best when grown on the dry side. During winter months plants should be watered only sparingly and not fed. When plants are in active growth during late spring and early summer a general purpose, well-balanced fertilizer should be administered at half strength. Leaf-tip dieback indicates excessive fertilizer. These species are particularly susceptible to a wide variety of fungal and bacterial infections if they are damaged. Consequently, it is recommended that these plants never be divided. Even under the best of conditions a majority of the asexual divisions will be lost. It is useful to cluster the *Brachypetalum* species together to ensure that they are not watered at the same rate as other types.

Subgenus *Parvisepalum*

The major problem in growing this group of species seems to be difficulty in producing flower buds. These plants tolerate temperatures that approach freezing but should be kept drier during the winter months. They also respond well to vigorous air movement. They can be divided into single fans of leaves, but individual growths take three or four years to mature and flower if conditions are not optimal. Some species in this section, for example *P. armeniacum* and *P. micranthum*, produce long runners that often emerge through the drainage holes at the bottom of the pot. If the leaves on those shoots are very large, the pot will have to be broken in order to free the new growth. A few growers have experimented with these plants, growing them in moss-lined baskets which allow the plant's stolons to emerge through the openings in the sides of the basket. However, it is very difficult to repot such plants. This group does respond well to frequent repotting. (HK)

Subgenus *Paphiopedilum*
Section *Coryopedilum*

The strap-leaved multifloral species, with the exception of *P. adductum* and *P. randsii*, are easy to maintain. Their special requirements include higher light intensities and a well-draining mix. These plants do not respond well to division and may need a number of years to recover, especially if the divisions comprise only a few growths. Much of the reluctance of multifloral species to flower has to do with recovery from transplant shock; therefore, special care should be taken in repotting not to damage the root system. However, all dead roots should be removed and damaged roots cut off before the damaged point. As with other repotted slipper orchids, all old potting medium should be washed off and plants repotted in barely damp potting mix. Repotted plants should not be watered for a few days to let damaged parts dry. Most of the larger-growing species need mixes with larger chunks of bark, and as this bark can last for several years they can often go for two or three years before repotting. Divisions should each have several strong mature growths. Single-growth divisions are known to have taken in excess of ten years to reach flowering size again.

Section *Paphiopedilum*

Most species in this group are among the easiest and most reliable to grow and flower. The smaller species such as *P. barbigerum*, *P. charlesworthii*, *P. fairrieanum*, *P. henryanum*, and possibly *P. helenae* need annual repotting and should not be broken into divisions with less than three or four fans of leaves. Most of the other species such as *P. insigne* and *P. villosum* are very vigorous plants and seem able to tolerate considerable amounts of abuse and still flower reliably.

Section *Barbata*

Most species of the *Barbata* group are very easy to grow. Those species difficult to maintain in cultivation include the closely related *P. bougainvilleanum*, *P. papuanum*, *P. violascens*, and *P. wentworthianum*. Another species that has proven difficult to maintain for a long period of time is *P. ciliolare*.

Growers who succeed with these species tend to grow them at lower light intensities but do not give them any special treatment. How these species grow in nature needs active research. Other *Barbata* species appear to offer no problems provided they are regularly repotted. If necessary, plants can be divided into small portions with one or two growths, but flower quality is improved if plants are maintained with several growths. In this group it is not unusual for the leading two or three growths to lack roots. Rootless leads will usually initiate their own roots if severed from the main plant. Rooting hormones applied to the cut surface of the rhizome may help. Such pieces must be anchored firmly in their new pots by tying a few pieces of wire to the rhizome and using the wire to hold the cutting firmly in place. Rootless pieces should have their leaves watered, but care should be taken that water does not accumulate and stand in the base of the leaves for extended periods of time.

Taxonomic literature

Albert, V. A. (1994). Cladistic relationships of the slipper orchids (Cypripedioideae: Orchidaceae) from congruent morphological and molecular data. *Lindleyana* 9, 115–32.

Albert, V. A. and Chase, M. W. (1992). *Mexipedium*: a new genus of slipper orchid (Cypripedioideae: Orchidaceae). *Lindleyana* 7, 172–6.

Albert, V. A. and Pettersson, B. (1994). Expansion of genus *Paphiopedilum* to include all conduplicate-leaved slipper orchids (Cypripedioideae: Orchidaceae). *Lindleyana* 9, 133–40.

Asher, J. (1980). A checklist for the genus *Paphiopedilum* for 1980–81. *Orchid Digest* 44, 175–85, 213–28.

Asher, J. (1981). A checklist for the genus *Paphiopedilum* for 1980–81. *Orchid Digest* 45, 15–26; 57–64.

Asher, J. (1985). A contribution to the study of *Paphiopedilum*, subgenus *Cochlopetalum*, with enumeration of the species within the subgenus, and a discussion of some modern techniques to evaluate and define species concepts. *Orchid Digest* 49, 15–30.

Atwood, J. T. (1984). The relationships of the slipper orchids (subfamily Cypripedioideae, Orchidaceae). *Selbyana* 7, 129–247.

Braem, G. (1988). *Paphiopedilum*. Brucke-Verlag, Kurt Schmersow, Hildesheim, Germany.

Brieger, F. G. (1973). 3. *Paphiopedilum*. In *Die Orchideen* (3rd edn) (ed Schlechter, R.), pp. 171–185. Paul Parey, Berlin.

Briquet, J. (1935). Appendix III: Nomina generica conservanda: 132. In *International Rules of Botanical Nomenclature*. G. Fischer, Jena.

Cox, A. V., Pridgeon, A. M., Albert, V. A., and Chase, M. W. (1997). Phylogenetics of the slipper orchids (Cypripedioideae: Orchidaceae): nuclear rDNA sequences. *Plant Systematics and Evolution* **208**, 197–223.

Cribb, P. J. (1987). *The genus* Paphiopedilum. Collingridge, London.

Cribb, P. J. (1997). *The genus* Cypripedium. Timber Press, Portland, Oregon.

Cribb, P. J. (1998). *The genus* Paphiopedilum (2nd edn). Natural History Publications, Kota Kinabalu, Sabah.

Dressler, R. (1981). *The orchids: natural history and classification*. Harvard University Press, Cambridge, Massachusetts.

Fowlie, J. A. (1966). An annotated checklist of the species of *Paphiopedilum*, 1966. *Orchid Digest* **30**, 307–13.

Karasawa, K. (1982). *The genus* Paphiopedilum. Karasawa, Hiroshima, Japan.

Karasawa, K. and Saito, K. (1982). A revision of the genus *Paphiopedilum* (Orchidaceae). *Bulletin of the Hiroshima Botanical Garden* **5**, 1–69.

Koopowitz, H. (1995). An annotated checklist of the genus *Paphiopedilum*. *Orchid Digest* **59**, 115–39.

Kränzlin, F. (1897). *Orchidacearum Genera et Species 1*. Mayer & Müller, Berlin.

Newton, G. D. and Williams, N. H. (1978). Pollen morphology of the Cypripedioideae and Apostasioideae (Orchidaceae). *Selbyana* **2**, 169–82.

Pfitzer, E. H. (1886). Morphologische Studien ueber die Orchideenbluethe. C. Winter, Heidelberg.

Pfitzer, E. H. (1894). Beitrage zur Systematik der Orchideen. Botanisches Jahrbücher für Systematik, Pflanzengeschichte und Pflanzengeographie **19**, 1–42.

Pfitzer, E. H. (1903). *Orchidaceae—Pleonandrae*. I. In *Das Pflanzenreich* (ed. A. Engler), IV, 50 (Heft 12) Orchidaceae Pleonandrae: 28–42.

Rafinesque, C. S. (1838). *Flora Telluriana* **4**, 45–7. H. Pobasco, Philadelphia, Pennsylvania.

Rickett, H. W. and Stafleu, F. A. (1959). Nomina generica conservanda et rejicienda Spermatophytorum. *Paphiopedilum. Taxon* **8**, 448–92.

Rolfe, R. A. (1896). The *Cypripedium* group. *Orchid Review* **4**, 327–34, 363–7.

Schlechter, R. (1927). *Die Orchideen* (2nd edn). Paul Parey, Berlin.

6. PHRAGMIPEDIUM

Phragmipedium Rolfe in *Orchid Rev*. **4**, 331 (1896), *nom. conserv.* Type species: *Cypripedium caudatum* Lindl. [= *Phragmipedium caudatum* (Lindl.) Rolfe].

Uropedium Lindl., *Orch. Linden*. 28 (1846). Type. *Uropedium lindenii* Lindl. [=*Phragmipedium lindenii* (Lindl.) L.O.Williams]
Uropedilum Lindl. emend. Pfitzer, *Entwurf Anordn. Orch*. 95 (1887), *orth. var*.
Phragmopedilum Rolfe emend. Pfitzer in *Bot. Jahrb. Syst*. **25**, 527 (1898), *orth. var*.
Phragmipedilum Rolfe in *Orchid. Rev*. **9**, 175 (1901), *orth. var*.

Derivation of name

From the greek '*phragma*', a fence or division, in allusion to the divisions in its trilocular ovary, and '*pedilon*', a sandal or slipper.

Description (Plate 15B–D, 16A–C; Fig. 6.1)

Small to large terrestrial, lithophytic or epiphytic *herbs* with elongate, fibrous roots arising from a short to elongate rhizome. Erect *shoots* leafy, clustered or less frequently well spaced, glabrous, the base enclosed by two to four sheathing sterile bracts, three- to several-leaved above. *Leaves* several, coriaceous, conduplicate, spreading or suberect, ligulate, elliptic, or oblong, obtuse to acute, often tridenticulate at apex, mid- to dark green on upper surface, lighter green below, glabrous, ciliate or not on the margins. *Inflorescence* terminal, occasionally branching, few- to many-flowered; *rachis* terete, hairy, glandular or glabrous; bracts conduplicate, elliptic, lanceolate, ovate or oblong, green, sometimes spotted or flushed or striped with purple, ciliate or not. *Flowers* deciduous, usually showy, concolorous or not, vernation imbricate; pedicel obscure to short; ovary trilocular, three-ribbed, glabrous or hairy. *Dorsal sepal* erect to hooded over lip, ovate, lanceolate, obovate or elliptic, obtuse, acute or acuminate, glabrous or pubescent on the outer surface, sometimes pubescent within at base, ciliate or not. *Lateral sepals* usually fused to form a concave synsepal that is more or less similar to the dorsal sepal, sometimes keeled on outer surface. *Petals*

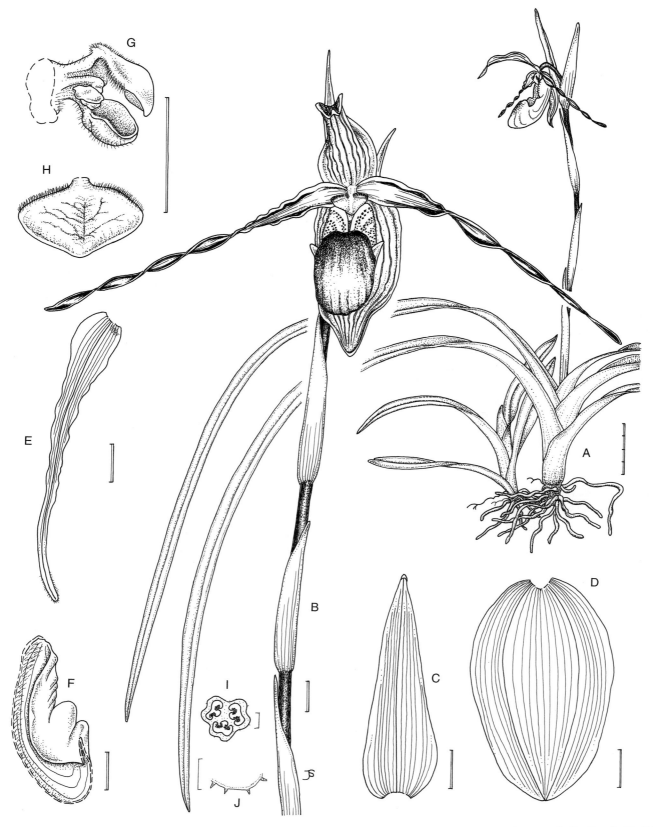

Fig. 6.1. *Phragmipedium longifolium*. A. Habit; B. Flower, front view; C. Dorsal sepal; D. Synsepal; E. Petal; F. Lip, longitudinal section; G. Column, side view; H. Staminode, front view; I. Ovary, transverse section; J. Hairs on ovary. Single bar = 1 mm, double bar = 1 cm. Drawn from *Hugh-Jones162*, colour slides of P. Taylor, and spirit material (cult. J. Kumisch) by Judi Stone.

free, spreading or pendent, flat, reflexed or spiral-
ing, elliptic, ovate, lanceolate, linear-lanceolate,
linear or oblanceolate, rounded, obtuse, acute or
acuminate at apex, often pubescent in basal half
within, usually ciliate. *Lip* deeply pouched and
inflated, slipper-shaped or urn-shaped, with more or
less pronounced incurved side lobes, rarely petaloid,
hairy within especially on lower surface, glabrous or
hairy on outer surface; front margin incurved or
not. *Column* short, stalked, porrect; anthers two,
bilocular, borne on short obtuse to acute filaments;
pollen powdery or viscid; staminode terminal on
column, sessile or shortly stalked, transversely reniform, oblong, ovate, obcordate or linear, flat, convex
or longitudinally conduplicate, glabrous to papillose
or finely pubescent, ciliate or not; stigma stalked,
dependent, tripartite, more or less papillose. *Capsule*
erect to pendent, three-ribbed, cylindrical to almost
ellipsoidal. **(PC)**

Fig. 6.2. Distribution map of *Phragmipedium*.

Distribution (Fig. 6.2)

Southern Mexico and Guatemala through Central
America to South America, south to Bolivia and
Brazil. A small genus of about 15 species.

Infrageneric treatment

Currently four sections are recognized (McCook
1989):

1. Section *Micropetalum* Hallier
Plants relatively small. Leaves thin-textured and
broad. Inflorescence often branching with flowers
blooming successively. Floral bracts short, fused on
basal margins. Flowers small and brightly coloured,
buds covered with multicellular hairs. Petals short
and broad. Lip lacking gibbous hollows at the base.
Two species and a variety.

2. Section *Platypetalum* Pfitzer
Plants large. Inflorescences often branched, with
flowers blooming successively. Flowers large; buds
covered with multicellular hairs. Petals spathulate.
Three closely allied species.

3. Section *Lorifolia* (Kränzl.) Pfitzer
Plants small to large. Inflorescence with flowers
blooming successively. Floral bracts equaling or

longer than the ovary. Flowers large. Petals elongate and tapering. Lip with tubercles on the
infolded margins of the claw; lip opening spurred.
Three species.

4. Section *Phragmipedium*
Plants large. Inflorescence with flowers blooming
simultaneously. Bracts with glands on adaxial surface,
marginally fused at base. Flowers large. Petals elongate, tapering and pendent. Lip rim pubescent, claw
face ribbed. Seven species. **(PC)**

Anatomy

Leaf

Möbius (1887) examined leaves of two species of
Paphiopedilum now included in *Phragmipedium*:
P. caudatum and *P. longifolium*. Stomata are always
abaxial (Solereder and Meyer 1930; Atwood 1984).
In contrast to *Paphiopedilum*, subsidiary cells are
present at least in *P. longifolium*, *P. lindenii*, and

P. sargentianum (N. H. Williams 1979) and in the primary hybrid *P.* × *sedenii* (Atwood 1984). Atwood and Williams (1979) reported that micropapillae and various sculpturing patterns occur on leaves of *P. schlimii*, *P. caricinum*, and *P. pearcei*; micropapillae differ in shape from those of *Paphiopedilum*. Anticlinal walls of epidermal cells are straight as in *Paphiopedilum* (*vis-à-vis* sinuous in *Cypripedium* and *Selenipedium*; Rosso 1966; Atwood 1984). Trichomes are absent from leaves (Rosso 1966). Chlorenchyma is differentiated into 1–2 palisade layers and spongy layers abaxially (Rosso 1966), but Atwood (1984) noticed only a slight difference in cell organization. As in *Paphiopedilum*, vascular bundles are in one row in transverse section, the midvein more prominent, surrounded by bundle sheaths of suberized fibres. Stegmata are absent from both leaf and stem (Solereder and Meyer 1930; Møller and Rasmussen 1984).

Stem

The axis of the plant comprises the leafy stem and rhizome. The leafy stem is bounded by a uniseriate epidermis lacking stomata (Rosso 1966). Stomata and trichomes are absent (Rosso 1966; Atwood 1984). Cortical parenchyma abuts both the epidermis and the sclerotic sheath of the vascular cylinder, which bears scattered vascular bundles (Solereder and Meyer 1930; Rosso 1966), which are amphivasal (Atwood 1984). The rhizome, highly condensed as in *Paphiopedilum*, has a uniseriate epidermis, often necrotic or sloughed, the cell walls impregnated with cutin or suberin. Cortical parenchyma is limited, bounded internally by a multiseriate endodermis with O-shaped thickenings (Rosso 1966). The vascular cylinder comprises a uniseriate pericycle and amphivasal vascular bundles embedded in ground parenchyma (Atwood 1984). Vessel elements are absent; tracheids have annular, spiral, and scalariform thickenings (Rosso 1966).

Root

The root is enclosed in a velamen of up to ten layers, although 4–7 layers are most common (Meinecke 1894; Rosso 1966; Pridgeon 1987). Unicellular root hairs are persistent on the outermost layer (Rosso 1966). Velamen is of the *Calanthe*-type according to the terminology of Porembski and Barthlott (1988). Long cells of the exodermis consistently have U-shaped thickenings

(Rosso 1966; Pridgeon 1987). Broadly laminate tilosomes have been reported for *P. boissierianum* and *P. caudatum* var. *lindenii* (Pridgeon *et al.* 1983). The innermost layer of the cortex, the endodermis, is uniseriate but biseriate locally in *P. caudatum* and *P. caudatum* var. *lindenii* (Rosso 1966) and perhaps also in *P. longifolium* (Meinecke 1894). Both Atwood (1984) and Rosso (1966) reported a multiseriate pericycle of sclerotic parenchyma cells. All endodermal walls may be thickened (O-shaped) or only the radial and inner tangential walls (U-shaped) depending on the taxon. Roots are 12–17-arch (Meinecke 1894, Rosso 1966). Vessels with simple perforation plates and metaxylem tracheids, both with scalariform pitting, are present in roots (Rosso 1966). **(AP)**

Palynology

The pollen of *Phragmipedium* taxa is consistently psilate (Schill and Pfeiffer 1977), with the exception of *P. longifolium* which is mostly fossulate (Newton and Williams 1978). Ultrastructural studies (Burns-Balogh and Hesse 1988) reveal a basically lamellate foot-layer, reduced columellae, and compact tectum for four species and *P.* × *Sedenii*, reaching extreme reduction of foot-layer and columellae in *Phragmipedium longifolium*. **(AP)**

Cytogenetics

Karyotypes for most species of *Phragmipedium* have been published (Karasawa 1979, 1980, 1986; Karasawa and Aoyama 1988). The somatic chromosome numbers for each of the sections are as follows: section *Phragmipedium* ($2n = 28$), section *Platypetalum* ($2n = 18$–22), section *Lorifolia* ($2n = 22$), and section *Micropetalum* ($2n = 24$–26, 28, or 30). Most of the chromosome complement in *Phragmipedium* are metacentrics or sub-metacentrics and some telocentrics, with many species exhibiting bimodal karyotypes where at least one pair of the larger chromosomes is metacentric (Karasawa 1980). If all telocentrics are paired to form metacentrics, a nearly uniform number of 18 is recovered in *Phragmipedium*, i.e. the total number of chromosome arms (i.e. *nombre fondamental*) of 34–37 is broadly conserved (Karasawa 1980). Extrapolations from phylogenetic analyses indicate that a diploid karyotype of 18 metacentric chromosomes is likely the plesiomorphic condition (apparently retained in *P. boissierianum*). **(AC)**

Phytochemistry

Some species of *Phragmipedium* have been studied for flavonoids and alkaloids. In a leaf survey of 142 species from 75 genera of Orchidaceae for presence of various classes of flavonoids, C. A. Williams (1979) found that *Phragmipedium sargentianum* contains flavone *C*-glycosides (= glycoflavones), the most common type of flavonoid in the family. However, the exact structures of the glycoflavones were not determined. No flavonol glycosides were detected in *P. sargentianum*. Stermitz *et al.* (1981) compared the alkaloids present in various *Phragmipedium* species by means of TLC. Two species in section *Phragmipedium*, *P. caudatum* and *P. warscewiczianum*, do not contain alkaloids, whereas two other species in this section, *P. wallisii* and *P. lindenii*, each contain one alkaloid (probably the same compound) in moderate concentrations. The same alkaloid plus an additional one are present in *P. czerwiakowianum* (section *Lorifolia*), both in high concentrations. The additional alkaloid is also present in another species belonging to section *Lorifolia*, *P. boissierianum*. Both compounds occur in high concentrations in *P. pearcei*, but in two other species belonging to this section, *P. ecuadorense* and *P. caricinum*, alkaloids seem to be absent. Finally, two or more different alkaloids are present in *P. sargentianum* and *P. kaieteurum* of section *Platypetalum*. Although the chemical structures of none of these alkaloids were determined, the data obtained indicate that study of the alkaloids in *Phragmipedium* species would be a valuable aid in the taxonomy and the identification of the various members of this genus. **(RG and NV)**

Phylogenetics

Phragmipedium has enjoyed great popularity with horticulturalists and has accordingly suffered (albeit to a lesser extent) from taxonomic inflation like *Paphiopedilum* (Table 6.1). The genus currently comprises four sections, *Micropetalum*, *Platypetalum*, *Lorifolia*, and *Phragmipedium* (Fig. 6.3).

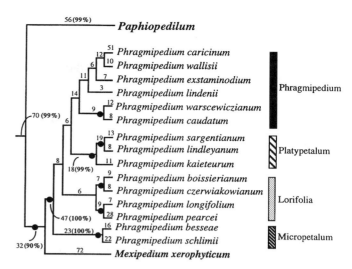

Fig. 6.3. One of the most-parsimonious trees from Cox *et al.* (1997a) showing cladistic relationships in the genus *Phragmipedium*. Sectional limits (*sensu* McCook 1989) are indicated by various shaded bars. Solid circles indicate clades strongly supported (node score >0.63) by the jackknife (Farris *et al.* 1996).

Most taxonomic schemes place section *Micropetalum* sister to the rest of *Phragmipedium*. Available data, both morphological and molecular, support these conclusions (Fig. 6.3). *Phragmipedium besseae* and *P. schlimii*, both of section *Micropetalum*, have inflated labella with involute lip margins, resembling subgenera *Parvisepalum* and *Brachypetalum* of *Paphiopedilum*, *Mexipedium*, and most species of *Cypripedium* and *Selenipedium*. Cribb (1987) has suggested this to be parallel adaptation to exploit similar pollinators; our results cannot confirm that this flower form is a plesiomorphic type, but they do not refute this or similar interpretations drawn by Albert (1994). This flower stereotype may be useful in attracting certain pollinators (mostly bees) and has subsequently been modified for attracting flies in both *Paphiopedilum* and *Phragmipedium*.

Phragmipedium species relationships inferred from DNA data (Cox *et al.* 1997a) closely parallel those in the monograph of the genus by McCook (1989), based primarily on morphological and isozyme data. McCook synonymized *P. kaieteurum* and *P. sargentianum* under *P. lindleyanum*, and molecular data do little to resolve this issue. The same is true for *P. czerwiakowianum*, which McCook treated as a synonym of *P. boissierianum*, and for *P. warscewiczianum*, which she treated as a synonym of *P. caudatum*. Data are unavailable for *P. klotzchianum*.

Table 6.1. Infrageneric treatments of *Phragmipedium* (revised from McCook 1989)

Pfitzer (1903)	Brieger (1973)	Garay (1979)	Atwood (1984)	McCook (1989)	Cox et al. (1997a)
Sect. Micropetalum	**Sect. Micropetalum**	**Sect. Micropetalum**	**Sect. Micropetalum**	**Sect. Micropetalum**	**Sect. Micropetalum**
P. schlimii	*P. schlimii*	*P. schlimii*	*P. schlimii*	*P. schlimii*	*P. schlimii*
			P. besseae	*P. besseae*	*P. besseae*
Sect. Platypetalum	**Sect. Platypetalum**	**Sect. Platypetalum**	**Sect. Platypetalum**	**Sect. Platypetalum**	**Sect. Platypetalum**
P. lindleyanum	*P. lindleyanum*	*P. lindleyanum*	*P. lindleyanum*	*P. lindleyanum*	*P. lindleyanum*
P. sargentianum	*P. sargentianum*	*P. sargentianum*	*P. sargentianum*		
		P. kaieteurum	*P. kaieteurum*		*P. kaieteurum*
Sect. Lorifolia	**Sect. Lorifolia**	**Sect. Lorifolia**		**Sect. Lorifolia**	**Sect. Lorifolia**
P. boissierianum	*P. boissierianum*	*P. boissierianum*	*P. boissierianum*	*P. boissierianum*	*P. boissierianum*
P. czerwiakowianum	*P. czerwiakowianum*	*P. czerwiakowianum*			
		P. reticulatum			
P. longifolium	*P. longifolium*	*P. longifolium*	*P. longifolium*	*P. longifolium*	*P. longifolium*
		P. dariense			
P. hartwegii	*P. hartwegii*	*P. hartwegii*			
		P. hinksianum			
		P. roezlii			
P. vittatum	*P. vittatum*	*P. vittatum*	*P. vittatum*	*P. vittatum*	*P. vittatum*
					P. pearcei
Sect. Phragmipedium		**Sect. Himantopetalum**		**Sect. Phragmipedium**	**Sect. Phragmipedium**
P. caricinum	*P. caricinum*	*P. caricinum*	*P. caricinum*	*P. caricinum*	*P. caricinum*
		P. ecuadorense	*P. ecuadorense*		
P. klotzschianum	*P. klotzschianum*	*P. klotzschianum*	*P. klotzschianum*	*P. klotzschianum*	*P. klotzschianum*
		P. pearcei	*P. pearcei*	*P. pearcei*	
Sect. Caudatum		**Sect. Phragmipedium**	**Sect. Caudatum**		
P. caudatum	*P. caudatum*	*P. caudatum*	*P. caudatum*	*P. caudatum*	*P. caudatum*
				P. exstaminodium	*P. exstaminodium*
		P. lindenii	*P. lindenii*	*P. lindenii*	*P. lindenii*
		P. wallisii	*P. wallisii*	*P. wallisii*	*P. wallisii*
		P. warszewiczianum	*P. warszewiczianum*		

Since the study of Cox *et al.* (1997*a*) ITS sequence data for *P. vittatum* have been collected, which place it sister *P. longifolium* in section *Lorifolia*. The only discrepancy between the findings of Cox *et al.* and McCook concerns the relationship of *P. pearcei*. McCook included this species in section *Phragmipedium*, but molecular data place *P. pearcei* in section *Lorifolia* sister to *Phragmipedium longifolium*, possibly the result of misidentification. **(AC)**

Ecology

Phragmipedium species are found in the tropics from southern Mexico to Bolivia and Brazil. Rarely found near sea level (*Phragmipedium longifolium*), they may be found as high as Machu Picchu in Peru (Vargas 1965) or close to 2500 m elevation. The southernmost species are found in Bolivia and São Paulo, Brazil, but apparently no species have been able to spread into the more temperate parts of South America. Some of the species such as *P. besseae* are found within narrow elevation bands (1300–1900 m), but *Phragmipedium longifolium* may be found from sea level to 1600 m elevation. We have noted that plants collected at 1600 m produce more lanky growth in heated greenhouses than do plants from near sea level.

A few species are epiphytic (some of the *P. caudatum* group), but most are terrestrial or lithophytic, particularly near waterfalls. The Central American populations of the *Phragmipedium caudatum* complex are largely epiphytic from Panama to Nicaragua but terrestrial or lithophytic on granite (Fowlie 1972) in South America. Dodson (personal communication) indicated that *P. lindenii* grows on lava. *Phragmipedium exstaminodium* is epiphytic in southern Mexico.

Most species grow terrestrially or lithophytically in crevices of volcanic rock, usually in association with river systems. *Phragmipedium longifolium* usually grows on landslides near rivers where they probably must recolonize before the canopy closes after a disturbance. Commonly, the substrate on which they are growing is dark and slippery, suggesting that the plants may benefit from a mat of cyanobacteria. Plants are sometimes seen invading road cuttings which apparently simulate landslides by rivers. Some populations in Panama, however, are known only on open rocks often in the centres of swift-running rivers where they receive sufficient light. In such a habitat the plants are dwarfed, sometimes producing an inflorescence less than 20 cm tall. These plants remain dwarfed in cultivation and flower poorly unless given full sunlight. They may be subjected to periodic flash flooding and become lost to raging torrents. Similarly, species such as *P. pearcei*, *P. boissierianum*, *P. hirtzii*, and *P. caricinum* grow near rivers. Perhaps the most specialized species is *P. klotzschianum* (Dunsterville 1982; Gorinsky 1972) which grows at the junction of Venezuela, Guyana, and Brazil. This species is found between 1300–1400 m elevation growing on sandstone by streams where plants have a constant supply of water to their roots. Fowlie (1986) reported that *P. sargentianum* grows in seeps over granite.

A few species grow on limestone or on serpentine. *Phragmipedium besseae* grows on vertical dripping cliffs often near waterfalls. Its precipitous habitat, generally avoided by plant collectors, probably explains why it remained undiscovered for so long. It seems to prefer growing on serpentine as the rock has a soapy texture and reacts to strong acid (Stig Dalström, personal communication).

Most *Phragmipedium* species can grow in fairly strong and sometimes direct sunlight, but more frequently the light is dappled by the surrounding vegetation. None of the species prefers deep shade, and the poor flowering of large plants growing in deep shade attests to their preference for stronger light.

Phragmipediums prefer high humidity of the rainforests in which they grow and where rainfall is frequent throughout the year. Some do not mind growing in areas where water is standing for short periods. In contrast, the *P. caudatum* complex will not tolerate standing water around their roots. **(JA)**

Pollination

Little is known of the pollination ecology in *Phragmipedium*. Most species have sequential flowering inflorescences except for the *P. caudatum* complex. Fowlie (1972) saw small bees alighting on flowers of *P. caudatum* but did not make detailed observations of pollination. McCook (personal communication) discovered syrphid flies visiting flowers of *P. caudatum*. Dodson and Gillespie (1967) reported both syrphids and small bees pollinating *P. longifolium*.

This orchid frequently reproduces vegetatively through elongated stolons, but seed set through pollination may be infrequent. There must be a similar story for *P. dalessandroi* Dodson and Gruss, but this species fails to produce stolons. Self-pollination sometimes occurs in *P. boissierianum* (*P. czerwiakowianum, sensu stricto*) and *P. schlimii* and is the normal mode of seed set in *P. lindenii*, which has a third stamen complete with filament that touches the stigma. Natural hybridization is suspected in an entity recently described as *P. richteri* (Roeth and Gruss 1994). This kind of plant showing moderately swollen auricles is known among populations of *P. boissierianum* and *P. pearcei* in Peru. The stature and moderately swollen auricles are intermediate among the other sympatric species. The problem with the hybrid hypothesis is that sometimes *P. richteri* is seen away from populations of the presumed parents. However, it is possible for the seed to become established far away from a mother plant. (JA)

Uses

Phragmipedium species and hybrids have been cultivated for more than a hundred years, but their appeal is markedly less than that of paphiopedilums. The earliest records of species in cultivation are probably those of *P. caudatum* and *P. lindenii* in the 1840s. The first artificial hybrid, *P. Dominianum*, a hybrid of *P. caricinum* and *P. caudatum*, was produced by Messrs. Veitch & Sons of Chelsea and Exeter in 1870. Hybridization continues and has received a boost from the introduction of the scarlet-flowered *P. besseae* and its 'lutino' variant into cultivation in the 1980s. The late Donald Wimber and the Eric Young Orchid Foundation produced spectacular colchicine-induced tetraploids that have increased the appeal of the genus and will drive forward breeding in the future.

Phragmipediums are grown as specimen plants, with little or no demand for their cut flowers.

No other uses have been reported. (PC)

Cultivation

Since this genus was placed under protection of Appendix I of CITES (Convention on International Trade in Endangered Species of Wild Fauna and Flora) the popularity of this group has sky-rocketed. Most of the species are relatively easy to grow and will succeed under similar conditions to paphiopedilums with only a few caveats. They will grow well in any of the paphiopedilum media, but dolomite should be omitted from the mix. Some growers have grown them well in a medium of granulated rockwool, charcoal, and perlite in equal quantities. Suggested light intensities are similar to those recommended for paphiopedilums.

Two of the most desirable species, *Phragmipedium besseae* and *P. schlimii*, have specific requirements. They are particularly sensitive to water quality and will signal their displeasure with leaf tip die-back. Both species, but particularly the latter, need water with very low salt content, and feeding should be at dilute levels. *Phragmipedium besseae* grows on steep cliffs by extending vertical rhizomes that can grow 15 cm or more. They often also do this in cultivation. Roots will usually not be produced from new growth until it is at least half-formed and may not develop if the growth is some distance from the pot. Sometimes divisions can be rooted, but these should not be removed from the mother plant until after the new growth has matured. Care should be taken in bending the new growths down to the medium as they can be quite brittle.

Most phragmipediums are adapted to grow in seepage areas and seem able to tolerate damper conditions than paphiopedilums. In fact, many growers have had great success placing potted phragmipediums in saucers of water, as the plants appear to grow more vigorously with their roots in standing water. Phragmipediums should be transplanted and repotted at regular intervals. Plants can be readily divided but should have at least three growths in each division. (HK)

Taxonomic literature

Albert, V. A. (1994). Cladistic relationships of the slipper orchids (Cypripedioideae: Orchidaceae) from congruent morphological and molecular data. *Lindleyana* 9, 115–32.

Albert, V. A. and Chase, M. W. (1992). *Mexipedium*: a new genus of slipper orchid (Cypripedioideae: Orchidaceae). *Lindleyana* 7, 172–6.

Albert, V. A. and Pettersson, B. (1994). Expansion of genus *Paphiopedilum* to include all conduplicate-leaved slipper orchids (Cypripedioideae: Orchidaceae). *Lindleyana* 9,133–40.

Atwood, J. T. (1984). The relationships of the slipper orchids (subfamily Cypripedioideae, Orchidaceae). *Selbyana* **7**, 129–247.

Atwood, J. T. and Dressler, R. L. (1999). Clarifications and new combinations in the *Phragmipedium caudatum* complex from Central America. *Selbyana* **19**, 245–8.

Briquet, J. (1935). Appendix III: Nomina generica conservanda: 132. In *International Rules of Botanical Nomenclature*. G. Fischer, Jena.

Brieger, F. G. (1973). 3. *Phragmipedium*. In *Die Orchideen* (3rd edn) (ed. R. Schlechter). Paul Parey, Berlin.

Cox, A. V., Pridgeon, A. M., Albert, V. A., and Chase, M. W. (1997). Phylogenetics of the slipper orchids (Cypripedioideae: Orchidaceae): nuclear rDNA sequences. *Plant Systematics and Evolution* **208**, 197–223.

Cribb, P. J. (1987). *The genus* Paphiopedilum. Collingridge, London.

Cribb, P. J. (1997). *The genus* Cypripedium. Timber Press, Portland, Oregon.

Dressler, R. (1981). *The orchids: natural history and classification*. Harvard University Press, Cambridge, Massachusetts.

Garay, L. A. (1979). The genus *Phragmipedium*. Orchid Digest **43**, 133–48.

Kränzlin, F. (1897). *Orchidacearum Genera et Species* 1. Mayer & Müller, Berlin.

McCook, L. (1989). *Systematics of* Phragmipedium *(Cypripedioideae; Orchidaceae)*. Ph.D. thesis. Cornell University, Ithaca, New York.

McCook, L. (1989). The genus *Phragmipedium*. *American Orchid Society Bulletin* **58**, 1095–100.

McCook, L. (1990). The genus *Phragmipedium*. *American Orchid Society Bulletin* **59**, 153–8.

Newton, G. D. and Williams, N. H. (1978). Pollen morphology of the Cypripedioideae and Apostasioideae (Orchidaceae). *Selbyana* **2**, 169–82.

Pfitzer, E. H. (1886). Morphologische Studien ueber die Orchideenbluethe. C. Winter, Heidelberg.

Pfitzer, E. H. (1894). Beitrage zur Systematik der Orchideen. *Botanisches Jahrbücher für Systematik, Pflanzengeschichte und Pflanzengeographie* **19**, 1–42.

Pfitzer, E. H. H. (1903). *Orchidaceae-Pleonandrae*. In *Das Pflanzenreich* (ed. A. Engler), IV, 50, 1–132. Engelmann, Leipzig.

Rickett, H. W. and Stefleu, F. A. (1959). Nomina generica conservanda et rejicienda Spermatophytorum. *Paphiopedilum. Taxon* **8**, 448–92.

Rolfe, R. A. (1896). The *Cypripedium* group. *Orchid Review* **4**, 327–34, 363–7.

Schlechter, R. (1927). *Die Orchideen* (2nd edn). Paul Parey, Berlin.

7. MEXIPEDIUM

Mexipedium V.A.Albert & M.W.Chase in *Lindleyana* 7 (3), 173 (1992). Type species: *Mexipedium xerophyticum* (Soto, Salazar & Hagsater) V.A.Albert & M.W.Chase.

Phragmipedium subgen. *Mexipedium* (V.A.Albert & M.W.Chase) V.A.Albert & Börge Pett. in *Lindleyana* 9 (2), 138.

Derivation of name

From Mexico, where this monotypic genus is found, and the Greek 'pedilon', a slipper or shoe.

Description (Plate 16D; Fig. 7.1)

Small lithophytic, colony-forming *herbs* with elongate, fibrous roots arising from an elongated rhizome; rhizome elongate, with 5–12 internodes between growths, covered by numerous sheaths. Erect *shoots* leafy, well-spaced, glabrous, the base enclosed by two to three sheathing sterile bracts, five- to eight-leaved above. *Leaves* coriaceous, conduplicate, spreading or suberect, ligulate or oblong, obtuse, mucronate at apex, clear green on upper surface, lighter green below, glabrous, ciliate or not on the margins. *Inflorescence* terminal, paniculate with one branch developing from the median sterile bract, 3- to 7-flowered, the flowers borne in succession; peduncle of 2 nodes, elliptic in cross-section, with a central conduplicate sterile bract; rachis abbreviated, hirsute; bracts imbricate, distichous, conduplicate, cymbiform, hairy, ciliate, the hairs multicellular. *Flowers* thin-textured, deciduous, lacking a scent, relatively small, white, densely brown-pubescent on the outer surface of the sepals, vernation valvate; pedicel obscure to short; ovary unilocular, three-ribbed, densely covered with multicellular hairs, placentation parietal. *Dorsal sepal* hooded over lip, elliptic, acute or subacute, pubescent on the outer surface, sometimes pubescent within at base, ciliate or not. *Lateral sepals* fused to form a concave synsepal that is pendent and suborbicular, obtuse or with two subacute tips, densely pubescent on outer surface. *Petals* free, incurved, oblong-ligulate, acute at apex, glabrous or ciliate at the base. *Lip* deeply pouched and inflated, slipper-shaped, subglobose, with very obscure incurved side

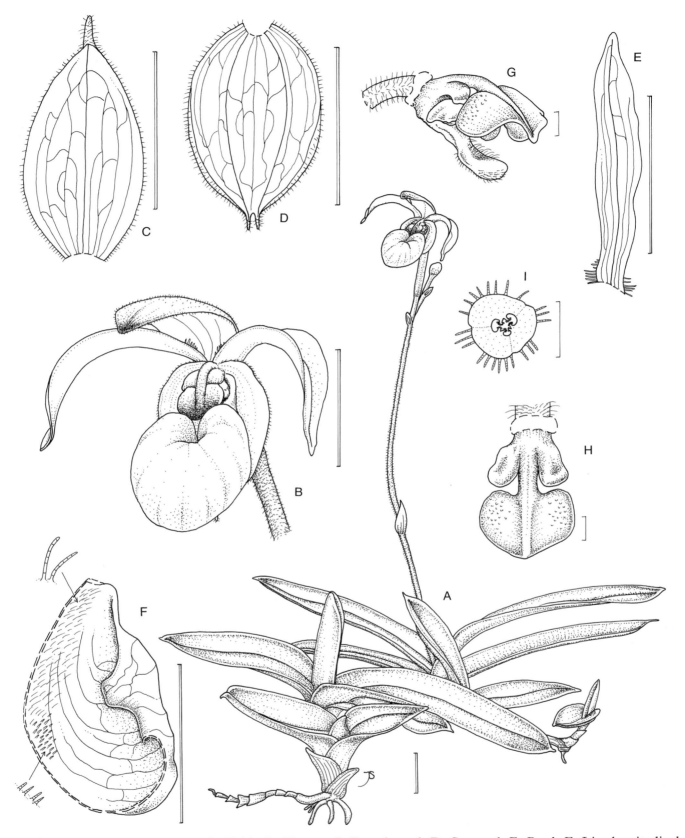

Fig. 7.1. *Mexipedium xerophyticum.* A. Habit; B. Flower; C. Dorsal sepal; D. Synsepal; E. Petal; F. Lip, longitudinal section; G. Column, side view; H. Staminode, front view; I. Ovary, transverse section. Single bar = 1 mm, double bar = 1 cm. Drawn from *Soto et al.* (1990) *59079* (isotype), colour slide 'Oaxaca' CBR/AOS, and Kew Spirit Collection no. 50896 by Judi Stone.

lobes, apical margins strongly incurved, glandular hairy within especially on lower surface, glabrous on outer surface, apparently lacking translucent windows. *Column* short, stalked, porrect; anthers two, bilocular, borne on short obtuse decurved filaments; pollinia granular, yellow; staminode terminal on column, shortly stalked, transversely oblong, obscurely three-lobed at the apex, convex, glabrous, with a hairy tuft on lower surface; stigma massive, stalked, dependent-curved, surface tripartite, the stalk bearing multicellular hairs. *Capsule* erect, three-ribbed, cylindrical. **(PC)**

Distribution (Fig. B.1)

Southern Mexico (Oaxaca) only. It is a very narrow endemic.

Anatomy

Nothing is known of the vegetative anatomy of *Mexipedium xerophyticum*, although it is likely to be similar to that of *Phragmipedium*. **(AP)**

Palynology

There are no reported observations of *Mexipedium* pollen. **(AP)**

Cytogenetics

The diploid complement of *Mexipedium xerophyticum* has 26 chromosomes, some of the pairs showing a degree of heteromorphy (Cox *et al.* 1997*b*). The karyotype probably comprises 20 metacentric/sub-metacentric and six telocentric chromosomes, although poorly defined centromeres of some small chromosomes makes it difficult to distinguish between metacentrics and telocentrics (See Figs VII.3, VII.4). The clearest feature of the karyotype is one pair of large metacentric chromosomes approximately twice the size of any other chromosome in this bimodal complement. Degrees of bimodality have been observed in many slipper orchid karyotypes (Karasawa 1980), although the number of chromosomes in each of the two quantitative chromosome groups may vary. In some cells of

the two roots of *Mexipedium* that were studied the two large metacentric chromosomes were dimorphic, the altered position of this centromere probably resulting from a local pericentric inversion. **(AC)**

Phytochemistry

Nothing is known of the phytochemistry of this species. **(RG and NV)**

Phylogenetics

Mexipedium xerophyticum was originally described as a *Phragmipedium* on the basis of four putative synapomorphies: (1) valvate sepal aestivation, (2) perianth epidermis with straight rather than sinuous cell walls, (3) fusion of the lateral lobes of the labellum, and (4) ventral synsepals that are larger than dorsal sepals (Albert and Chase 1992; Atwood 1984). However, within *Phragmipedium* only the first of these characters is consistent (Albert and Chase 1992). The known distribution of *Mexipedium* is limited to the state of Oaxaca, Mexico, which does not overlap the ranges of *Phragmipedium* species occurring farther south in southern Mexico (Chiapas), Mesoamerica, and South America.

Description of *Phragmipedium xerophyticum* (Soto *et al.* 1990) highlighted a number of internal conflicts in generic delimitation of the conduplicate genera. Albert and Pettersson (1994) attempted to resolve these conflicts by expanding the concept of *Paphiopedilum* to embrace *Phragmipedium*, *Mexipedium*, and *Paphiopedilum*, although this suggestion has not been generally adopted. Available data unambiguously place *Mexipedium xerophyticum* sister to *Phragmipedium*. Such an arrangement receives significant support from the jackknife in Cox *et al.* (1997*a*), and a similar topology has been recovered from an analysis of plastid *rbcL* sequences (Albert and Chase 1992).

A principal point of discussion concerns whether *Mexipedium* warrants generic rank. Although similarities between *Mexipedium* and *Phragmipedium* may be as significant as their differences, current evidence demonstrates a degree of divergence between the two that is comparable to that between conduplicate- and plicate-leaved genera (Figs B.2, B.3). Accordingly, the division of Cypripedioideae into five genera appears appropriate, although strong supporting arguments based on DNA data *alone* are insufficient to justify generic status. *Mexipedium* is surely one of the most interesting slipper orchids for many reasons, and it may be pivotal in trying to understand not only the systematics but also the biogeography of the subfamily. **(AC)**

Ecology

The species occurs in warm humid regions on karst outcrops surrounded by woods of pines, sweet gum (*Liquidambar*), and 'encinos'. The vegetation of the karst outcrops comprises stunted trees of *Plumeria rubra*, *Bursera simaruba*, and *Pseudobombax ellipticum* with occasional plants of *Beaucarnia*, *Yucca*, *Agave*, and *Acanthocereus*. *Mexipedium xerophyticum* forms colonies on the limestone rocks by vegetative means. It is nowhere common in its very restricted distribution. Recent fires in the region may have brought it to the verge of extinction (G. Salazar, personal communication). **(PC)**

Pollination

Although *M. xerophyticum* chiefly reproduces vegetatively by elongate stolons, numerous capsules and a pair of small bees were observed by Soto *et al.* (1990). Small bees are suspected as the pollinators of these small, fragrant flowers. **(JA)**

Cultivation

Mexipedium xerophyticum succeeds under a wide variety of cultural situations. It appears to grow happily in a well-draining *Paphiopedilum* mix. The species, however, has a few cultural requirements that need to be addressed. It produces thin, bract-covered horizontal runners that can reach 20 cm or more before turning up to produce the next shoot. In a few clones, short rhizomes only a few centimetres long are produced, but those plants are rare. Roots are initiated from the rhizome near the base of the leaves. As each cluster of leaves matures, it will produce additional rhizomes so that eventually a mature plant can have a network of rhizomes connecting a large number of growths. Each growth needs its own flower pot, and as adjacent shoots are some distance apart a single plant can occupy a large number of containers. To accommodate this habit, most enthusiasts tend to grow and flower plants of this species in small-diameter but relatively deep pots. Containers 5–6 cm wide and 10 cm deep appear to be adequate. Divisions or seedlings are centred in the pot. The rhizomes are thin, and care needs to be taken not to break them. Although rhizomes can be severed once a shoot has rooted, it seems that multi-growth interconnected plants are more vigorous, and it is recommended that rhizomes be kept intact. Feeding and watering follow general recommendations for *Paphiopedilum*. **(HK)**

Taxonomic literature

Albert, V. A. (1994). Cladistic relationships of the slipper orchids (Cypripedioideae: Orchidaceae) from congruent morphological and molecular data. *Lindleyana* **9**, 115–32.

Albert, V. A. and Chase, M. W. (1992). *Mexipedium*: a new genus of slipper orchid (Cypripedioideae: Orchidaceae). *Lindleyana* **7**, 172–6.

Albert, V. A. and Pettersson, B. (1994). Expansion of genus *Paphiopedilum* to include all conduplicate-leaved slipper orchids (Cypripedioideae: Orchidaceae). *Lindleyana* **9**, 133–40.

Soto, M. A., Salazar, G. A., and Hágsater, E. (1990). *Phragmipedium xerophyticum*, una nueva especie del sureste de Mexico. *Orquídea (México)* **12**, 1–10.

Glossary

Abaxial the surface of an organ directed away from the main axis, in general the lower surface of a leaf, outer surface of a sepal or petal, etc.

Acrotonic referring to the attachment of the rostellum or viscidium to the apex of the anther.

Actinomorphic radially symmetrical; regular.

Adaxial the surface of an organ directed toward the main axis, in general the upper surface of a leaf, inner surface of a sepal or petal, etc.

Adventitious applied to roots that do not rise from the radicle but from the stem, etc.; also applied to embryo-like structures in a seed that arise from outside the true embryo and often abort.

Agenous a form of stomatal development in which the guard-cell mother-cell is surrounded by protoderm cells which do not divide (agene cells).

Aestivation the arrangement (folding, interleaving) of flower parts in the bud, e.g. valvate, imbricate.

Allogamy cross-fertilization or outbreeding.

Allopolyploid a hybrid between two (or more) different species, possessing two (or more) unlike sets of chromosomes from each parent; *cf.* **autopolyploid**.

Alveolate deeply pitted, like a honeycomb.

Amphistomaty the presence of stomata on both surfaces of an organ such as a leaf.

Amphivasal referring to a type of vascular bundle in which xylem surrounds the phloem.

Amyloplast a type of plastid which stores starch.

Anatropous applied to an ovule which is bent parallel to its funiculus or stalk.

Androecium the assemblage of stamens in a flower.

Aneuploid having more or fewer than an exact multiple of the haploid set of chromosomes.

Anther that part of the stamen in which pollen is produced by microsporogenesis.

Anticlinal referring to a cell wall or the plane of cell division perpendicular to the nearest surface.

Aperture a depressed area of the pollen grain through which the pollen tube emerges, characterized by a localized thinning of the exine.

Apomictic referring to reproduction without fertilization, or asexual reproduction.

Apomorphy a derived character state.

Auricle a small lobe or ear.

Autapomorphy a derived character state that occurs in only the clade under study and is therefore uninformative in a more comprehensive cladistic context.

Autogamy self-fertilization.

Autopolyploid an organism with two (or more) sets of chromosomes derived from the same species; *cf.* **allopolyploid**.

Axile a type of placentation in which the ovules are in the middle of the ovary in the angles formed by the septa.

B-chromosome relatively inactive heterochromatin in a nucleus, usually smaller than normal chromosomes, numerically inconstant, and generally of limited taxonomic use.

Baculate   rod-shaped; used for instance to describe exine structure in pollen grains.

Basic number   the minimum haploid chromosome number in a species series.

Basitonic   referring to the attachment of the rostellum or viscidium at the base of the anther.

Bitegmic   having two integuments surrounding the nucellus and developing into outer and inner seed coats.

Bivalent   either a chromosome which has duplicated to form two sister chromatids or a pair of duplicated homologous chromosomes held together at meiosis.

Calcareous   composed chiefly of calcium carbonate, or growing on limestone or chalky soil.

Calceiform   slipper-shaped.

Calcicole   a plant that thrives in lime-rich substrate such as chalk, limestone, or calcareous sand.

Callus   a waxy, fleshy, or other protuberance, as on a labellum.

Calymmate   a condition of pollen tetrads or polyads in which the ektexine/sexine of the component monads is continuous around the entire pollen unit.

Calyx   the assemblage of sepals in a flower.

Capitulum   a dense inflorescence ('head') composed of sessile flowers on a receptacle.

Capsule   a dry fruit splitting along one or more sutures, as in most Orchidaceae.

Carpel   the ovule-bearing organ of the flower, collectively the gynoecium.

Casparian strip   a suberized/lignified band in primary cell walls, especially applied to the endodermis of plant roots.

Caudicles   extensions of tissue derived from the anther and connecting the pollinia to stipe or, in orchids without a stipe, directly to the viscidium.

Centric fission   the production of two telocentric chromosomes from one metacentric chromosome by splitting at the centromeric region.

Chalazal   referring to the region of the ovule opposite the micropyle, where the integuments fuse with the ovule stalk.

Chlorenchyma   parenchyma tissue containing chloroplasts.

Chloroplast   a cell organelle in all green plants which contains DNA, chlorophyll, and other pigments involved in photosynthesis; a green plastid.

Chromocentre   a tightly coiled chromosome or portion thereof in the interphase nucleus of a cell.

Chromosome   a rod-shaped body composed of DNA and histones and other proteins, responsible for transmission of genetic information from cell to cell and generation to generation and also for control of cell function and development.

Clade   a monophyletic group.

Cladistics   a method of systematics used to reconstruct phylogenies of organisms and to construct classifications which are based on the identification of clades and their interrelationships.

Cladogram   a branching diagram of taxa showing relationships as defined by synapomorphies (derivative or novel states).

Cleistogamy   the condition of having flowers that never open and are self-pollinated.

Clinandrium   the portion of the orchid column underneath the anther; the 'anther bed'.

Colpate   furrowed, as in pollen grains.

Columella (pl. **columellae**)   columnar structural elements of the exine of some pollen grains.

Column   an organ of the orchid flower representing the fusion of filaments and style.

Column foot   an extension at the base of the column in some orchids, to which the labellum is attached.

Conduplicate   folded once longitudinally down the middle.

Convolute   rolled during development, as some orchid leaves.

Coriaceous   of a leathery texture, as of some orchid leaves.

Corm   a condensed stem, generally underground.

Corolla   the assemblage of petals of a flower.

Cortex   the region of ground tissue between the epidermis and vascular tissues in a stem or root.

Cuticle   a hydrophobic layer composed of cutin on the outer wall of epidermal cells.

Cutin   a wax-like fatty compound impervious to water and making up the cuticle.

Cymbiform   boat-shaped.

Cyme   a broad, determinate, repeatedly branching inflorescence, with the oldest flowers at the end of the branch.

Dehiscent   a spontaneously opening along certain lines to expose reproductive structures, as in orchid anthers, capsules or seeds, or the dropping of leaves.

Diploid   having two sets of chromosomes, designated '2*n*'.

Distichous   in two ranks or rows on opposite sides of an axis, as in leaf arrangement.

Dune slack   a dip between dunes where water accumulates.

Dysploid   referring to differing basic chromosome numbers in populations or species (often confused with **aneuploid,** in which there are single chromosome changes, i.e. monosomy or trisomy).

Edaphic   pertaining to condition of soil or substrate.

Ektexine   the outermost stratum of the exine of pollen grains, comprising tectum, columellae, and foot layer.

Elaters   spiral thickenings or hairs which help to disperse spores or seeds, often by hygroscopic action.

Embryo sac   the female gametophyte of flowering plants containing the egg cell which is later fertilized in the formation of the embryo.

Endexine   the innermost stratum of the exine of pollen grains.

Endodermis   the innermost layer of the cortex in roots and some stems of seed plants.

Endosperm   the nutritive tissue formed in the embryo sac, which is often absorbed by other tissues by the time of germination; absent in nearly all orchids.

Endothecial thickening   secondary thickening in a wall layer of the anther.

Entire   smooth and continuous; said of leaves and leaf-like organs in which the margins are not broken by teeth.

Epiphyte   a plant growing on another as its substrate, but not parasitic.

Excrescence   an outgrowth, as from the epidermis.

Exine   outer wall of pollen grains, composed of sporopollenin and often intricately structured and sculptured.

Exodermis   the cell layer immediately subjacent to the velamen of orchid roots and developmentally the outermost layer of the cortex.

Extraxylary fibres   sclerenchyma fibres appearing outside the xylem in a plant organ.

Fen   an open plant community on generally alkaline or neutral wet peat, characterized by tall herbaceous plants such as reeds.

Filament   the stalk supporting an anther.

Foot layer   the innermost layer of the ektexine of pollen grains.

Fossulate   with minute grooves, applied to the sculpturing of pollen exine.

Foveolate   with minute pores or pits, applied to the sculpturing of pollen exine.

Funiculus   the stalk of an ovule.

Fusiform   spindle-shaped, as some pseudobulbs.

Gametic   pertaining to gametes (egg and sperm cells).

Gametophyte   the haploid, gamete-forming phase in alternation of generations in plants.

Genome   the genetic chromosomal complement of an organism or cell; also refers to the circular DNA molecules found in plastids and mitochondria.

Glabrous   smooth or hairless.

Gynoecium   the assemblage of carpels in a flower.

Gynostemium   *see* **column**.

Hamulate   having small, hook-like processes.

Hamulus   a type of stipe representing the recurved apex of the rostellum.

Haploid   having one set of chromosomes, designated '*n*'.

Hemiperigenous   a form of stomatal development in which the guard-cell mother-cell is surrounded by agene cells and perigene cells.

Heteroblastic   having pseudobulbs of a single internode.

Heterobrochate   with reticulate sculpturing, such as pollen exine.

Heterochromatin   regions of densely staining chromatin that are highly condensed in the interphase nucleus and not transcribed.

Hexaploid   having six sets of paired chromosomes, designed '6*n*'.

Homoblastic   having pseudobulbs of a several internodes.

Homology   traits attributable to inheritance from a common ancestor rather than from independent origins.

Homoplasy   resemblance not attributable to inheritance from a common ancestor but to convergence, parallelism, or reversal.

Hypodermis   a layer or layers of cells beneath the epidermis and distinguishable from underlying layers.

Hypophysis   the uppermost cell of the suspensor of the embryo from which the root and root cap are derived.

Hypostase   the proliferation of the nucellus at the chalazal end of the embryo sac, often partly lignified or suberized.

Hypostomaty   the condition of having stomata only on the lower surface of the leaf.

Idioblast   a cell in a tissue that differs from others in form, size, or contents.

Imbricate   overlapping; referring, for example, to sepals in the flower bud or to leaves on the stem.

Incumbent   lying upon or bending downwards, as the anther of many orchids during development.

Indehiscent   not opening spontaneously.

Indumentum   a hairy, pubescent, scaly, or waxy covering of an organ such as a leaf or sepal.

Intectate   without a tectum, with reference to pollen exine.

Integument   outer cell layer(s) surrounding the nucellus of the ovule, developing into the seed coat.

Internode   the region of an axis between to nodes.

Interphase   the period in the cell cycle between one mitosis or meiosis and the next, during which DNA is replicated.

Intine   the cellulosic inner wall of a pollen grain, below the exine.

Introrse   turned or faced inward toward the axis; in anthers, opening toward the centre of the flower.

Involute   rolled inward or toward the upper side, as some leaves.

Isozyme *or* **isoenzyme**   any one of several different forms of some enzymes.

ITS   the internal transcribed spacers of 18S–26S nuclear ribosomal DNA, characterized by tandem repeat structure and high copy number.

Karyotype a representation of the chromosome complement in a cell, with the mitotic chromosomes arranged in pairs in order of size and shape.

Labellum the median, modified petal of an orchid flower also known as the **lip**.

Laciniate deeply slashed into narrow divisions or lobes.

Laevigate smooth, with reference to the sculpture of pollen exine.

Ligulate tongue- or strap-shaped.

lip *see* **labellum**.

Lithophyte a plant growing on rocks.

Locule a chamber or cavity, as that bearing ovules in an ovary or pollen in an anther.

Marcescent withering, but persisting.

Massula (pl. **massulae**) a mass or packet of pollen grains in sectile pollinia.

Maximum likelihood a method of phylogenetic inference applied in terms of the probability that a proposed model of evolution and hypothesized history could explain observed data, often but not exclusively used with DNA sequence data.

Megasporogenesis in angiosperms, the formation of megaspores containing the embryo sac.

Mentum a chin-like projection in some orchid flowers, formed by the fusion of the lateral sepals and base of the lip/column.

Mesogene any cell derived by division of a meristemoid (that is not itself a meristemoid) before the meristemoid divides into two guard cells.

Mesoperigenous a form of stomatal development in which the guard-cell mother-cell is surrounded by both mesogene and perigene cells.

Mesophyll the usually photosynthetic parenchyma cells between the upper and lower epidermis of a leaf.

Meristemoid a cell formed by an unequal division of a protodermal cell.

Metacentric referring to a chromosome with the centromere in the middle.

Metaphase a stage in mitosis or meiosis when the chromosomes are aligned along the equator of the cell, with the centromeres along the spindle equator.

Metaxylem that part of the primary xylem that differentiates after protoxylem and before secondary xylem (if formed).

Micropylar referring to the pore or opening in the integuments of the ovule through which the pollen tube enters, or to the homologous opening in the seed coat; also the distal region of the nucellus.

Microsporogenesis in angiosperms, the formation of microspores, which become the male gametophytes (germinated pollen grain and pollen tube).

Monad a single cell, rather than a **tetrad**, resulting from meiosis; a single pollen grain.

Monophyletic referring to all descendents derived from a common ancestor or to a group or taxon containing all the descendents of a single progenitor.

Monopodial referring to a growth habit in which new leaves develop from the same meristem or growing point as all previous leaves; *cf.* **sympodial**.

Multiseriate arranged in several parallel rows.

Multivalent a structure formed by the association of more than two chromosomes during meiosis, characterizing some polyploids.

Mycorrhiza the association of fungi and roots of higher plants, often termed a symbiosis.

Neighbor-joining a tree-building algorithm using a matrix of corrected[*] pairwise distances to produce phylogenetic estimates ([*]based on a specific model of DNA sequence evolution).

Nexine in Erdtman's system of pollen wall stratification, the inner layer of the exine.

Node the region on an axis (stem, rhizome, etc.) to which other plant parts (e.g. leaves) are attached.

Nombre fondamental   fundamental number; the total number of chromosome arms in the complete complement of a species.

Nucellus   the tissue that surrounds the megaspore (and embryo sac) in seed plants.

Nucleoside   a group of compounds consisting of a purine or pyrimidine base linked to the sugar ribose or deoxyribose, e.g. adenosine.

Nucleotide   a phosphate ester of a nucleoside, either a purine or pyrimidine base linked to ribose or deoxyribose phosphates; one of the basic repeating units in DNA or RNA (e.g. adenine triphosphate).

Operculate   having a lid; said of anthers when the anther wall shrivels and is shed after anthesis, leaving the pollinia exposed.

Osmophore   a gland, usually floral, which produces fragrances that attract pollinators.

Ovary   the part of the carpel which contains the ovules and becomes the fruit.

Ovule   the unit of the ovary which contains the egg cell and becomes the seed.

Palisade mesophyll   leaf chlorenchyma characterized by elongate cells arranged with their long axes perpendicular to the surface.

Panicle   a branching inflorescence.

Papilla   a rounded projection from a portion of an epidermal cell; a bump or protuberance.

Paraphyletic   referring to a taxon or group not including all descendents of a common ancestor.

Parenchyma   a plant tissue composed of thin-walled, relatively unspecialized cells with one or more physiological functions.

Parietal   a type of placentation in which the ovules are in rows on the ovary wall or its extensions.

Parsimony   a method of phylogenetic inference emphasizing the criterion of the simplest and most economical explanation to explain the observed variation in data.

Pedicel   the stalk of an individual flower.

Pedicellate   supported by a stalk.

Peduncle   the stalk of a solitary flower or an entire inflorescence.

Pellicle   a thin, flexible outer layer.

Pentaploid   having five sets of chromosomes, designated '$5n$'.

Perforation plate   a part of the wall of a vessel element (xylem) that is perforated.

Perianth   the sepals and petals of a flower.

Pericentric inversion   a chromosome inversion in which the breaks are on either side of the centromere and include it.

Periclinal   referring to a cell wall or the plane of cell division parallel to the nearest surface.

Pericycle   that part of the stele of a root (one or more layers) between the endodermis and phloem, giving rise to lateral roots.

Perigene cell   a daughter cell formed by one or more divisions in a protodermal cell adjacent to the guard-cell mother-cell of a stomatal complex.

Petal   any of the whorl of flower parts generally just inside the sepals, usually colourful and showy.

Phenetics   an approach to classification based on overall similarity of phenotypic characters, not necessarily reflecting phylogenetic relationships.

Phloem   the food-conducting tissue of vascular plants.

Phototaxis   movement in response to light.

Phylogenetics   an approach to the study of evolution and classification based on reconstruction of relationships or evolutionary history of organisms (i.e. cladistics).

Pilate   like a hair.

Pith   ground tissue (usually parenchyma) in the centre of a stem or root.

Placenta a region(s) in the ovary to which ovules are attached.

Placentation the manner in which ovules are attached in an ovary.

Plesiomorphy an ancestral character state.

Plicate folded or pleated.

Plumule the embryonic shoot of a plant.

Polar nuclei one or more (generally two) nuclei in the central cell of a mature embryo sac.

Pollen microspores with highly structured walls, produced by microsporogenesis in anthers and giving rise to the male gametophyte.

Pollinarium (pl. **pollinaria**) the functional unit of pollen transfer in orchid pollination, consisting of two or more pollinia (sometimes with caudicles), often a stalk or stipe, and viscidium.

Pollinium (pl. **pollinia**) a coherent mass of pollen grains.

Polyphyletic referring to a group or taxon not sharing an exclusive common ancestor (i.e. derived from different ancestors).

Polyploid having more than two complete sets of chromosomes.

Polytomy three or more branches diverging from one point on a cladogram (also polychotomy).

Porate with a pore or aperture, as in pollen grains.

Proembryo the embryo in its early stages of development, before differentiation of main body and suspensor.

Protocorm the ephemeral structure resulting from the germinated orchid seed and from which the first true shoot and root differentiate.

Protoderm the meristematic tissue that produces the epidermis of an organ.

Pseudobulb the variously thickened portion of an aerial orchid stem.

Psilate smooth, with reference to the surface of a pollen grain.

Raceme an indeterminate inflorescence of stalked flowers.

Rachis that portion of an inflorescence above the peduncle, bearing flowers

Radicle the embryonic root.

rbcL the plastid gene that codes for the large subunit of the enzyme ribulose 1,5-bisphosphate carboxylase/oxygenase, which fixes carbon dioxide in photosynthesis.

Reniform kidney-shaped.

Restriction site mapping plotting of the positions at which particular restriction endonucleases cleave a genome; usually referring to the circular plastid genome or the tandemly repeating 18S–26S region of the nuclear ribosomal DNA.

Resupinate referring to an orchid flower with the labellum lowermost.

Revolute rolled backward; with margin rolled toward lower side, as some leaves.

Rhizoid a filamentous outgrowth which functions like a root.

Rhizome the indeterminate axis or system of axes of many plants, such as sympodial orchids, which successively give rise to new shoots and flowers, often horizontal or underground.

Robertsonian change a change in chromosome structure due to centric fission or centric fusion that does not disrupt the *nombre fondamental*.

Rostellum part of the median stigma lobe of orchid flowers.

Saprophyte a plant which derives its nourishment, in whole or part, from dead organic matter (often used incorrectly for achlorophyllous plants parasitic on fungi).

Scabrate rough, or with a covering of stiff hairs or points.

Scalariform arranged in a ladder-like fashion, as the pitting in some xylem tracheary elements.

Sclerenchyma a cell with thickened, usually lignified secondary walls; a tissue of such cells functioning primarily in support or protection.

Sclerification   the process of being changed into sclerenchyma.

Sectile   referring to pollinia comprising several 'packets' connected by elastoviscin.

Secund   directed to one side only, as flowers on an inflorescence.

Seep   a spring out of which water or oil oozes.

Semitectate   referring to the exine of a pollen grain with a discontinuous 'roof' supported by columellae, giving a reticulate appearance to the surface.

Sepal   one of the outermost whorl of flower parts, often as colourful and showy as the petals in orchids.

Septum   a partition.

Serrate   toothed, as the margin of some leaves and leaflike organs.

Sexine   in Erdtman's system of pollen wall stratification, the outer layer of the exine.

Silica bodies   conical or spherical masses of silica contained in stegmata of many plants including some orchids.

Somatic   of cell types, structures, and processes in an individual other than those giving rise to gametes.

Spike   an indeterminate inflorescence with sessile flowers.

Spiranthosomes   specialized amyloplasts characterizing orchids in tribe Cranichideae, among others.

Spongy mesophyll   leaf chlorenchyma characterized by loosely arranged cells and conspicuous intercellular spaces.

Sporophyte   the diploid or asexual, (usually) spore-forming phase in alternation of generations in plants; the dominant phase in all seed plants.

Sporopollenin   a cyclic alcohol composing the outer wall of pollen grains and spores, highly resistant to decay.

Spur   a saccate or tubular extension of the labellum (or other floral parts) in many orchids, often containing nectar.

Stamen   the pollen-bearing organ of the flower, composed of anther and filament, collectively the androecium.

Staminodium *or* **staminode**   a sterile stamen.

Stegmata   cells lining sclerenchyma fibres and containing silica bodies.

Stele   a morphological region of the stem or root containing vascular tissues and associated ground tissues.

Stigma   the upper region of the carpel that is receptive to pollen grains.

Stipes (pl. **stipites**)   a pollinium stalk derived from the rostellum; also called **stipe**.

Stolon   a creeping stem that can root and form a new plantlet.

Stoma (pl. **stomata**)   an epidermal pore of many plant parts surrounded by guard cells and functioning in gas exchange.

Suberin   a fatty substance found in endodermal cell walls and cork.

Subsidiary cells   epidermal cells associated with stomata and distinguishable from guard cells either by development or morphology.

Sulcate   with a groove or furrow, as some pollen grains.

Suspensor   an extension at the base of the embryo.

Symplesiomorphy   a shared, ancestral character.

Sympodial   referring to a growth habit in which new shoots arise successively from axillary buds of a rhizome.

Synapomorphy   a shared, derived character.

Synergid   one of two cells at the micropylar end of the embryo sac and associated with the egg cell.

Synsepal   a floral part formed by the partial or complete fusion of two or more sepals.

Systematics   the science of classification based on natural relationships and the study of variation and evolution of taxa.

Tabula infrastigmatica   fleshy structure on either the base of the column or at the juncture of the lip and column.

Taxon (pl. **taxa**)   any taxonomic group, e.g. species, genus, or tribe.

Taxonomy   the science of identification and nomenclature of organisms.

Tectate   referring to the exine of a pollen grain with a generally continuous 'roof' supported by columellae.

Tegula   a pollinium stalk consisting of the modified rostellar epidermis and possibly subtending layers of cells.

Telocentric   referring to a chromosome with a terminal centromere.

Tenuinucellate   a type of nucellar organization in which the embryo sac is surrounded by only an epidermis and a few layers of cells.

Terete   pencil-like; round in cross-section.

Tessellated   with a chequered pattern.

Testa   the seed coat.

Tetracytic   referring to stomata surrounded by four subsidiary cells.

Tetrad   as applied to pollen, four grains attached and shed as a unit.

Tetraploid   having four complete sets of chromosomes, designated '$4n$'.

Theca (pl. **thecae**)   a set of two of the four locules of the anther.

Tilosome   a lignified excrescence from the cell wall of the innermost velamen layer of roots adjacent to the thin-walled passage cells of the exodermis.

Torus   the receptacle of a flower.

Tracheid   a tracheary element of the xylem without perforation plates.

Trichome   a plant hair.

Triploid   having three complete sets of chromosomes, designated '$3n$'.

Tuber   a thickened, usually underground storage root or stem.

Tubercle   a small, rounded protuberance; a nodule on some roots.

Ultramafic   referring to soils low in available calcium, potassium, and phosphorus but high in iron and nickel, considered too toxic for many plants.

Umbel   a more or less flat-topped inflorescence in which the pedicels arise from the same point.

Uniseriate   occurring in a single row or layer.

Unitegmic   having only one integument surrounding the nucellus.

Urceolate   urn-shaped.

Valvate   meeting by the edges without overlapping.

Velamen   a tissue that arises from root dermatogen, consisting of dead cells at maturity and bordered internally by an exodermis.

Vernation   the arrangement of leaves in a bud.

Verrucose   covered with wart-like projections.

Vessel members   one of the cellular components of a vessel, a tracheary element with perforation plates.

Viscidium (pl. **viscidia**)   the sticky portion of the rostellum which is often connected to pollinia.

Xylem   the water-conducting tissue in vascular plants.

Zygomorphic   bilaterally symmetrical.

References

Abe, K. (1968). Contributions to the embryology of the family Orchidaceae, II. Development of the embryo sac in *Pogonia japonica* Reichb.f. *Science Reports of the Tohoku University. Series IV, Biology*, **34**, 59–65.

Abe, K. (1972). Contributions to the embryology of the family Orchidaceae, VII. A comparative study of the embryo sac. *Science Reports of the Tohoku University. Series IV, Biology*, **36**, 179–201.

Ackerman, J. D. and Williams, N. H. (1980). Pollen morphology of the tribe Neottieae and its impact on the classification of the Orchidaceae. *Grana*, **19**, 7–18.

Ackerman, J. D. and Williams, N. H. (1981). Pollen morphology of the Chloraeinae (Orchidaceae: Diurideae) and related subtribes. *American Journal of Botany*, **68**, 1392–402.

Adams, H. (1959). Aspects of variation in the Orchidaceae. In *The orchids: a scientific survey* (ed. C. L. Withner), pp. 73–154. Wiley, New York.

Afzelius, K. (1916). Zur Embryosackentwicklung der Orchideen. *Svensk Botanisk Tidskrift*, **10**, 183–227.

Afzelius, K. (1954). Embryo sac development in *Epipogium aphyllum*. *Svensk Botanisk Tidskrift*, **48**, 513–20.

Aiton, W. (1789). *Hortus Kewensis*. G. Nichol, London.

Albert, V. A. (1990). *In situ*, fluorochrome-mediated visualization of nuclear and cytoplasmic DNA. II. Extra-embryonal nuclei in *Cypripedium acaule* seeds: Persistent evidence of endosperm failure. *Lindleyana*, **5**, 151–7.

Albert, V. A. (1994). Cladistic relationships of the slipper orchids (Cypripedioideae: Orchidaceae) from congruent morphological and molecular data. *Lindleyana*, **9**, 115–32.

Albert, V. A. and Chase, M. W. (1992). *Mexipedium*: a new genus of slipper orchid (Cypripedioideae: Orchidaceae). *Lindleyana*, **7**, 172–6.

Albert, V. A. and Pettersson, B. (1994). Expansion of genus *Paphiopedilum* Pfitzer to include all conduplicate-leaved slipper orchids (Cypripedioideae: Orchidaceae). *Lindleyana*, **9**, 133–9.

Amici, G. B. (1824). Observations microscopiques sur diverses espéces de plantes. *Annales des Sciences Naturelles, Paris*, **2**, 41–70, 211–48.

Amici, G. B. (1830). Note sur le mode d'action du pollen sur le stigmate. Extrait d'une lettre d'Amici a Mirbel. *Annales des Sciences Naturelles, Paris*, **21**: 329–32.

Amici, G. B. (1847). Sur la fécondation des Orchidees. *Annales des Sciences Naturelles, Botanique*. Sér. 3, **7/8**, 193–205.

Aoyama, M. and Karasawa, K. (1997). A karyomorphological study on *Phragmipedium xerophyticum*, Orchidaceae. *Chromosome Science*, **1**, 73–6.

Arber, A. (1925). *Monocotyledons: a morphological study*. Cambridge University Press.

Arditti, J. (1984). An history of orchid hybridization, seed germination and tissue culture. *Botanical Journal of the Linnean Society*, **89**, 359–89.

Arditti, J. (1992). *Fundamentals of orchid biology*. Wiley, New York.

Arditti, J. and Ernst, R. (1993). *Micropropagation of orchids*. Wiley, New York.

Arditti, J., Michaud, J. D., and Healey, P. L. (1979). Morphometry of orchid seeds. I. *Paphiopedilum* and native California and related species of *Cypripedium*. *American Journal of Botany*, **66**, 1128–37.

Arditti, J., Michaud, J. D., and Healey, P. L. (1980). Morphometry of orchid seeds. II. Native California and related species of *Calypso*, *Cephalanthera*, *Corallorhiza*, and *Epipactis*. *American Journal of Botany*, **67**, 347–60.

Asher, J. (1980). A checklist for the genus *Paphiopedilum* for 1980–81. *Orchid Digest*, **44**, 175–85, 213–28.

Asher, J. (1985). A contribution to the study of *Paphiopedilum*, subgenus *Cochlopetalum*, with enumeration of the species within the subgenus, and a discussion of some modern techniques to evaluate and define species concepts. *Orchid Digest*, **49**, 15–30.

Atwood, J. T. (1984). The relationships of the slipper orchids (subfamily Cypripedioideae, Orchidaceae). *Selbyana*, **7**, 129–247.

Atwood, J. T. (1985). Pollination of *Paphiopedilum rothschildianum*: brood-site deception. *National Geographic Research*, Spring 1985, 247–54.

Atwood, J. T. (1986). The size of the Orchidaceae and the systematic distribution of epiphytic orchids. *Selbyana*, **9**, 171–86.

Atwood, J. T. and Williams, N. H. (1979). Surface features of the adaxial epidermis in the conduplicate-leaved Cypripedioideae (Orchidaceae). *Botanical Journal of the Linnean Society*, **78**, 141–56.

Atwood, J. T. and Dressler, R. L. (1999). Clarifications and new combinations in the *Phragmipedium caudatum* complex from Central America. *Selbyana*, **19**, 245–8.

Averyanov, L. V., Hiep, N. T., Loc, P. K., and Huyen, D. D. (1996a). Endangered Vietnamese paphiopedilums. I. *Paphiopedilum helenae*. *Orchids*, **65**, 1062–8.

Averyanov, L. V., Hiep, N. T., Loc, P. K., and Huyen, D. D. (1996b). Rediscovery of the rare slipper orchid *Paphiopedilum delenatii* Guillaum. in Vietnam. *New Orchids*, **80**, 22–5.

Backlund, A. and Bremer, K. (1998). To be or not to be—principles of classification and monotypic families. *Taxon*, **47**, 391–400.

Baker, R. K. (1972). *Foliar anatomy of the Laeliinae (Orchidaceae)*. Ph.D. thesis. Washington University, St. Louis, Missouri.

Bänziger, H. (1996). The mesmerizing wart: the pollination strategy of the epiphytic lady slipper orchid *Paphiopedilum villosum* (Lindl.) Stein (Orchidaceae). *Botanical Journal of the Linnean Society*, **121**, 59–90.

Baranov, P. (1915). Recherchen sur le développement du sac embryonnaire chez les *Spiranthes australis* Lindl. et *Serapias pseudocordigera* Morie. *Bulletin de la Société Imperiale des Naturalistes de Moscou (New series)*, **29**, 74–92.

Barkman, T. J. (1993). *Cladistic and phenetic relationships within* Cypripedium *(Orchidaceae) inferred from floral fragrance-compound data*. M.S. thesis. Michigan State University, East Lansing, Michigan.

Barthlott, W. (1976a). Morphologie der Samen von Orchideën im Hinblick auf taxonomische und funktionelle Aspekte. In *Proceedings of the 8th World Orchid Conference* (ed. K. Senghas), pp. 444–445. Deutsche Orchideen Gesellschaft, Frankfurt.

Barthlott, W. (1976b). Struktur und Funktion des *Velamen radicum* der Orchideen. In *Proceedings of the 8th World Orchid Conference* (ed. K. Senghas), pp. 438–443. Deutsche Orchideen Gesellschaft, Frankfurt.

Barthlott, W. and Ziegler, B. (1981). Mikromorphologie der Samenschalen als systematisches Merkmal bei Orchideen. *Berichte der Deutschen Botanischen Gesellschaft*, **94**, 267–73.

Bateman, R. M., Pridgeon, A. M., and Chase, M. W. (1997). Phylogenetics of subtribe Orchidinae (Orchidoideae, Orchidaceae) based on nuclear ITS sequences. 2. Infrageneric relationships and reclassification to achieve monophyly of *Orchis sensu stricto*. *Lindleyana*, **12**, 113–41.

Bauer, F. A. and Lindley, J. (1830–1838). *Illustrations of orchidaceous plants*. Ridgway and Sons, London.

Baumann, H., Kunkele, S., and Lorenz, R. (1989). Die nomenklatorischen Typen der von Linnaeus veröffentlichen Namen europäischer Orchideen. *Mitteilungen Arbeitskreis Heimische Orchideen Baden-Württemberg*, **21**, 452.

Beer, J. G. (1863). *Beiträge zur Morphologie und Biologie der Familie der Orchideen*. Carl Gerold's Sohn, Wien.

Bell, A. R. (1991). *Plant form: an illustrated guide to flowering plant morphology*. Oxford University Press.

Belling, J. (1924). Detachment (elimination) of chromosomes in *Cypripedium acaule*. *Botanical Gazette*, **78**, 458–60.

Bennett, M. D. and Leitch, I. J. (1995). Nuclear DNA amounts in angiosperms. *Annals of Botany*, **76**, 113–76.

Bentham, G. (1881). Notes on Orchideae. *Journal of the Linnean Society (Botany)*, **18**, 281–360.

Bentham, G. (1883). Orchideae. In *Genera plantarum* (ed. G. Bentham and J. D. Hooker), vol. 3, pp. 460–636. L. Reeve, London.

Bernard, N. (1899). Sur la germination du *Neottia nidus-avis*. *Comptes Rendus Hebdomadaires des Séances de l'Académie des Sciences. Paris*, **1437**, 483–5.

Bernard, N. (1902). Ètudes sur la tubérisation. *Revue Générale de Botanique*, **14**, 5–101.

Bernard, N. (1909). L'évolution dans la symbiose. *Annales des Sciences Naturelles, Botanique*, **9**, 1–196.

Blume, C. L. (1825a). *Tabellen en Platen voor de Javaansche Orchideën*. Ter Lands Drukkerij, Batavia.

Blume, C. L. (1825b). Bijdragen tot de kennis der Javansche orchideën. In *Bijdragen tot de flora van Nederlandsch Indië*, pp. 260–434. Ter Lands Drukkerij, Batavia.

Boorsma, W. G. (1902). Onderzoek naar de plantenstoffen van Nederlands-Indie, Vol. 4: Mededelingen mit Lands Plantentuin 52. Batavia, Dutch Indies.

Bown, D. (1995). *Royal Horticultural Society encyclopedia of herbs and their uses*. Dorling Kindersley, London.

Braem, G. J. (1988). Paphiopedilum: *Eine Monographie aller Frauenschuh-Orchideen der asiatischen Tropen und Subtropen*. Kurt Schmersow, Hildesheim.

Brandham, P. E. and Johnson, M. A. T. (1977). Population cytology of structural and numerical chromosome variants in the *Aloineae (Liliaceae)*. *Plant Systematics and Evolution*, **128**, 105–22.

Brandham, P. E. and Johnson, M. A. T. (1982). Polyploidy and chromosome interchange in *Aloe* L. from Somalia. *Kew Bulletin*, **37**, 387–95.

Bremer, K. (1988). The limits of amino acid sequence data in angiosperm phylogenetic reconstruction. *Evolution*, **42**, 795–803.

Brieger, F. G. (1971). Die Unterscheidung der Unterfamilien und Triben. In *Rudolf Schlechter: Die Orchideen 3. völlig neubearbeitete auflage* (ed. F. Brieger, R. Maatsch, K. Senghas). I/A, Lieferung 3, pp. 147–158. Paul Parey, Berlin.

Brieger, F. G. (1973). Unterfamilie: Cypripedioideae. In *Die Orchideen* (ed. R. Schlechter), pp. 161–198. Paul Parey, Berlin.

Brieger, F. G. (1976). On the orchid system: general principles and the distinction of subfamilies. In *Proceedings of the 8th World Orchid Conference* (ed. K. Senghas), pp. 488–504. Deutsche Orchideen Gesellschaft, Frankfurt.

Briquet, J. (1935). Appendix III: Nomina generica conservanda: 132. In *International Rules of Botanical Nomenclature*. G. Fischer, Jena.

Brown, R. (1810). Prodromus florae Novae Hollandiae et Insulae van-Diemen. J. Johnson, London.

Brown, R. (1813). *Cypripedium*. In *Hortus Kewensis*, (2nd edn) (ed. W. Aiton), 5, 220–2.

Brown, R. (1831). Observations on the organs and modes of fecundation in Orchideae and Asclepiadeae. Richard Taylor, London.

Brown, W. H. (1909). The embryo sac of *Habenaria*. *Botanical Gazette*, 48, 241–50.

Brummitt, R. K. (1992). *Vascular plant families and genera*. Royal Botanic Gardens, Kew.

Brummitt, R. K. (1997). Taxonomy versus cladonomy, a fundamental controversy in biological systematics. *Taxon*, 46, 723–34.

Brummitt, R. K. and Powell, C. E. (1992). *Authors of plant names*. Royal Botanic Gardens, Kew.

Burgeff, H. (1932). *Saprophytismus und symbiose: studien an tropischen Orchideen*. G. Fischer, Jena.

Burgeff, H. (1959). Mycorrhiza of orchids. In *The orchids: a scientific survey* (ed. C. L. Withner), pp. 361–95. Wiley, New York.

Burns-Balogh, P. (1983). A theory on the evolution of the exine in Orchidaceae. *American Journal of Botany*, 70, 1304–12.

Burns-Balogh, P. and Funk, V. (1986). A phylogenetic analysis of the Orchidaceae. *Smithsonian Contributions to Botany*, 61, 1–79.

Burns-Balogh, P. and Hesse, M. (1988). Pollen morphology of the cypripedioid orchids. *Plant Systematics and Evolution*, 158, 165–82.

Butzin, F. (1971). Die Namen der supragenerischen Einheiten der Orchidaceae. *Willdenowia*, 6, 301–40.

Cameron, K. M. (1996). *Phylogenetic relationships of the vanilloid orchids: an integration of molecular, morphological, and anatomical data*. Ph.D. thesis. University of North Carolina, Chapel Hill, North Carolina.

Cameron, K. M. and Chase, M. W. (1998). Seed morphology of the vanilloid orchids. *Lindleyana*, 13, 148–69.

Cameron, K. M., Chase, M. W., Whitten, W. M., Kores, P. J., Jarrell, D. C., Albert, V. A., Yukawa, T., Hills, H. G., and Goldman, D. H. (1999). A phylogenetic analysis of the Orchidaceae: evidence from *rbcL* nucleotide sequences. *American Journal of Botany*, 86, 208–24.

Capesius, I. and Barthlott, W. (1975). Isotopen-Markierungen und Raster-elektronenmikroskopische Untersuchungen des Velamen radicum der Orchideen. *Zeitschrift für Pflanzenphysiologie*, 75, 436–48.

Caputo, P. S., Cozzolino, G. L., Moretti, A., and Stevenson, D. W. (1996). Karyology and phylogeny of some Mesoamerican species of *Zamia* (Zamiaceae). *American Journal of Botany*, 83, 1513–20.

Carlson, M. C. (1954). Megasporogenesis and development of the embryo sac of *Cypripedium parviflorum*. *Botanical Gazette*, 107, 107–14.

Carnevali, G. and de Carnevali, I. R. (1993). New or noteworthy orchids for the Venezuelan flora IX. *Novon*, 3, 102–25.

Case, F. H. (1987). *Orchids of the western Great Lakes region*. Cranbrook Institute of Science, Bloomfield Hills, Michigan.

Case, M. A. (1993). High levels of allozyme variation within *Cypripedium calceolus* (Orchidaceae) and low levels of divergence among its varieties. *Systematic Botany*, 18, 663–77.

Case, M. A. (1994). Extensive variation in the levels of genetic diversity and degree of relatedness among five species of *Cypripedium* (Orchidaceae). *American Journal of Botany*, 81, 175–84.

Catling, P. M. (1982). Breeding systems of northeastern North American *Spiranthes* (Orchidaceae). *Canadian Journal of Botany*, 60, 3017–39.

Catling, P. (1983). Autogamy in eastern Canadian Orchidaceae: a review of current knowledge and some other considerations. *Naturaliste Canadien*, 110, 37–45.

Catling, P. M. (1990). Auto-pollination in the Orchidaceae. In *Orchid biology: reviews and perspectives*, Vol. 5 (ed. J. Arditti), pp. 121–158. Timber Press, Portland, Oregon.

Chase, M. W. (1986). A reappraisal of the oncidioid orchids. *Systematic Botany*, 11, 477–91.

Chase, M. W. and Cox, A. V. (1998). Gene sequences, collaboration, and analysis of large data sets. *Australian Systematic Botany*, 11, 215–29.

Chase, M. W. and Hills, H. G. (1992). Orchid phylogeny, flower sexuality, and fragrance-seeking: evidence from variation in chloroplast DNA among subtribes Catasetinae and Cyrtopodiinae. *BioScience*, 42, 43–9.

Chase, M. W. and Olmstead, R. G. (1988). Isozyme number in subtribe Oncidiinae (Orchidaceae): an evaluation of polyploidy. *American Journal of Botany*, **75**, 1080–5.

Chase, M. W. and Palmer, J. D. (1992). Floral morphology and chromosome number in subtribe Oncidiinae (Orchidaceae): evolutionary insights from a phylogenetic analysis of chloroplast DNA restriction site variation. In *Molecular systematics of plants* (ed. P. S. Soltis, D. E. Soltis, and J. J. Doyle), pp. 324–339. Chapman and Hall, London.

Chase, M. W. and Palmer, J. D. (1997). Leapfrog radiation in floral and vegetative traits among twig epiphytes in the orchid subtribe Oncidiinae. In *Molecular evolution and adaptive radiation* (ed. T. J. Givnish and K. J. Sytsma), pp. 331–352. Cambridge University Press.

Chase, M. W. and Pippen, J. S. (1988). Seed morphology in the Oncidiinae and related subtribes (Orchidaceae). *Systematic Botany*, **13**, 313–23.

Chase, M. W. and Pippen, J. S. (1990). Seed morphology and phylogeny in subtribe Catasetinae (Orchidaceae). *Lindleyana* **5**, 126–34.

Chase, M. W., Cameron, K. M., Hills, H. G., and Jarrell, D. (1994). DNA sequences and phylogenetics of the Orchidaceae and other monocots. In *Proceedings of the 14th World Orchid Conference* (ed. A. M. Pridgeon), pp. 61–73. HMSO, Edinburgh.

Chase, M. W., Stevenson, D. W., Wilkin, P., and Rudall, P. J. (1995). Monocot systematics: a combined analysis. In *Monocotyledons: systematics and evolution* (ed. P. Rudall, P. J. Cribb, D. F. Cutler, and C. J. Humphries), pp. 685–730. Royal Botanic Gardens, Kew.

Chatin, A. (1856). Anatomie des plantes aériennes de l'ordre des Orchidées. 1. Mémoire: Anatomie des racines. *Mémoires de la Société des Sciences Naturelle de Cherbourg*, **4**, 5–18.

Chen, S. C. (1965). A primitive new orchid genus *Tangtsinia* and its meaning in phylogeny. *Acta Phytotaxonomica Sinica*, **10**, 193–206.

Chen, S. C. (1978). *Sinorchis*—a primitive new genus of Orchidaceae from China. *Acta Phytotaxonomica Sinica*, **16**, 82–5.

Chen, S. C. (1979). On *Diplandrorchis*, a very primitive and phylogenetically significant new genus of Orchidaceae. *Acta Phytotaxonomica Sinica*, **17**, 1–6.

Chen, S. C. and Tsi, Z. H. (1984). On *Paphiopedilum malipoense sp. nov.*—an intermediate form between *Paphiopedilum* and *Cypripedium*. *Acta Phytotaxonomica Sinica*, **22**, 119–24.

Chen, S. C. and Xi, Y. Z. (1987). Chinese cypripediums, with a discussion on the classification of the genus. In *Proceedings of the 12th World Orchid Conference* (ed. K. Saito and R. Tanaka), pp. 141–146. Twelfth World Orchid Conference, Inc., Tokyo.

Chesselet, P. C. M. H. (1989). *Systematic implications of leaf anatomy and palynology in the Disinae and Coryciinae*. M.S. thesis. University of Cape Town. Cape Town, South Africa.

Clements, M. A. (1988). Orchid mycorrhizal associations. *Lindleyana*, **3**, 73–86.

Clements, M. A. (1995). *Reproductive biology in relation to phylogeny of the Orchidaceae especially the tribe Diurideae*. Ph.D. thesis. The Australian National University, Canberra.

Clements, M. A. (1996). Reproductive biology in relation to phylogeny of the Orchidaceae especially the tribe Diurideae (Abstract). *Journal and Proceedings of the Royal Society of New South Wales*, **129**(1/2),83.

Clifford H. T. and Smith, W. K. (1969). Seed morphology and classification of Orchidaceae. *Phytomorphology*, **19**, 133–9.

Cocucci, A. E. and Jensen, W. A. (1969). Orchid embryology: the mature megagametophyte of *Epidendrum scutella*. *Kurtziana* **5**, 23–38.

Correll, D. (1950). *Native orchids of North America*. Chronica Botanica, Waltham, Massachusetts.

Cox, A. V. (1995). *The utility of 5S r-DNA in phylogenetic reconstructions: development of the polymerase chain reaction in plant systematics*. Ph.D. thesis. University of Reading. Reading, England.

Cox, A. V., Pridgeon, A. M., Albert, V. A., and Chase, M. W. (1997*a*). Phylogenetics of the slipper orchids (Cypripedioideae: Orchidaceae): nuclear rDNA sequences. *Plant Systematics and Evolution*, **208**, 197–223.

Cox, A. V., Pridgeon, A. M., and Johnson, M. A. T. (1997*b*). Cytological characterization of *Mexipedium xerophyticum* (Cypripedioideae: Orchidaceae). *Lindleyana*, **12**, 162–5.

Cox, A. V., Abdelnour, G. J., Bennett, M. D., and Leitch, I. J. (1998). Genome size and karyotype evolution in the slipper orchids (Cypripedioideae: Orchidaceae). *American Journal of Botany*, **85**, 681–7.

Cozzolino, S., Aceto, S., Caputo, P., Gaudio, L., and Nazzaro, R. (1998). Phylogenetic relationships in *Orchis* and some related genera: an approach using chloroplast DNA. *Nordic Journal of Botany*, **18**, 79–87.

Cribb, P. (1987). *The genus* Paphiopedilum. Collingridge, London.

Cribb, P. (1997). *The genus* Cypripedium. Timber Press, Portland, Oregon.

Cribb, P. (1998*a*). John Lindley. *Orchids*, **67**, 268–73.

Cribb, P. (1998*b*). *The genus* Paphiopedilum, (2nd edn). Natural History Publications, Kota Kinabalu, Sabah, and Royal Botanic Gardens, Kew.

Cronquist, A. (1981). *An integrated system of classification of flowering plants*. Columbia University Press, New York.

Cronquist, A. (1987). A botanical critique of cladism. *Botanical Review*, **53**, 1–52.

Curry, K. J., McDowell, L. M., Judd, W. S., and Stern, W. L. (1991). Osmophores, floral features, and systematics of *Stanhopea* (Orchidaceae). *American Journal of Botany*, **78**, 610–23.

Dahlgren, R. (1977). A system of classification of the angiosperms to be used to demonstrate the distribution of characters. *Botaniska Notiser*, **128**, 119–47.

Dahlgren, R. (1980). A revised system of classification of the angiosperms. *Botanical Journal of the Linnean Society*, **80**, 91–124.

Dahlgren, R. and Clifford, H. T. (1982). *The monocotyledons: a comparative study*. Academic Press, London.

Dannenbaum, C., Wolter, M., and Schill, R. (1989). Stigma morphology of the orchids. *Botanisches Jahrbücher für Systematik, Pflanzengeschichte und Pflanzengeographie*, **110**, 441–60.

Darwin, C. (1862). *On the various contrivances by which British and foreign orchids are fertilised by insects*. John Murray, London.

Darwin, C. (1877). *The various contrivances by which orchids are fertilised by insects*, (2nd edn). D. Appleton, New York.

Daumann, E. (1933). Über die 'Scheinnektarien' von *Parnassia palustris* und anderer Blütenarten. *Jahrbuch für Wissenschaftliche Botanik*, **77**, 104–49.

Daumann, E. (1968). Zur Bestaubbungsoekologie von *Cypripedium calceolus*. *Oesterreichische Botanische Zeitschrift*, **115**, 434–46.

Davis, G. L. (1966). *Systematic embryology of the angiosperms*. Wiley, New York.

Delpino, F. (1867). *Sugli apparecchi della fecondazione nelle piante antocarpee (fanerogame), sommario di osservazioni fatte negli anni 1865–1866*. Cellini, Firenze.

Delpino, F. (1873). Ulteriori osservazioni sulla dicogamia nel regno vegetale. *Atti della Società Italiana di Scienze Naturali*, **16**, 200.

DeWolf, G. P. (1956) Primary classification in the Orchidaceae—a critique. *Taxon*, **5**, 46–53.

Dodson, C. H. (1957). Studies in *Oncidium*. I. *Oncidium pusillum* and its allies. *American Orchid Society Bulletin*, **26**, 170–2.

Dodson, C. H. and Gillespie, R. J. (1967). *The biology of the orchids*. Benson Printing, Nashville, Tennessee.

Donoghue, M. J., Olmstead, R. G., Smith, J. F., and Palmer, J. D. (1992). Phylogenetic relationships of Dipsacales based on *rbc*L sequences. *Annals of the Missouri Botanical Garden*, **79**, 333–45.

Douzery, E. J. P., Pridgeon, A. M., Kores, P., Kurzweil, H., Linder, H. P., and Chase, M. W. (In press). Molecular phylogenetics of Diseae (Orchidaceae): a contribution from nuclear ribosomal ITS sequences. *American Journal of Botany*.

Dressler, R. L. (1960). A review—on the evolution of the Orchidaceae. *American Orchid Society Bulletin*, **29**, 759–60.

Dressler, R. L. (1974). Classification of the orchid family. In *Proceedings of the 7th World Orchid Conference* (ed. M. Ospina), pp. 259–279. Editorial Bedout, Medellín, Colombia.

Dressler, R. L. (1976). How to study orchid pollination without any orchids. In *Proceedings of the 8th World Orchid Conference* (ed. K. Senghas), pp. 534–537. Deutsche Orchideen Gesellschaft, Frankfurt.

Dressler, R. L. (1979). The subfamilies of the Orchidaceae. *Selbyana*, **5**, 197–206.

Dressler, R. L. (1981). *The orchids: natural history and classification*. Harvard University Press, Cambridge, Massachusetts.

Dressler, R. L. (1983). Classification of the Orchidaceae and their probable origin. *Telopea*, **2**, 413–24.

Dressler, R. L. (1986). Recent advances in orchid phylogeny. *Lindleyana*, **1**, 5–20.

Dressler, R. L. (1989*a*). The improbable *Selenipedium chica*. *American Orchid Society Bulletin*, **58**, 549–52.

Dressler, R. L. (1989*b*). Rostellum and viscidium: divergent definitions. *Lindleyana*, **4**, 48–9.

Dressler, R. L. (1989*c*). The vandoid orchids: a polyphyletic grade? *Lindleyana*, **4**, 89–93.

Dressler, R. L. (1990*a*). The Neottieae in orchid classification. *Lindleyana*, **5**, 102–9.

Dressler, R. L. (1990*b*). The Spiranthoideae: grade or subfamily? *Lindleyana*, **5**, 110–16.

Dressler, R. L. (1990*c*). The major clades of the Orchidaceae-Epidendroideae. *Lindleyana*, **5**, 117–25.

Dressler, R. L. (1993). *Phylogeny and classification of the orchid family*. Cambridge University Press.

Dressler, R. L. and Chase, M. W. (1995). Whence the orchids? In *Monocotyledons: systematics and evolution* (ed. P. J. Rudall, P. J. Cribb, D. F. Cutler, and C. J. Humphries), pp. 217–226. Royal Botanic Gardens, Kew.

Dressler, R. L. and Dodson, C. H. (1960). Classification and phylogeny in the Orchidaceae. *Annals of the Missouri Botanical Garden*, **47**, 25–67.

du Petit-Thouars, L.-M. A. A. (1809). Extrait de trois mémoires lus à la première classe de l'institut, sur l'historie des plantes orchidées des îles australes d'Afrique. *Noveau Bulletin des Sciences, par la société philomatique*, **1**, 314–19.

du Petit-Thouars, L.-M. A. A. (1822). Histoire particulière des plantes orchidées recueillies sur les trois iles australes d'Afrique, de France, de Bourbon et de Madagascar. Paris.

Dumée, M. P. (1910). Quelques observations sur l'embryon des Orchidées. *Bulletin de la Société Botanique de France*, **57**, 83–7.

Dunsterville, G. C. K. (1982). Hunting *Phragmipedium klotzscheanum*—an agony in eight fits. *American Orchid Society Bulletin*, **51**, 709–12.

Duvall, M. R., Clegg, M. T., Chase, M. W., Clark, W. D., Kress, W. J., Hills, H. G., Eguiarte, L. E., Smith, J. F., Gaut, B. S., Zimmer, E. A., and Learn, G. H. (1993). Phylogenetic hypotheses for the monocotyledons constructed from *rbc*L sequence data. *Annals of the Missouri Botanical Garden*, **80**, 607–19.

Erdtman, G. (1944). Pollen morphology and plant taxonomy. II. Notes on some monocotyledonous pollen types. *Svensk Botanisk Tidskrift*, **38**, 163–8.

Erdtman, G. (1952). *Pollen morphology and plant taxonomy. Angiosperms*. Almqvist & Wiksell, Stockholm.

Faber, F. C. von. (1904). *Beitrag zur vergleichenden Anatomie der Cypripedilinae*. Ph.D. thesis. University of Heidelberg. Druck von C. Grüninger, Stuttgart.

Fabre, J.-H. (1856). De la germination des Ophrydées et de la nature de leurs tubercules. *Annales des Sciences Naturelles, Botanique*, sér. 4, **5**, 163–86 + 1 pl.

Farris, J. S., Albert, V. A., Källersjö, M., Lipscomb, D., and Kluge, A. G. (1996). Parsimony jackknifing outperforms neighbor-joining. *Cladistics*, **12**, 99–124.

Faegri, K. and van der Pijl, L. (1971). *The principles of pollination ecology*. 2nd ed. Pergamon Press, Oxford.

Fedorov, A. (ed.). (1969). *Chromosome numbers of flowering plants*. Academy of Sciences of the USSR, Leningrad.

Felsenstein, J. (1985). Confidence limits on phylogenies: an approach using the bootstrap. *Evolution*, **39**, 783–91.

Felsenstein, J. (1993). 'PHYLIP (Phylogeny Inference Package),' version 3.5, Department of Genetics, University of Washington, Seattle.

Fockens, J. W. (1857). *Über die Luftwurzeln der Gewächse*. Ph.D. thesis. University of Göttingen. Universitäts-Buchdruckerei von E. A. Huth, Göttingen.

Foldats, E. (1969). *Flora de Venezuela: Orchidaceae*, Part 1. Edicion Especial del Instituto Botanico, Caracas.

Fowlie, J. A. (1966). An annotated checklist of the species of *Paphiopedilum*, 1966. *Orchid Digest*, **30**, 307–13.

Fowlie, J. A. (1972). In search of *Phragmipedium caudatum*. *Orchid Digest*, **36**, 47–8.

Fowlie, J. A. (1977). Malaya revisited: part 12. *Orchid Digest*, **41**, 190–7.

Fowlie, J. A. (1986). In Brazil: Part XXXIII. *Orchid Digest*, **50**, 105–9.

Fowlie, J. A. (1989a). China: awash in the Bitter Sea, Part. II. *Paphiopedilum concolor* on rocky limestone summits of southwestern Guangxi Province. *Orchid Digest*, **53**, 2–24.

Fowlie, J. A. (1989b). China: awash in the Bitter Sea, part III. The habitat of *Paphiopedilum micranthum* among limestone on the southeast border of Guizhou Plateau. *Orchid Digest*, **53**, 132–9.

Fowlie, J. A. (1990a). China: awash in the Bitter Sea. Part IV. The habitat of *Paphiopedilum emersonii* on limestone concretions of southeast of the Guizhou Plateau. *Orchid Digest*, **54**, 41–7.

Fowlie, J. A. (1990b). China: awash in the Bitter Sea. Part V. *Paphiopedilum esquirolei* on cliffs over limestone caves in Guangxi. *Orchid Digest*, **54**, 137–42.

Francini, E. (1931). Ricerche embryologische e cariologiche sul genere 'Cypripedium' s.l. *Nuovo Giornale Botanico Italiano*, **38**, 154–212.

Fredrickson, M. (1990). Embryological study of *Herminium monorchis* (Orchidaceae) using confocal scanning laser microscopy. *American Journal of Botany*, **77**, 123–7.

Fredrickson, M. (1991). An embryological study of *Platanthera bifolia* (Orchidaceae). *Plant Systematics and Evolution*, **174**, 213–20.

Fredrickson, M. (1992). The development of the female gametophyte of *Epipactis* (Orchidaceae) and its inference for reproductive ecology. *American Journal of Botany*, **79**, 63–8.

Fredrickson, M. and Franksson, O. (1988). Confocal scanning laser microscopy, a new technique used in an embryological study on *Dactylorhiza maculata* (Orchidaceae). *Nordic Journal of Botany*, **8**, 369–74.

Freudenstein, J. V. (1991). A systematic study of endothecial thickenings in the Orchidaceae. *American Journal of Botany*, **78**, 766–81.

Freudenstein, J. V. (1998). Paraphyly, ancestors, and classification —a response to Sosef and Brummitt. *Taxon*, **47**, 95–104.

Freudenstein, J. V., Senyo, D. M. and Chase, M. W. (1998). Evolution and phylogenetic significance of a mithchondrial intron in Orchidaceae. *American Journal of Botany*, **85**, 129–30 [Abstract].

Freudenstein, J. V. and Rasmussen, F. N. (1996). Pollinium development and number in the Orchidaceae. *American Journal of Botany*, **83**, 813–24.

Freudenstein, J. V. and Rasmussen, F. N. (1997). Sectile pollinia and relationships in the Orchidaceae. *Plant Systematics and Evolution*, **205**, 125–46.

Freudenstein, J. V. and Rasmussen, F. N. (1999). What does morphology tell us about orchid relationship?—a cladistic analysis. *American Journal of Botany*, **86**, 225–48.

Friedman, W. E. (1990). Double fertilization in *Ephedra*, a nonflowering seed plant: its bearing on the origin of angiosperms. *Science*, **247**, 951–4.

Friedman, W. E. (1992). Double fertilization in nonflowering seed plants and its relevance to the origin of flowering plants. *International Review of Cytology*, **140**, 319–55.

Friedman, W. E. (1994). The evolution of embryogeny in seed plants and the developmental origin and early history of endosperm. *American Journal of Botany*, **81**, 1468–86.

Fuchs, A. and Ziegenspeck, H. (1925). Bau und Form der Wurzeln der einheimischen Orchideen in Hinblick auf ihre Aufgaben. *Botanisches Archiv*, **12**, 290–379.

Fuchs, A. and Ziegenspeck, H. (1926*a*). Entwicklungsgeschichte der Axen der einheimischen Orchideen und ihre Physiologie und Biologie. I. Teil. *Cypripedium, Helleborine, Limodorum, Cephalanthera. Botanisches Archiv*, **14**, 165–260.

Fuchs, A. and Ziegenspeck, H. (1926*b*). Entwicklungsgeschichte der Axen der einheimischen Orchideen und ihre Physiologie und Biologie. II. Teil. *Listera, Neottia, Goodyera. Botanisches Archiv*, **16**, 360–413.

Fuchs, A. and Ziegenspeck, H. (1927*a*). Die Dactylorchisgruppe der Ophrydineen. *Botanisches Archiv*, **19**, 163–274.

Fuchs, A. and Ziegenspeck, H. (1927*b*). Entwicklungsgeschichte der Axen der einheimischen Orchideen und ihre Physiologie und Biologie. III. Teil. *Botanisches Archiv*, **18**, 378–475.

Fuchs, A. and Ziegenspeck, H. (1927*c*). Entwicklung, Axen und Blatter. IV. Teil. *Botanisches Archiv*, **20**, 275–422.

Garay, L. A. (1960). On the origin of the Orchidaceae. *Botanical Museum Leaflets Harvard University*, **19**, 57–96.

Garay, L. A. (1972). On the origin of the Orchidaceae, II. *Journal of the Arnold Arboretum*, **53**, 202–15.

Garay, L. A. (1976). The history of orchids prior to the Renaissance —codices. In *Proceedings of the 8th World Orchid Conference* (ed. K. Senghas), pp. 510–515. Deutsche Orchideen Gesellschaft, Frankfurt.

Garay, L. A. (1978). *Selenipedium*. In *Flora of Ecuador* (ed. G. Harling and B. Sparre), **9**, 12–13. University of Göteborg, Stockholm.

Garay, L. A. (1979). The genus *Phragmipedium*. *Orchid Digest*, **43**, 133–48.

Garay, L. A. (1986). Olim Vanillaceae. *Botanical Museum Leaflets Harvard University*, **30**, 223–36.

Geitler, L. (1956). Zur Fortpflanzungsbiologie, Embryologie und mechanistischen Deutung der Embryogenese von *Epipogium aphyllum. Osterreichische Botanische Zeitschrift*, **103**, 312–35.

Goldblatt, P. (1981). *Index to plant chromosome numbers 1975–1978.* Monographs in Systematic Botany, Vol. 5. Missouri Botanical Garden, St. Louis, Missouri.

Goldblatt, P. (1984). *Index to plant chromosome numbers 1979–1981.* Monographs in Systematic Botany, Vol. 8. Missouri Botanical Garden, St. Louis, Missouri.

Goldblatt, P. (1985). *Index to plant chromosome numbers 1982–1983.* Monographs in Systematic Botany, Vol. 13. Missouri Botanical Garden. St. Louis, Missouri.

Goldblatt, P. (1988). *Index to plant chromosome numbers 1984–1985.* Monographs in Systematic Botany, Vol. 23. Missouri Botanical Garden, St. Louis, Missouri.

Goldblatt, P. and Johnson, D. E. (1990). *Index to plant chromosome numbers 1986–1987.* Monographs in Systematic Botany, Vol. 30. Missouri Botanical Garden, St. Louis, Missouri.

Goldblatt, P. and Johnson, D. E. (1991). *Index to plant chromosome numbers 1988–1989.* Monographs in Systematic Botany, Vol. 40. Missouri Botanical Garden, St. Louis, Missouri.

Goldblatt, P. and Johnson, D. E. (1994). *Index to plant chromosome numbers 1990–1991.* Monographs in Systematic Botany, Vol. 51. Missouri Botanical Garden, St. Louis, Missouri.

Goldblatt, P. and Johnson, D. E. (1996). *Index to plant chromosome numbers 1992–1993.* Monographs in Systematic Botany, Vol. 58. Missouri Botanical Garden, St. Louis, Missouri.

Gorinsky, P. (1972). Habitat notes from a collection in Guyana of *Phragmipedium klotzscheanum. Orchid Digest*, **36**, 151–4.

Gray, A. (1862). Fertilization of orchids through the agency of insects. *American Journal of Science and Arts*, **34**, 420–9.

Griesbach, R. and Asher, J. (1983). Orchids of the Boundary Waters Canoe Area of northern Minnesota. In *North American terrestrial orchid symposium 2* (ed. E. H. Plaxton), pp. 65–80. Michigan Orchid Society, Livonia, Michigan.

Gruss, O. and J. Roeth. (1994). *Phragmipedium richteri* Roeth et Gruss—ein neue Art der Gattung *Phragmipedium*, Sektion *Himantopetalum* (Hallier) Garay auf Peru. *Die Orchidee*, **45**(3), back cover.

Guignard, J. A. (1886). Insects and orchids. *Annual Report of the Entomological Society of Ontario*, **16**, 39–48.

Guttenberg, H. von. (1926). Die Bewegungsgewebe. In *Handbuch der Pflanzenanatomie* (ed. K. Linsbauer), Vol. 5, Sec. 1, Pt. 2. Gebrüder Borntraeger, Berlin.

Hagerup, O. (1947). The spontaneous formation of haploid, polyploid and aneuploid embryos in some orchids. *Kongelige Dansk Videnskabernes Selskabs, Biologiske Meddelelser*, **20**, 1–22.

Hágsater, E. (1984). *Cypripedium dickinsonianum* Hágsater, a new species from Chiapas, Mexico. *Orquídea (México)*, **9**, 198–213.

Haig, D. (1990). New perspectives on the angiosperm female gametophyte. *Botanical Review*, **56**, 236–74.

Hallé, N. (1977). Orchidacées. *Flore de Nouvelle Caledonie et dependances*, fascicle 8. Muséum National d'Histoire Naturelle, Paris.

Hallier, H. (1896). Über *Paphiopedilum amabile* und die Hochgebirgsflora des Berges K'Lamm in West Borneo nebst einer Übersicht über die Gattung *Paphiopedilum*. *Annales du Jardin Botanique de Buitenzorg*, **14**, 18–52.

Hatch, E. D. (1954). Primary classification in the Orchidaceae. *Ceiba*, **4**, 222–4.

Hausen, B. M. (1979). New allergenic quinones in orchids. *Archives of Dermatology Research*, **264**, 102–3.

Hausen, B. M. (1980). Allergic contact dermatitis to quinones in *Paphiopedilum haynaldianum* (Orchidaceae). *Archives of Dermatology (USA)*, **116**, 327–8.

Hennig, W. (1966). *Phylogenetic systematics*. University of Illinois Press, Urbana.

Hering, L. (1900). Zur Anatomie der monopodialen Orchideen. *Botanisches Centralblatt*, **84**, 1–11, 35–45, 73–81, 113–22, 145–52, 177–84, tables I–III.

Heslop-Harrison, J. (1968). Synchronous pollen mitosis and the formation of the generative cell in massulate orchids. *Journal of Cell Science* **3**, 457–66.

Hesse, M., Burns-Balogh, P., and Wolff, M. (1989). Pollen morphology of the 'primitive' epidendroid orchids. *Grana*, **28**, 261–78.

Heusser, K. (1915). Die Entwicklung der generativen Organe von *Himantoglossum hircinum* Spr. (*Loroglossum hircinum* Rich.). *Beihefte zum Botanischen Centralblatt*, **32**, 218–77.

Hildebrand, F. (1863). Die Fruchtbildung der Orchideen, ein Beweis für die doppelte Wirkung des Pollens. *Botanische Zeitung*, **21**, 329–33, 337–45.

Hillis, D. M. (1996). Inferring complex phylogenies. *Nature*, **383**, 130–13.

Hofmeister, W. (1847). Untersuchungen des Vorgangs bei der Befruchtung der Oenotheren. *Botanische Zeitung*, **5**, 785–92.

Holman, R. T. (1983). Studies on the fragrances of North American terrestrial orchids. In *North American Terrestrial Orchid Symposium 2* (ed. E. H. Plaxton), pp. 32–40. Michigan Orchid Society, Livonia, Michigan.

Hsu, C.-C. (1972). Preliminary chromosome studies on the vascular plants of Taiwan. V, Cytotaxonomy of some monocotyledons. *Taiwania*, **17**, 48–65.

Hünecke, G. (1904). *Zur Anatomie der Pleurothallidinae*. Ph.D. thesis. University of Heidelberg. Hörning & Berkenbusch, Heidelberg.

Irmisch, T. (1850). *Zur Morphology der monokotylischen Knollen- und Zwiebelgewächse*. G. Reimer, Berlin.

Jacquet, P. (1994). History of orchids in Europe, from antiquity to the 17th century. In *Orchid biology: reviews and perspectives*, Vol. 6 (ed. J. Arditti), pp. 33–102. Wiley, New York.

Jansen, R. K., Michaels, H. J., Wallace, R. S., Kim, K.-J., Keeley, S. C., Watson, L. E., and Palmer, J. D. (1992). Chloroplast DNA variation in the Asteraceae: phylogenetic and evolutionary implications. In *Molecular systematics of plants* (ed. P. S. Soltis, D. E. Soltis, and J. J. Doyle), pp. 252–279. Chapman and Hall, London.

Jeyanayahgy, S. and Rao, A. N. (1966). Flower and seed development in *Bromheadia findlaysoniana*. *Bulletin of the Torrey Botanical Club*, **93**, 97–103.

Johansen, D. A. (1950). *Plant embryology: embryogeny of the Spermatophyta*. Chronica Botanica, Waltham, Massachusetts.

Johri, B. M. (1984). *Embryology of angiosperms*. Springer-Verlag, Berlin.

Johri, B. M., Ambegaokor, K. B., and Srivastava, P. S. (1992). *Comparative embryology of angiosperms*, 2 vols. Springer-Verlag, Berlin.

Jones, K. (1966). Chromosomes of orchids. I. *Polystachya. Kew Bulletin*, **20**, 357–9.

Jones, K., Lim, K. Y., and Cribb, P. J. (1982). The chromosomes of orchids. VII. *Dendrobium. Kew Bulletin*, **37**, 221–7.

Jones, R. N. and Rees, H. (1982). *B chromosomes*. Academic Press, London.

Jorapur, S. M. (1980). Chromosome number reports, LXVIII. *Taxon*, **29**, 546.

Joshi, A. C. (1933). Perennation and vegetative propagation in *Zeuxine sulcata* Lindley. *Journal of the Indian Botanical Society*, **12**, 20–3.

Judd, W. S., Stern, W. L., and Cheadle, V. I. (1993). Phylogenetic position of *Apostasia* and *Neuwiedia* (Orchidaceae). *Botanical Journal of the Linnean Society*, **113**, 87–94.

Jussieu, A. L. de. (1789). *Genera plantarum*. Herrisant, Paris.

Karasawa, K. (1979). Karyomorphological studies in *Paphiopedilum*, Orchidaceae. *Bulletin of the Hiroshima Botanical Garden*, **2**, 1–149.

Karasawa, K. (1980). Karyomorphological studies in *Phragmipedium*, Orchidaceae. *Bulletin of the Hiroshima Botanical Garden*, **3**, 1–9.

Karasawa, K. (1981). A revision of chromosome number in some hybrids of *Paphiopedilum*. *Bulletin of the Hiroshima Botanical Garden*, **4**, 1–8.

Karasawa, K. (1982). *The genus* Paphiopedilum. Karasawa, Hiroshima, Japan.

Karasawa, K. (1986). Karyomorphological studies on nine taxa of *Paphiopedilum*. *Bulletin of the Hiroshima Botanical Garden*, 8, 23–42.

Karasawa, K. and Aoyama, M. (1986). Karyomorphological studies on *Cypripedium* in Japan and Formosa. *Bulletin of the Hiroshima Botanical Garden*, 8, 1–22.

Karasawa, K. and Aoyama, M. (1988). Karyomorphological studies on two species of *Paphiopedilum*. *Bulletin of the Hiroshima Botanical Garden*, 10, 1–6.

Karasawa, K. and Saito, K. (1982). A revision of the genus *Paphiopedilum* (Orchidaceae). *Bulletin of the Hiroshima Botanical Garden*, 5, 1–69.

Karasawa, K. and Tanaka, R. (1980). C-banding study on centric fission in the chromosomes of *Paphiopedilum*. *Cytologia*, 45, 97–102.

Karasawa, K. and Tanaka, R. (1981). A revision of chromosome numbers in some hybrids of *Paphiopedilum*. *Cytologia*, 45, 97–102.

Karasawa, K., Aoyama, M., and Kamimura, T. (1997). Karyomorphological studies on five rare species of *Paphiopedilum*, Orchidaceae. *Annals of the Tsukuba Botanical Garden*, 16, 29–39.

Kashyap, S. K. and Mehra, P.N. (1983). Cytological investigations on West Himalayan orchids, tribe Orchideae. II. Several genera. *Cytologia*, 48, 647–57.

Kimura, C. (1971). Embryological studies of *Galeola septentrionalis* Reichb. f. *Science Reports of the Tohoku University. Series IV, Biology*, 35, 253–8.

Klier, K., Leoschke, M. J., and Wendel, J. F. (1991). Hybridization and introgression in white and yellow ladyslipper orchids (*Cypripedium candidum* and *C. pubescens*). *Journal of Heredity*, 82, 305–18.

Koopowitz, H. (1995). An annotated checklist of the genus *Paphiopedilum*. *Orchid Digest*, 59, 115–39.

Kores, P. J., Cameron, K. M., Molvray, M., and Chase, M. W. (1997). The phylogenetic relationships of Orchidoideae and Spiranthoideae (Orchidaceae) as inferred from *rbc*L plastid sequences. *Lindleyana*, 12, 1–11.

Kränzlin, F. (1897–1904). *Orchidacearum Genera et Species*. Mayer & Müller, Berlin.

Kränzlin, F. (1901). *Cypripedium* L. In *Orchidacearum Genera et Species* Vol. 1 (ed. F. Kränzlin), pp. 11–86. Mayer & Müller, Berlin.

Kugust, K. (1966). Hybridizing with *Oncidium*. In *Proceedings of the 5th World Orchid Conference* (ed. L. R. De Garmo), pp. 45–52. Harvard University Press, Cambridge, Massachusetts.

Kurnar, M. and Manilal, K. S. (1988). Floral anatomy of *Apostasia odorata* and the taxonomic status of Apostasioids (Orchidaceae). *Phytomorphology*, 38, 159–62.

Kurpko, S., Israelstam, G. F., and Martinovic, B. (1954). Embryo sac development and chromosome number in *Vanilla roscheri* from Inhaca Island. *South African Journal of Science*, 51, 115–17.

Kurzweil, H. (1987a). Developmental studies in orchid flowers I: epidendroid and vandoid species. *Nordic Journal of Botany*, 7, 427–42.

Kurzweil, H. (1987b). Developmental studies in orchid flowers II: orchidoid species. *Nordic Journal of Botany*, 7, 443–51.

Kurzweil, H. (1988). Developmental studies in orchid flowers III: neottioid species. *Nordic Journal of Botany*, 8, 271–82.

Kurzweil, H. (1993). Developmental studies in orchid flowers IV: cypripedioid species. *Nordic Journal of Botany*, 13, 423–30.

Kurzweil, H., Linder, H. P., Stern, W. L., and Pridgeon, A. M. (1995). Comparative vegetative anatomy and classification of Diseae (Orchidaceae). *Botanical Journal of the Linnean Society*, 117, 171–220.

Lavarack, P. S. (1971). Numerical methods in the taxonomy of the orchids. In *Proceedings of the 6th World Orchid Conference* (ed. M. J. G. Corrigan), pp. 55–61. Sydney.

Lawler, L. (1984). Ethnobotany of the Orchidaceae. In *Orchid biology: reviews and perspectives*, Vol. 3 (ed. J. Arditti), pp. 27–149. Cornell University Press, Ithaca, New York.

Lee, T. C. (1975). *Paphiopedilum purpuratum* (Lindl.) Pfitz. *Orchid Digest*, 39, 192–4.

Leitgeb, H. (1865). *Die Luftwurzeln der Orchideen. Denkschriften Kaiserlichen Akademie der Wissenschaften.* Mathematisch-Naturwissenschaftlichen Klasse (Wien), 24, 179–222.

Lidén, M., Oxelman, B., Backlund, A., Andersson, L., Bremer, B., Eriksson, R., Moberg, R., Nordal, I., Persson, K., Thulin, M., and Zimmer, B. (1997). Charlie is our darling. *Taxon*, 46, 735–8.

Lim, K. Y. (1985). The chromosomes of orchids at Kew. 3, Miscellaneous species. *American Orchid Society Bulletin*, 54, 1234–5.

Lim, K. Y. and Jones, K. (1982). The chromosomes of orchids. VI. *Bulbophyllum*. *Kew Bulletin*, 37, 217–19.

Linder, H. P. and Kurzweil, H. (1994). The phylogeny and classification of the Diseae (Orchidoideae: Orchidaceae). *Annals of the Missouri Botanical Garden*, 81, 687–713.

Lindley, J. (1826). *Orchidearum sceletos*. R. Taylor, London.

Lindley, J. (1830–1840). *Genera and species of orchidaceous plants*. J. Ridgway, London.

Lindley, J. (1833). *Nixus plantarum*. Ridgway and Sons, London.

Lindley, J. (1836). *A natural system of botany*, (2nd edn). Longman, London.

Lindley, J. (1837–1841). *Sertum Orchidaceum*. Ridgway and Sons, London.

Lindley, J. (1840). *Cypripedium*. In *Genera and species of orchidaceous plants*, pp. 525–32. Ridgways, London.

Lindley, J. (1847). *The vegetable kingdom*, (2nd edn). Bradbury & Evans, London.

Lindley, J. (1852–1859). *Folia orchidacea*. J. Matthews, London.

Link, H. F. (1824). *Elementa philosophiae botanicae*. Haude & Spenersche, Berlin.

Link, H. F. (1851). Bemerkungen über den Bau der Orchideen. *Abhandlungen der Königlichen Akademie der Wissenschaften zu Berlin*, **1849**, 103–27 + Table. IV.

Linnaeus, C. (1753). *Species plantarum*. L. Salvius, Stockholm.

Linnaeus, C. (1762–1763). *Species plantarum*, (2nd edn). Salvius, Stockholm.

Lopes, M. A. and Larkins, B. A. (1993). Endosperm origin, development, and function. *The Plant Cell*, **5**, 1383–99.

Luer, C. A. (1975). *The native orchids of the United States and Canada excluding Florida*. New York Botanical Garden.

McCook, L. (1989). *Systematics of Phragmipedium (Cypripedioideae; Orchidaceae)*. Ph.D. thesis. Cornell University, Ithaca, New York.

McCormick, S. (1993). Male gametophyte development. *The Plant Cell*, **5**, 1265–75.

MacDougal, D. T. (1894). On the poisonous influence of *Cypripedium spectabile* and *Cypripedium pubescens*. *Minnesota Botanical Studies, Bulletin*, **9**, part I, 32–6.

MacDougal, D. T. (1895). Poisonous influence of various species of *Cypripedium*. *Minnesota Botanical Studies, Bulletin*, **9**, part VII, 450–1.

Maheshwari, P. (1950). *An introduction to the embryology of angiosperms*. McGraw-Hill, New York.

Maheshwari, S. C. (1955). The occurrence of bisporic embryo sacs in angiosperms—a critical review. *Phytomorphology*, **5**, 67–79.

Manning, J. C. and Linder, H. P. (1992). Pollinators and evolution in *Disperis* (Orchidaceae), or why are there so many species? *South African Journal of Science*, **88**, 38–49.

Mansfeld, R. (1937a). Ueber das system der Orchidaceae. *Blumea* suppl. 1, 25–37.

Mansfeld, R. (1937b). Über das System der Orchidaceae-Monandrae. *Notizblatt des Botanischen Gartens und Museums zu Berlin-Dahlem*, **13**, 666–76.

Mansfeld, R. (1955). Über die Verteilung der Merkmale innerhalb der Orchidaceae-Monandrae. *Flora*, **142**, 65–80.

Martínez, A. (1985). The chromosomes of orchids. VIII. Spiranthinae and Cranichidinae. *Kew Bulletin*, **40**, 139–47.

Mayr, E. (1988). *Towards a new philosophy of biology*. Harvard University Press, Cambridge, Massachusetts.

Meinecke, E. P. (1894). Beiträge zur Anatomie der Luftwurzeln der Orchideen. *Flora*, **78**, 133–203 + 2 plates.

Meyen, F. J. F. (1837). *Neues System der Pflanzenphysiologie*. Haude & Spenersche, Berlin.

Möbius, M. (1887). Über den anatomischen Bau der Orchideenblätter und dessen Bedeutung für das System dieser Familie. *Jahrbücher (Pringsheim) für wissenschaftliche Botanik*, **18**, 530–07.

Møller, J. D. and Rasmussen, H. (1984). Stegmata in Orchidales: character state distribution and polarity. *Botanical Journal of the Linnean Society*, **89**, 53–76.

Molvray, M. and Kores, P. J. (1995). Character analysis of the seed coat in Spiranthoideae and Orchidoideae, with special reference to the Diurideae (Orchidaceae). *American Journal of Botany*, **82**, 1443–54.

Moore, R. J. (1970). *Index to plant chromosome numbers for 1968*. Regnum Vegetabile, Vol. 68. Utrecht, the Netherlands.

Moore, R. J. (1971). *Index to plant chromosome numbers for 1969*. Regnum Vegetabile, Vol. 77. Utrecht, The Netherlands.

Moore, R. J. (1972). *Index to plant chromosome numbers for 1970*. Regnum Vegetabile, Vol. 84. Utrecht, The Netherlands.

Moore, R. J. (1973). *Index to plant chromosome numbers 1967–1971*. Regnum Vegetabile, Vol. 90. Utrecht, The Netherlands.

Moore, R. J. (1974). *Index to plant chromosome numbers for 1972*. Regnum Vegetabile, Vol. 91. Utrecht, The Netherlands.

Moore, R. J. (1977). *Index to plant chromosome numbers for 1973/74*. Regnum Vegetabile, Vol. 96. Utrecht, The Netherlands.

Moreau, L. (1913). Étude anatomique des Orchidées in pseudo-bulbes des pay chauds et de quelques autres espèces tropicales de plantes a tubercules. *Revue Générale de Botanique*, **25**, 503–48.

Morris, F. and Eames, E. A. (1929). *Our wild orchids*. Scribners, New York.

Müller, H. (1868). Beobachtungen an westfalischen Orchideen. *Verhandlungen des Naturhistorischen Vereins der preussischen Rheinlande und Westfalens*, 25, 1–62.

Müller, H. (1869). Über die Anwendung der Darwinischen Theorie auf Blumen und blumenbesuchende Insekten. *Verhandlungen des Naturhistorischen Vereins der preussischen Rheinlande und Westfalens*, **26**, *Corr. Bl.*, 43–66.

Müller, H. (1873). *Die Befruchtung der Blumen durch Insekten*. Engelmann, Leipzig.

Nagashima, T. (1989). Embryogenesis, seed formation and immature seed germination in vitro on *Ponerorchis graminifolia* Reichb.f. *Journal of the Japanese Society for Horticultural Science*, **58**, 187–94.

Nakata M. and Hashimoto, T. (1983). Karyomorphological studies on species of *Pleurothallis*, Orchidaceae. *Annals of the Tsukuba Botanical Garden*, **2**, 11–32.

Nandi, O., Chase, M. W., and Endress, P. K. (1998). A combined cladistic analysis of angiosperms using *rbc*L and non-molecular data sets. *Annals of the Missouri Botanical Garden*, **85**, 137–212.

Natesh, S. and Rau, M. A. (1984). The embryo. In *Embryology of angiosperms* (ed. B. M. Johri), pp. 377–443. Springer-Verlag, Berlin.

Newton, G. D. and Williams, N. H. (1978). Pollen morphology of the Cypripedioideae and the Apostasioideae (Orchidaceae). *Selbyana*, **2**, 169–82.

Neyland, R. and Urbatsch, L. E. (1995). A terrestrial origin for the Orchidaceae suggested by a phylogeny inferred from *ndh*F chloroplast gene sequences. *Lindleyana*, **10**, 244–51.

Neyland, R. and Urbatsch, L. E. (1996). Phylogeny of subfamily Epidendroideae (Orchidaceae) inferred from *ndh*F chloroplast gene sequences. *American Journal of Botany*, **83**, 1195–206.

Nilsson, L. A. (1979). Anthecological studies on the lady's slipper, *Cypripedium calceolus*. *Botaniska Notiser*, **132**, 329–49.

Nishimura, G. (1991). Comparative anatomy of cotyledonous orchid seedlings. *Lindleyana*, **6**, 140–6.

Oesterberg, J. A. (1883). Bidrag til kaennedom af perikarpets anatomi och kaerlstrengforloebet I blomman hos Orchideerne. *Kongl. Svenska Vetenskapsakademiens Handlingar*, **1883**, 47–62.

Olsson, O. (1967). Embryological studies in the Orchidaceae. The genus *Hetaeria*. *Svensk Botanisk Tidskrift*, **61**, 33–42.

Ornduff, R. (1968). *Index to plant chromosome numbers for 1966*. Regnum Vegetabile, Vol. 55. Utrecht, The Netherlands.

Ornduff, R. (1969). *Index to plant chromosome numbers for 1967*. Regnum Vegetabile, Vol. 59. Utrecht, The Netherlands.

Oudemans, C. A. J. A. (1861). Über den Sitz der Oberhaut bei den Luftwurzeln der Orchideen. *Veroffentlicht durch die Konigliche Akademie der Wissenschaften zu Amsterdam*, **9**, 1–31 + 3 pl.

Pace, L. (1907). Fertilization in *Cypripedium*. *Botanical Gazette*, **44**, 353–74.

Pastrana, M. D. and Santos, J. K. (1931). A contribution to the life history of *Dendrobium anosmum* Lindley. *Natural and Applied Science Bulletin, University of the Philippines*, **1**, 133–44.

Perner, H. (1997). Cultivation. In *The genus* Cypripedium (ed. P. Cribb), pp. 60–96. Timber Press, Portland Oregon.

Pfitzer, E. (1882). *Grundzüge einer vergleichen Morphologie der Orchideen*. Carl Winter's Universitätsbuchhandlung, Heidelberg.

Pfitzer, E. H. H. (1886). *Morphologische Studien über die Orchideenblüthe*. C. Winter, Heidelberg.

Pfitzer, E. H. H. (1887). *Entwurf einer natürlichen Anordnung der Orchideen*. Carl Winter, Heidelberg.

Pfitzer, E. H. H. (1889). Orchidaceae. In *Die natürlichen Pflanzenfamilien* Vol. 2 (ed. A. Engler and K. Prantl), **6**, pp. 52–224.

Pfitzer, E. H. H. (1894). Beitrage zur Systematik der Orchideen. *Botanisches Jahrbücher für Systematik, Pflanzengeschichte und Pflanzengeographie*, **19**, 1–42.

Pfitzer, E. H. H. (1903). Orchidaceae-Pleonandrae. In *Das Pflanzenreich* Vol. 4 (ed. A. Engler), **50**, pp. 1–132. Engelmann, Leipzig.

Pijl, L. van der and Dodson, C. H. (1966). *Orchid flowers: their pollination and evolution*. University of Miami Press, Coral Gables, Florida.

Poddubnaya-Arnoldi, V. A. (1960). Study of fertilisation in the living material of some angiosperms. *Phytomorphology*, **20**, 185–98.

Poddubnaya-Arnoldi, V. A. (1967). Comparative embryology of the Orchidacaee. *Phytomorphology*, **17**, 312–20.

Pohl, F. (1931). Ölüberzüge verschiedener Pflanzenorgane besonders der Blüte. *Jahrbuch für wissenschaftliche Botanik*, **70**, 565–655.

Porembski, S. and Barthlott, W. (1988). Velamen radicum morphology and classification of Orchidaceae. *Nordic Journal of Botany*, **8**, 117–37.

Prakash, N. and Lee-Lee, A. (1972). Life history of a common Malaysian orchid *Spathoglottis plicata*. *Phytomorphology*, **23**, 9–17.

Pridgeon, A. M. (1982). Diagnostic anatomical characters in the Pleurothallidinae (Orchidaceae). *American Journal of Botany*, **69**, 921–38.

Pridgeon, A. M. (1987). The velamen and exodermis of orchid roots. In *Orchid biology: reviews and perspectives*, Vol. 4 (ed. J. Arditti), pp. 139–192. Cornell University Press, Ithaca, New York.

Pridgeon, A. M. (1994). Systematic leaf anatomy of Caladeniinae. *Botanical Journal of the Linnean Society*, **114**, 31–48.

Pridgeon, A. M. and Chase, M. W. (1995). Subterranean axes in tribe Diurideae (Orchidaceae): morphology, anatomy, and systematic significance. *American Journal of Botany*, **82**, 1473–95.

Pridgeon, A. M. and Chase, M. W. (1998). Phylogenetics of subtribe Catasetinae (Orchidaceae) from nuclear and chloroplast DNA sequences. In *Proceedings of the 15th World Orchid Conference* (ed. C. E. B. Pereira), pp. 275–281. Naturalia Publications, Turriers, France.

Pridgeon, A. M. and Stern, W. L. (1983). Ultrastructure of osmophores in *Restrepia* (Orchidaceae). *American Journal of Botany*, **70**, 1233–43.

Pridgeon, A. M. and Stern, W. L. (1985). Osmophores of *Scaphosepalum* (Orchidaceae). *Botanical Gazette*, **146**, 115–23.

Pridgeon, A. M., Stern, W. L., and Benzing, D. H. (1983). Tilosomes in roots of Orchidaceae: morphology and systematic occurrence. *American Journal of Botany*, **70**, 1365–77.

Pridgeon, A. M., Bateman, R. M., Cox, A. V., Hapeman, J. R., and Chase, M. W. (1997). Phylogenetics of subtribe Orchidinae (Orchidoideae, Orchidaceae) based on nuclear ITS sequences. 1. Intergeneric relationships and polyphyly of *Orchis sensu lato*. *Lindleyana*, **12**, 89–109.

Proctor, M. and Yeo, P. (1973). *The pollination of flowers*. Collins, London.

Prosina, M. N. (1930). Über die vom *Cypripedium*-Typus abweichende Embryosackentwicklung von *Cypripedium guttatum* Sw. *Planta*, **12**, 532–44.

Rafinesque, C. S. (1819). Criosanthes. *Journal de physique, de chimie, d'histoire naturelle et des arts*, **89**, 102.

Rafinesque, C. S. (1828). *Medical flora; or Manual of the medical botany of the United States of North America*. Atkinson & Alexander, Philadelphia.

Rafinesque, C. S. (1838). *Sacodon*. In *Flora Telluriana* **4**, 46. H. Pobasco, Philadelphia, Pennsylvania.

Rao, A. N. (1967). Flower and seed development in *Arundina graminifolium*. *Phytomorphology*, **20**, 291–300.

Rao, A. N. and Chua, L. G. (1978). Fruit and seed development in certain local orchids. In *Proceedings of the symposium on orchidology, Singapore 1978* (ed. A. G. Alphonso, P. Y. Seng, A. N. Rao, T. E. Soon, and G. C. Jin), pp. 24–42. The Orchid Society of South-East Asia, Singapore.

Rao, P. R. M. and Rao, K. M. (1983). Embryology of *Liparis viridiflora*. *Acta Botanica Indica*, **11**, 288–34.

Rasmussen, F. N. (1982). The gynostemium of the neottioid orchids. *Opera Botanica*, **65**, 1–96.

Rasmussen, F. N. (1985). Orchids. In *The families of the monocotyledons: structure, evolution, and taxonomy* (ed. R. M. T. Dahlgren, H. T. Clifford, and P. F. Yeo), pp. 247–274. Springer-Verlag, Berlin.

Rasmussen, F. N. (1986). On the various contrivances by which pollinia are attached to viscidia. *Lindleyana*, **1**, 21–32.

Rasmussen, H. N. (1981). The diversity of stomatal development in Orchidaceae subfamily Orchidoideae. *Botanical Journal of the Linnean Society*, **82**, 381–93.

Rasmussen, H. N. (1986). The vegetative architecture of orchids. *Lindleyana*, **1**, 42–50.

Rasmussen, H. N. (1995). *Terrestrial orchids: from seed to mycotrophic plant*. Cambridge University Press.

Rauh, W., Barthlott, W., Ehler, W., and Ehler, N. (1975). Morphologie und Funktion der Testa staubförmiger Flugsamen. *Botanische Jahrbücher der für Systematik, Pflanzengeschichte und Pflanzengeographie*, **96**, 353–74.

Raunkiaer, C. (1895–1899). *De Danske blomsterplanters naturhistorie*, Vol. 1. De enkimbladede—Orchidaceae, Skudbyging og voksemaade. Blomstring, pp. 305–382. Gyldendalske Boghandels Forlag, Copenhagen.

Reichenbach, H. G. (1852). *De pollinis orchidearum genesi ac structura et de orchideis in artem ac systema redigendis*. Ph.D. thesis. F. Hofmeister, Leipzig.

Reichenbach, H. G. (1854–1900). *Xenia orchidacea*. 3 vols. Brockhaus, Leipzig.

Reichenbach, H. G. (1885). Ueber das System der Orchideen. In *Bulletin du Congrès international de botanique et d'horticulture*, pp. 39–58. Imprimerie de l'Académie Impériale des Sciences, St. Petersbourg.

Reveal, J. L. and Hoogland, R. D. (1991). Validation of three family names in the Magnoliophyta. *Bulletin du Muséum national d'Histoire naturelle, Paris, 4e série*, **13**, *section B, Adansonia*, 91–3.

Rheede Tot Drakenstein, H. van. (1678–1693). *Hortus Malabaricus*. 12 vols. Someren & Dyck, Amsterdam.

Richard, L. C. (1817). *De orchideis europaeis annotationes, praesertim ad genera dilucidanda spectantes*. A. Belin, Paris.

Richard, L. C. (1818). De Orchideis Europaeis Annotationes. *Mémoires du Muséum d'Histoire Naturelle. Paris*, **4**, 23–61.

Rickett, H. W. and Stafleu, F. A. (1959). Nomina generica conservanda et rejicienda Spermatophytorum. *Paphiopedilum*. *Taxon*, **8**, 448–92.

Ridley, H. N. (1888). Notes on self-fertilisation and cleistogamy in orchids. *Journal of the Linnean Society of London* (*Botany*), **24**, 389–95.

Rogers, T., Bohan, G., and Koopowitz, H. (1998). An initial assessment of the wild populations of *Paphiopedilum sanderianum* (Rchb.f.) Stein and their reproductive fitness. In *Proceedings of the 15th World Orchid Conference* (ed. C. E. B. Pereira), pp. 415–421. Naturalia Publications, Turriers, France.

Rolfe, R. A. (1890). A morphological and systematic review of the Apostasieae. *Journal of the Linnean Society, Botany* **25**, 211–43.

Rolfe, R. A. (1896). The *Cypripedium* group. *Orchid Review*, **4**, 327–34.

Rolfe, R. A. (1909). The evolution of the Orchidaceae. *Orchid Review*, **17**, 129–32, 193–6, 289–92, 353–6.

Rolfe, R. A. (1910). The evolution of the Orchidaceae. *Orchid Review*, **18**, 33–6, 97–9, 129–32, 162–6, 289–94, 321–5.

Rolfe, R. A. (1911). The evolution of the Orchidaceae. *Orchid Review*, 19, 68–9, 289–92.

Rolfe, R. A. (1912). The evolution of the Orchidaceae. *Orchid Review*, 20, 204–7, 225–8, 260–4.

Rosso, S. W. (1966). The vegetative anatomy of the Cypripedioideae (Orchidaceae). *Journal of the Linnean Society* (*Botany*), 59, 309–41 + 5 pl.

Rudall, P. J., Chase, M. W., Cutler, D. F., Rusby, J., and de Bruijn, A. Y. (1998). Anatomical and molecular systematics of Asteliaceae and Hypoxidaceae. *Botanical Journal of the Linnean Society*, 127, 1–42.

Rumphius, G. E. (1741–1750). *Herbarium Amboinense.* 6 vols. Changuion and Uytwerf, Amsterdam.

Rüter, B. and Stern, W. L. (1994). An assessment of quantitative features of velamen stratification and protoxylem strands in roots of Orchidaceae. *Lindleyana*, 9, 219–25.

Saitou, N. and Nei, M. 1987. The neighbor-joining method: a new method for reconstructing phylogenetic trees. *Molecular Biology and Evolution*, 4, 406–25.

Sargant, E. (1900). Recent work on the results of fertilization in angiosperms. *Annals of Botany*, 14, 689–712.

Sarojini Manon, V., Seenui, S., Decruse, S. W., and Gangaprasad, A. (1995). preliminary study on the habitat, distribution and population density of *Paphiopedilum druryi* (Bedd.) Stein at Agasthyamalai in southern India. *Journal of the Orchid Society of India*, 9, 55–67.

Sasikumar, B. (1975). Polystele in terrestrial orchids. *Science and Culture*, 41, 216–17.

Savina, G. L. (1965). Development of sexual elements in the course of fertilization in some species of *Orchis. Botanische Zeitung*, 50, 96–102.

Savina, G. L. (1978). Certain pecularities in the embryology of orchids. *Proceedings of the Indian National Science Academy. Part B, Biological Sciences*, 44, 141–5.

Savolainen, V., Chase, M. W., Morton, C. M., Hoot, S. B., Soltis, D. E., Bayer, C., Fay, M. F., de Bruijn, A., Sullivan, S., and Qiu, Y.-L. (In press). Phylogenetics of flowering plants based upon a combined analysis of plastid *atp*B and *rbc*L gene sequences. *Annals of the Missouri Botanical Garden.*

Schacht, H. (1856). Lehrbuch der Anatomie und Physiologie der Gewächse. G. W. F. Müller, Berlin.

Schill, R. (1978). Palynologische Untersuchungen zur systematischen Stellung der Apostasiaceae. *Botanisches Jahrbücher für Systematik, Pflanzengeschichte und Pflanzengeographie*, 99, 353–62.

Schill, R. and Pfeiffer, W. (1977). Untersuchungen an Orchideenpollinien unter besonderer Beruecksichtigung ihrer Feinskulpturen. *Pollen et Spores*, 19, 5–118.

Schlechter, R. (1911). Die Polychondreae (Neottiinae Pfitz.) und ihre systematische Einteilung. *Botanische Jahrbücher für Systematik, Pflanzengeschichte und Pflanzengeographie*, 45, 375–410.

Schlechter, R. (1911–14). Die Orchidaceen von Deutsch–Neu-Guinea. *Repertorium specierum novarum regni vegetabilis. Beiheft*, 1, 1–1079

Schlechter, R. (1914–15). *Die Orchideen.* Paul Parey, Berlin.

Schlechter, R. (1918). Versuch einer natürlichen Neuordning der afrikanischen angraekoiden Orchidaceen. *Beihefte zum Botanischen Centralblatt*, 36, 62–181.

Schlechter, R. (1920). Versuch einer systematische Neuordnung der Spiranthinae. *Beihefte zum Botanischen Centralblatt*, 37, 454.

Schlechter, R. (1923–28). Figuren-Atlas zu den Orchidaceen von Deutsch–Neu-Guinea. *Repertorium specierum novarum regni vegetabilis. Beiheft*, 21, 1–372.

Schlechter, R. (1926). Das System der Orchidaceen. *Notizblatt des Botanischen Gartens und Museums zu Berlin-Dahlem*, 9, 563–91.

Schlechter, R. (1927). *Die Orchideen*, (2nd edn). Paul Parey, Berlin.

Schlechter, R. (1970). *Die Orchideen.* 3. (ed. F. Brieger, R. Maatsch, and K. Senghas). Paul Parey, Berlin.

Schleiden, M. J. (1845–46). *Grundzüge der wissenschaftlichen Botanik*, 2 vols. Verlag von Wilhem Engelmann, Leipzig.

Schleiden, M. J. (1849). *Principles of scientific botany* (trans. E. Lankester from 2nd German edn). Longman, Brown, Green, and Longmans, London.

Schmalle, H. W. and Hausen, B. M. (1979). A new sensitizing quinone from lady slipper (*Cypripedium calceolus*). *Naturwissenschaften*, 66, 527–8.

Schwarzacher, T., Ambros, P., and Schweizer, D. (1980). Application of Giemsa banding to orchid karyotype analysis. *Plant Systematics and Evolution*, 134, 293–7.

Schweinfurth, C. (1959). Key to the orchids— Appendix I. In *The orchids: a scientific survey* (ed. C. L. Withner), pp. 511–28. Wiley, New York.

Seidenfaden, G. (1992). The orchids of Indochina. *Opera Botanica*, 114, 1–502.

Seidenfaden, G. and Wood, J. J. (1992). *The orchids of Peninsular Malaysia and Singapore.* Olsen & Olsen, Fredensborg, Denmark.

Seshagiriah, K. H. (1941). Morphological studies in Orchidaceae. I. *Zeuxine sulcata* Lindley. *Journal of the Indian Botanical Society*, 20, 357–65.

Sharp, L. W. (1912). The orchid embryo sac. *Botanical Gazette*, 54, 373–85.

Sheahan, M. C. and Chase, M. W. (1996). A phylogenetic analysis of Zygophyllaceae R. Br. based on morphological, anatomical, and *rbc*L DNA sequence data. *Botanical Journal of the Linnean Society*, **122**, 279–300.

Sheviak, C. (1983). United States terrestrial orchids. In *North American terrestrial orchid symposium 2* (ed. E. H. Plaxton), pp. 49–60. Michigan Orchid Society, Livonia, Michigan.

Siddall, M. E. and Kluge, A. G. (1997). Probabilism and phylogenetic inference. *Cladistics*, **13**, 313–36.

Siebe, M. (1903). *Ueber den anatomischen Bau der Apostasiinae*. Ph.D. thesis. University of Heidelberg. Hörning & Berkenbusch, Heidelberg.

Simon, W. (1968). Chromosome numbers and B chromosomes in *Listera*. *Caryologia*, **21**, 181–9.

Sinotô, Y. (1962). Chromosome numbers in *Oncidium* alliance. *Cytologia*, **27**, 306–13.

Slyusarenko, A. G. (1981). *Cypripedium* L.—the lady's slippers. *American Orchid Society Bulletin*, **50**, 776–80.

Solereder, H. and Meyer, F. J. (1930). *Systematic anatomy of the monocotyledons. Vol. 6. Microspermae* (trans. A. Herzberg and ed. B. Golek). Israel Program for Scientific Translations, Jerusalem.

Soltis, D. E., Hibsch-Jetter, C., Soltis, P. S., Chase, M. W., and Farris, J. S. (1997). Molecular phylogenetic relationships among angiosperms: an overview based on *rbc*L and 18S rDNA sequences. In *Evolution and diversification of land plants* (ed. K. Iwatsuki and P. H. Raven), pp. 157–178. Springer-Verlag, Tokyo.

Soltis, D. E., Soltis, P. S., Mort, M., Chase, M. W., Savolainen, V., Hoot, S. B., and Morton, C. M. (1998). Inferring complex phylogenies using parsimony: an empirical approach using three large DNA data sets for angiosperms. *Systematic Biology*, **47**, 32–42.

Sood, S. K. (1984a). An embryological study of *Neottia listeroides*, a saprophytic orchid in India. *Journal of Plant Anatomy and Morphology*, **1**, 69–75.

Sood, S. K. (1984b). Embryo development in *Goodyera biflora* Hook.f. (Orchidaceae). *Current Science*, **53**, 1247–9.

Sood, S. K. (1986). Gametogenesis, integuments initiation and embryogeny in three species of *Habenaria* (Orchidaceae, Orchideae). *Proceedings of the Indian Academy of Sciences*, **96**, 487–94.

Sood, S. K. (1988). Development of gametophytes, embryogeny and pericarp in *Goodyera repens* (Orchidaceae, Neottieae). *Proceedings of the Indian Academy of Sciences*, **98**, 149–56.

Sood, S. K. (1992). Embryology of *Malaxis saprophyta*, with comments on the systematic position of *Malaxis* (Orchidaceae). *Plant Systematics and Evolution*, **179**, 95–105.

Sood, S. K. and Rao, P. R. M. (1986). Gametophytes, embryogeny and pericarp of *Microstylis wallichii* Lindl. (Orchidaceae). *Botanical Magazine Tokyo*, **99**, 351–9.

Sood, S. K. and Rao, P. R. M. (1988). Studies in the embryology of the diandrous orchid *Cypripedium cordigerum* (Cypripedieae, Orchidaceae). *Plant Systematics and Evolution*, **160**, 159–68.

Sosef, M. S. M. (1997). Hierarchical models, reticulate evolution and the inevitability of paraphyletic taxa. *Taxon*, **46**, 75–85.

Soto, M. A., Salazar, G. A., and Hágsater, E. (1990). *Phragmipedium xerophyticum*, una nueva especie del sureste de Mexico. *Orquídea (México)*, **12**, 1–10.

Souéges, R. (1934–39). Exposés d'embryologie et de morphologie végétales. Volumes I–X. Paris.

Sprenger, M. (1904). *Ueber den anatomischen Bau der Bolbophyllinae*. Ph.D. thesis. University of Heidelberg. J. Hörning, Heidelberg.

Staudermann, W. (1924). Die Haare der Monokotylen. *Botanisches Archiv*, **8**, 105–84.

Stebbins, G. L. (1971). *Chromosomal evolution in higher plants*. Edward Arnold, London.

Steele, W. H. (1995). Growing *Cypripedium reginae* from seed. *American Orchid Society Bulletin*, **64**, 382–91.

Steiner, K. E. (1989). The pollination of *Disperis* (Orchidaceae) by oil-collecting bees in southern Africa. *Lindleyana*, **4**, 164–83.

Stermitz, F. R., Fink, N. H., and Hegedus, L. S. (1981). A comparison of alkaloid content in some *Phragmipedium* species. *American Orchid Society Bulletin*, **50**, 1346–52.

Stern, W. L. (1997a). Vegetative anatomy of subtribe Habenariinae (Orchidaceae). *Botanical Journal of the Linnean Society*, **125**, 211–27.

Stern, W. L. (1997b). Vegetative anatomy of subtribe Orchidinae (Orchidaceae). *Botanical Journal of the Linnean Society*, **124**, 121–36.

Stern, W. L. and Warcup, J. H. (1994). Root tubercles in apostasiad orchids. *American Journal of Botany*, **81**, 1571–5.

Stern, W. L., Curry, K. J., and Pridgeon, A. M. (1987). Osmophores of *Stanhopea* (Orchidaceae). *American Journal of Botany*, **74**, 1323–31.

Stern, W. L., Cheadle, V. I., and Thorsch, J. (1993a). Apostasiads, systematic anatomy, and the origins of the Orchidaceae. *Botanical Journal of the Linnean Society*, **111**, 411–55.

Stern, W. L., Morris, M. W., Pridgeon, A. M., and Dressler, R. L. (1993b). Comparative vegetative anatomy and systematics of Spiranthoideae. *Botanical Journal of the Linnean Society*, **113**, 161–97.

Stevens, P. F. (1984). Metaphors and typology in the development of botanical systematics 1690–1960, or the art of putting new wines in old bottles. *Taxon*, **33**, 169–211.

Stevens, P. F. (1986). Evolutionary classification in botany, 1960–1985. *Journal of the Arnold Arboretum*, **67**, 313–39.

Stevens, P. F. (1987). Pattern and process: phylogenetic reconstruction in botany. In *Biological metaphor and cladistic classification—an interdisciplinary perspective* (ed. H. M. Hoenigswald and L. F. Wiener), pp. 155–79. University of Pennsylvania Press, Philadelphia.

Stevens, P. F. (1997). How to interpret botanical classifications—suggestions from history. *Bioscience*, **47**, 243–50.

Stewart, J. and Stearn, W. T. (1993). *The orchid paintings of Franz Bauer*. Herbert Press in association with The Natural History Museum, London.

Stoutamire, W. (1963). Terrestrial orchid seedlings. *Australian Plants*, **2**, 119–22.

Stoutamire, W. P. (1967). Flower biology of the lady's slippers (Orchidaceae: *Cypripedium*). *Michigan Botanist*, **3**, 107–19.

Stoutamire, W. (1974). Terrestrial orchid seedlings. In *The orchids: scientific studies* (ed. C. L. Withner), pp. 101–128. Wiley, New York.

Summerhayes, V. S. (1951). *Wild orchids of Britain*. Collins, London.

Summerhayes, V. S. (1968). *Wild orchids of Britain*, (3rd edn). Collins, London.

Swamy, B. G. L. (1941). The development of the male gamete in *Cymbidium bicolor* Lindl. *Proceedings of the Indian Academy of Sciences, Section B* **14**, 454–60.

Swamy, B. G. L. (1942). Female gametophyte and embryogeny in *Cymbidium bicolor* Lindl. *Proceedings of the Indian Academy of Sciences, Section B* **15**, 194–201.

Swamy, B. G. L. (1943a). Embryology of Orchidaceae. *Current Science*, **12**, 13–17.

Swamy, B. G. L. (1943b). Gametogenesis and embryogeny of *Eulophia epidendraea* Fischer. *Proceedings of the National Institute of Sciences of India*, **9**, 59–65.

Swamy, B. G. L. (1944). Embryo sac and the embryo of *Satyrium nepalense* Don. *Journal of the Indian Botanical Society*, **23**, 66–70.

Swamy, B. G. L. (1945). Embryo sac and fertilization in *Cypripedium spectabile*. *Botanical Gazette*, **107**, 291–5.

Swamy, B. G. L. (1946a). Embryology of *Habenaria*. *Proceedings of the National Institute of Sciences of India*, **12**: 413–26.

Swamy, B. G. L. (1946b). The embryology of *Zeuxine sulcata* Lindl. *New Phytologist*, **45**, 132–6.

Swamy, B. G. L. (1947). On the life-history of *Vanilla planifolia*. *Botanical Gazette*, **108**, 449–56.

Swamy, B. G. L. (1948a). Agamospermy in *Spiranthes cernua*. *Lloydia*, **11**, 149–62.

Swamy, B. G. L. (1948b). The embryology of *Epidendrum prismatocarpum*. *Bulletin of the Torrey Botanical Club*, **75**: 245–9.

Swamy, B. G. L. (1948c). Vascular anatomy of orchid flowers. *Botanical Museum Leaflets Harvard University*, **13**: 61–95.

Swamy, B. G. L. (1949a). Embryological studies in the Orchidaceae. I. Gametophytes. *American Midland Naturalist*, **41**, 184–201.

Swamy, B. G. L. (1949b). Embryological studies in the Orchidaceae. II. Embryogeny. *American Midland Naturalist*, **41**, 202–5.

Swartz, O. (1800a). Afhandling om orchidernes slaegter och deras systematiska indelning. *Kongl. Vetenskaps Academiens Nya Handling-ar*, **21**, 115–39.

Swartz, O. (1800b). Orchidernes slaegter och arter up ställde. *Kongl. Vetenskaps Academiens Nya Handlingar*, **21**, 202–54.

Swartz, O. (1805). Genera et Species Orchidearum. *Neues Journal für die Botanik*, **1**, 1–108.

Szlachetko, D. L. (1991). Thelymitroideae, a new subfamily within Orchidaceae. *Fragmenta Floristica et Geobotanica*, **36**, 33–49.

Szlachetko, D. L. (1995). Systema Orchidalium. *Fragmenta florisitica et geobotanica supplementum*, **3**, 1–152.

Tanaka, R. and Kamemoto, H. (1974). List of chromosome numbers in species of the Orchidaceae. In *The orchids: scientific studies* (ed. C. L. Withner), pp. 411–483. Wiley, New York.

Tanaka, R. and Kamemoto, H. (1984). Chromosomes in orchids: counting and numbers. In *Orchid biology: reviews and perspectives*, Vol. 3 (ed. J. Arditti), pp. 323–410. Cornell University Press, Ithaca, New York.

Tohda, H. (1967). An embryological study of *Hetaeria sikokiana*, a saprophytic orchid in Japan. *Science Reports of the Tohoku University. Series 4, Biology*, **33**, 83–95.

Tohda, H. (1968). Development of the embryo of *Bletilla striata*. *Science Reports of the Tohoku University. Series 4, Biology*, **34**, 125–31.

Tohda, H. (1971a). Development of the embryo of *Lecanorchis japonica*. *Science Reports of the Tohoku University. Series 4, Biology*, **35**, 245–51.

Tohda, H. (1971b). Development of the embryo of *Orchis aristata*. *Science Reports of the Tohoku University. Series 4, Biology*, **35**, 239–43.

Tohda, H. (1974a). Embryological study on *Gymnadenia cucullata*. *Science Reports of the Tohoku University. Series 4, Biology*, **37**, 71–7.

Tohda, H. (1974b). Embryological study on *Gymnadenia camtschatica*. *Science Reports of the Tohoku University. Series 4, Biology*, **37**, 79–87.

Tohda, H. (1974c). Development of the embryo of *Pogonia* (Orchidaceae). *Science Reports of the Tohoku University. Series 4, Biology*, **37**, 89–93.

Tohda, H. (1983). Seed morphology in Orchidaceae. 1. *Dactylorchis, Ponerorchis, Chondradenia,* and *Galeorchis. Science Reports of the Tôhoku Imperial University, Series 4, Biology,* 38, 253–68.

Tohda, H. (1985). Seed morphology in Orchidaceae. II. Tribe Cranichideae. *Science Reports of the Tôhoku Imperial University, Series 4, Biology,* 39: 21–43.

Tohda, H. (1986). Seed morphology in Orchidaceae. III. Tribe Neottieae. *Science Reports of the Tôhoku Imperial University, Series 4, Biology,* 39: 103–19.

Tominski, P. (1905). *Die Anatomie des Orchideenblattes in ihrer Abhängigkeit von Klima und Standort.* Ph.D. thesis. Friedrich-Wilhelms-University. E. Eberling, Berlin.

Toscano de Brito, A. L. V. (1994). *Systematic studies in the subtribe Ornithocephalinae (Orchidaceae).* Ph.D. thesis. University of Reading. Reading, England.

Treub, M. (1879). Notes sur l'embryogénie de quelques Orchidées. *Verhandeingen der Koninklijke Nederlandsche Akademie van Wetenschappen, Afdeeling Natuurkunde,* 19, 1–50.

Troll, W. (1951). Botanische Notizen II. *Akademie der Wissenschaften und der Literatur. Abhandlungen der mathematisch-naturwissenschaftlichen Klasse. Mainz & Wiesbaden,* 1951 (2), 81–117.

Tsi, Z., Luo, Y., Cribb, P. J., McGough, N., Siu, G. and Chau, Z. (1999). A preliminary report on the population size, ecology, and conservation status of some *Paphiopedilum* species (Orchidaceae) in southwest China. *Lindleyana,* 14, 12–23.

Vargas, C. (1965). Orchids of Machu Pichu (Peru). *American Orchid Society Bulletin,* 34, 960–3.

Vermeulen, P. (1947). *Studies on dactylorchids.* F. Schotarius and Jens, Utrecht, The Netherlands.

Vermeulen, P. (1966). The system of the Orchidales. *Acta Botanica Neerlandica,* 15, 224–53.

Veyret, Y. (1955a). Embryogénie des Orchidacées. Développement de l'embryon chez le *Vanilla fragrans* (Salis.) Ames. *Comptes Rendus Hebdomadaires des Séances de l' Académie des Sciences,* 241, 1409–11.

Veyret, Y. (1955b). Embryogénie des Orchidacées. Développement de l'embryon chez l'*Orchis longibracteata* Biv. *Comptes Rendus Hebdomadaires des Séances de l' Académie des Sciences,* 241, 1828–31.

Veyret, Y. (1956a). Embryogénie des Orchidacées. Développement de l'embryon chez l'*Orchis laxiflora* Lam. *Bulletin de la Société Botanique de France,* 102, 121–6.

Veyret, Y. (1956b). Embryogénie des Orchidacées. Développement de l'embryon chez le *Limodorum abortivum* Sw. *Comptes Rendus Hebdomadaires des Séances de l' Académie des Sciences,* 242, 545–7.

Veyret, Y. (1956c). Embryogénie des Orchidacées. Les lois du développement chez le *Goodyera repens* R. Br. *Bulletin de la Société Botanique de France,* 103, 577–81.

Veyret, Y. (1956d). Embryogénie des Orchidacées. Développement de l'embryon chez le *Neottia nidus-avis* Rich. *Comptes Rendus Hebdomadaires des Séances de l' Académie des Sciences,* 243, 1549–51.

Veyret, Y. (1956e). Embryogénie des Orchidacées. Développement de l'embryon chez le *Cephalanthera ensifolia* Rich. *Comptes Rendus Hebdomadaires des Séances de l' Académie des Sciences,* 244, 930–3.

Veyret, Y. (1957). Embryogénie des Orchidacées. Développement de l'embryon chez l'*Eulophia oedoplectron* Summ. *Comptes Rendus Hebdomadaires des Séances de l' Académie des Sciences,* 244, 2017–20.

Veyret, Y. (1958a). Embryogénie des Orchidacées. Les lois du développement chez le *Coelogyne parishii* Hook. *Comptes Rendus Hebdomadaires des Séances de l' Académie des Sciences,* 247, 656–9.

Veyret, Y. (1958b). Embryogénie des Orchidacées. Les lois du développement chez l'*Angraecum distichum* Lindl. *Bulletin de la Société Botanique de France,* 105, 5–6, 227–30.

Veyret, Y. (1961a). Embryogénie des Orchidacées. Développement de l'embryon chez le *Platycoryne paludosa* Rolfe. *Comptes Rendus Hebdomadaires des Séances de l' Académie des Sciences,* 253, 1116–18.

Veyret, Y. (1961b). Embryogénie des Orchidacées. Développement de l'embryon chez le *Masdevallia veitchiana* Reich. *Comptes Rendus Hebdomadaires des Séances de l' Académie des Sciences,* 253, 2580–2.

Veyret, Y. (1965). Embryogénie comparée et blastogénie chez les Orchidaceae-Monandrae. O.R.S.T.O.M., Paris.

Veyret, Y. (1972). Ètudes embryologiques dans le genre *Cynorkis* (Orchidaceae). *Adansonia (series 2),* 12, 389–402.

Veyret, Y. (1974). Development of the embryo and the young seedling stages of orchids. In *The orchids: scienitific studies* (ed. C. L. Withner), pp. 223–265. Wiley, New York.

Vij, S. P. and Shekhar, N. (1985). Chromosome number reports LXXXVII. *Taxon,* 34, 346–51.

Vij, S. P., Shekhar, N., and Kuthiala, R. (1981). Chromosome number reports LXXI. *Taxon,* 30, 512.

Vijayaraghavan, M. R., Shukla, A. K., and Gupta, V. (1987). Discontinuities in the microsporocyte wall and microsporogenesis in *Zeuxine strateumatica* (Lindl.) Schltr. *Journal of the Orchid Society of India,* 1, 75–80.

Vogel, E. F. de. (1969). Monograph of the tribe Apostasieae (Orchidaceae). *Blumea,* 17, 313–50.

Vogel, S. (1962). Duftdrüsen im Dienste der Bestäubung: Über Bau und Funktion der Osmophoren. *Abhandlung der Mathematische-Naturwissenschaftlichen Klasse,* 10, 598–763.

Vogel, S. (1974). *Ölblumen und Ölsammelnde Bienen.* Akademie der Wissenschaften und der Literatur, Mainz, No. 7: 281–547.

Vogel, S. (1981). Bestäubungs Konzepte der monokotylen und ihre Ausdruck im System. *Berichte der deutschen botanischen Gesellschaft,* **94,** 667–75.

Wagner, W. H., Jr. (1980). Orgin and philosophy of the Groundplan-divergence Method of cladistics. *Systematic Botany,* **5,** 173–93.

Warcup, J. H. (1985). *Rhizanthella gardneri* (Orchidaceae), its rhizoctonia endophyte and close association with *Melaleuca unicinata* (Myrtaceae) in Western Australia. *New Phytologist,* **99,** 273–80.

Webster, A. D. (1886). On the growth and fertilisation of *Cypripedium calceolus. Transactions and Proceedings of the Botanical Society Edinburgh,* **16,** 357–60.

Weltz, M. (1897). *Zur Anatomie der monandrischen sympodialen Orchideen.* Ph.D. thesis. University of Heidelberg. J. Hörning, Heidelberg.

White, J. H. (1907). On polystely in roots of Orchidaceae. *University of Toronto Studies, Biological Series,* **6,** 163–79.

Whitten, W. M., Chase, M. W., and Stern, W. L. (1996). Molecular systematics of Stanhopeinae (Orchidaceae). *American Journal of Botany,* **83,** 201[Abstract].

Wiley, E. O. (1979). An annotated Linnean hierarchy, with comments on natural taxa and competing systems. *Systematic Zoology,* **28,** 308–37.

Wiley, E. O. (1981). *Phylogenetics: the theory and practice of phylogenetic systematics.* Wiley, New York.

Williams, C. A. (1979). The leaf flavonoids of the Orchidaceae. *Phytochemistry,* **18,** 803–13.

Williams, N. H. (1979). Subsidiary cells in the Orchidaceae: their general distribution with special reference to development in the Oncidieae. *Botanical Journal of the Linnean Society,* **78,** 41–66.

Williams, N. H. and Broome, C. R. (1976). Scanning electron microscope studies of orchid pollen. *American Orchid Society Bulletin,* **45,** 699–707.

Willis, J. C. (1973). *A dictionary of the flowering plants and ferns.* 8th edn (revised H. K. Airy Shaw). Cambridge University Press.

Wimber, D. (1989). In *Odontoglossum* (ed. L. Bockemühl), pp. 319–320. Brücke-Verlag Kurt Schmersow, Hildesheim, Germany.

Wirth, M. and Withner, C. L. (1959). Embryology and development in the Orchidaceae. In *The orchids: a scientific survey* (ed. C. L. Withner), pp. 155–188. Wiley, New York.

Wit, H. C. D. de. (1977). Orchids in Rumphius' Herbarium Amboinense. In *Orchid biology: reviews and perspectives,* Vol. 1 (ed. J. Arditti), pp. 47–94. Cornell University Press, Ithaca, New York.

Wolter, M. and Schill, R. (1986). Ontogenie von Pollen, Massulae, und Pollinien bei den Orchideen. *Tropische und subtropische Pflanzenwelt,* **56,** 1–93.

Wood, J. J. and Cribb, P. J. (1994). *A checklist of the orchids of Borneo.* Royal Botanic Gardens, Kew.

Wood, J. J., Beaman, R. S., and Beaman, J. H. (1993). *The plants of Mt. Kinabalu. Vol. 2. Orchids.* Royal Botanic Gardens, Kew.

Yeung, E. C. (1987). Development of pollen tube and accessory structures in orchids. In *Orchid biology: reviews and perspectives,* Vol. 4 (ed. J. Arditti), pp. 193–226. Cornell University Press, Ithaca.

Yeung, E. C. (1989). Embryology of *Epidendrum ibaguense.* I. Ovule development. *Canadian Journal of Botany,* **67,** 2219–26.

Yeung, E. C. and Law, S. K. (1992). Embryology of *Calypso bulbosa.* II. Embryo development. *Canadian Journal of Botany,* **70,** 461–8.

Yeung, E. C. and Meinke, D. W. (1993). Embryogenesis in angiosperms: development of the suspensor. *The Plant Cell,* **5,** 1371–81.

Yukawa, T., Kurita, S., Nishida, M., and Hasebe, M. (1993). Phylogenetic implications of chloroplast DNA restriction site variation in the subtribe Dendrobiinae (Orchidaceae). *Lindleyana,* **8,** 211–21.

Yukawa, T., Cameron, K. M., and Chase, M. W. (1996). Chloroplast DNA phylogeny of subtribe Dendrobiinae (Orchidaceae): insights from a combined analysis based on *rbc*L DNA sequences and restriction site variation. *Journal of Plant Research,* **109,** 169–76.

Zavada, M. S. (1983). Comparative morphology of monocot pollen and evolutionary trends of apertures and wall structures. *Botanical Review,* **49,** 331–79.

Zavada, M. S. (1990). A contribution to the study of pollen wall ultrastructure of orchid pollinia. *Annals of the Missouri Botanical Garden,* **77,** 785–801.

Zhu, Y. C. (1989). Orchidaceae. In *Plantae Medicinales Chinae bor.-Orient,* pp. 182–185.

Ziegenspeck, H. (1928). Orchidaceae. In *Lebensgeschichte der Blütenpflanzen Mitteleuropas: spezielle Ökologie der Blütenpflanzen Deutschlands* (ed. E. O. O. Kirchner, E. Loew, and C. Schröter). E. Ulmer, Stuttgart.

Ziegenspeck, H. (1936). Orchidaceae. In *Lebesgeschichte der Blütenpflanzen Mitteleuropas* (ed. W. Walgerin and C. Schröter), Vol. 1, Pt. 4, pp. 1–840. E. Ulmer, Stuttgart.

Ziegler, B. (1981). *Mikromorphologie der Orchideensamen unter Berücksichtigung taxonomischer Aspekte.* Ph.D. thesis. Ruprecht Karls Universität, Heidelberg.

Zörnig, H. (1903). *Beiträge zur Anatomie der Coelogyninen.* Ph.D. thesis. University of Heidelberg. Wilhelm Engelmann, Leipzig.

INDEX TO SCIENTIFIC NAMES

*Page numbers in **boldface** refer to relevant illustrations of taxa. For embryo types and chromosome numbers of specific taxa, see alphabetized lists on pages 46–7 and 70-80, respectively.*

Aceras 88
Acianthinae 36
Acianthus 57
   A. macroglossa, seed **63**
Acranthae 5, 7, 9, 23
Acrotonae 5, 7, 9, 10
Aerangis clavigera **Pl. 6C**
Anacamptis 82–3, 88
Aphyllorchis queenslandica, embryo **48**
Aporostylis bifolia, embryo **49**
Apostasia 6, 19, 22, 25, 26, 27, 31, 33, 34, 37, 38, 45, 47, 57, 64, 94, 95, 96, 100-4
   A. latifolia 102
   A. nipponica 102
   A. nuda 100, 102, 104
   A. odorata 94, 100, 102
   A. parvula 102
   A. wallichii **101**, 102, 104, **Pl. 9B**
   sect. *Adactylus* 100, 102
   sect. *Apostasia* 102
Apostasiaceae 6, 10, 33, 39
Apostasieae 4, 9
Apostasioideae 5, 6, 7, 19, 26, 28, 29, 33, 37, 64, 93, 94-104
Arethuseae 4, 5, 6, 20, 36, 57
Arthrochilus dockrillii, embryo **49**
Aspasia principissa, column **Pl. 7A–D**

Basitonae 5, 7, 8, 10
Brassavola nodosa, capsule **Pl. 8C**
Bulbophyllinae 25, 55
Bulbophyllum
   B. lasiochilum **Pl. 3B**
   B. lepidum **Pl. 5C**

Caladenia
   C. flava **Pl. 1D**
   C. gemmata 57

Caladeniinae 34
Calopogon tuberosus var. *simpsonii*, chromosomes **71**
Calypso 55
Catasetinae 20, 87
Catasetum 20
   pollinarium **Pl. 8A**
Cattleya 82
   C. amethystoglossa 18, 20, **Pl. 6A**
Cephalanthera 45
   C. erecta, seed **62**
Chloraea alpina, endothecial thickenings **27**
Chloraeinae 34, 81
Cochlioda 82, 83, 84
Coeloglossum 88
Coelogyne
   C. fragrans, chromosomes **72**
   C. pandurata **Pl. 4C**
Coelogyninae 25
Convolutae 7
Corallorhiza striata, endothecial thickenings **27**
Coryciinae 27, 34
Corymborkis veratrifolia **Pl. 3D**
Cranichideae 19, 23, 34, 61, 64, 66
Cranichis ciliata, spiranthosomes **28**
Cycnoches 20
Cymbidium 54
   C. canaliculatum, embryo **54**
   C. lowianum 18, 20, **Pl. 6B**
   C. rectum **Pl. 4A**
Cypripediaceae 6, 10, 39
Cypripedieae 4, 5, 6, 9
Cypripedioideae 5, 9, 19, 23, 25, 26, 27, 29, 33, 34, 37, 39, 57, 61, 64, 81, 87, 93, 105-64

Cypripedium 25, 34, 106, 108, 110, 111, 112, 114–32
   C. acaule 116, 124, 125, 126, 127, **Pl. 11C**
   C. arietinum 115, 116, 124, 126, **Pl. 11A**
   C. bardolphianum, leaf transection **118**
   C. calceolus 105, 114, 115, 124, 125, 126, 127, 130, 131, **Pl. 10B**
   C. californicum 124
   C. candidum 124, 125, 126, 127
   C. cordigerum 44
   C. debile 116, 125, 127, **Pl. 12A**
   leaf transection **118**
   C. dickinsonianum 127
   C. elegans 124
   C. fasciculatum 116, 123
   C. flavum 125, **Pl. 10A**
   C. formosanum 124, 128, 130, 131, **Pl. 11B**
   C. forrestii 125
   C. guttatum 116, 127, **Pl. 11D**
   C. henryi 125, 128
   C. himalaicum 124
   C. irapeanum 27, 115, 123, 128, 134, 135, **Pl. 9C**
   C. japonicum 116
   C. kentuckiense 124, 131
   C. lichiangense 124, 125, 127
   trichome, leaf **118**
   C. macranthos 115, **116**, 124, 125, 127, 128, 130, 131
   C. margaritaceum 22, 116, 123, 124, 125, 127, **Pl. 12B**
   C. molle 124
   C. montanum 127
   C. parviflorum
      var. *parviflorum* 124, 127, 129, 131

      var. *pubescens* 125, 126, 127, 128, 129, 131, **Pl. 10C**
   C. passerinum 127
   C. reginae 115, 124, 125, 126, 127, 131, **Pl. 9D**
   C. shanxiense 124, 125
   C. subtropicum 115
   C. tibeticum 124, **Pl. 10D**
   C. ventricosum 130
   sect. *Acaulia* 116
   sect. *Bifolia* 116, 119
   sect. *Criosanthes* 116
   sect. *Cypripedium* 115, 119
   sect. *Enantiopedilum* 116
   sect. *Flabellinervia* 116, 119
   sect. *Irapeana* 115
   sect. *Obtusipetala* 115
   sect. *Retinervia* 116, 119
   sect. *Subtropica* 115
   sect. *Trigonopedilum* 116

Dactylorhiza 83, 88
Dendrobiinae 55
Dendrobium 55
   D. speciosum **Pl. 1C**
Diandrae 3, 4, 5, 7, 8, 81
Diceratosteleae 28
Dimorphorchis rossii **Pl. 8B**
Diplandrorchis 21
Disa 34, 65
   D. filicornis, endothecial thickenings **27**
   D. rungweensis, seed **61**
   D. uniflora **Pl. 2A**
Diseae 19, 24, 28, 34, 61, 64, 88
Disinae 27, 34
Diurideae 6, 10, 19, 24, 28, 34, 36, 38, 49–51, 57, 61, 66, 81
Dryadella edwallii, seed **64**
Duckeella 65
Duplicatae 7

Epidendranthe 6, 10
Epidendreae 4, 5, 6, 9, 10,
    20, 23, 54
Epidendroideae 5, 6, 7, 10,
    19, 20, 25, 26, 28, 29,
    33, 37, 61, 65–6, 81, 93
Epipactis 21, 45
    E. palustris 44
Epipogieae 23
Epipogium 54–5
Epistephium 4, 22
    E. laxiflorum, seed **65**
Erythrodeae 19

Gastrodia sesamoides, embryo
    48
Gastrodieae 19, 29, 36
Genoplesium apostasioides,
    embryo **50**
Genyorchis 20
Glomeria flammula **Pl. 5B**
Gomphichis valida, endothecial
    thickenings **27**
Goodyerinae 52, 61
Gymnadenia 88

Habenariinae 28, 51
Heteroblastae 7
Homoblastae 7

Isotria 65

Kerosphaereae 5, 9, 10, 23

Laelia 82
Laeliinae 19, 28, 87
Lecanorchis 34, 35, 36, 65
Leochilus oncidioides, pollen
    36
Liparis 54
Listera ovata, embryo **53**
Lockhartia 82
    L. oerstedii **Pl. 5A**
Lyperanthus 57

Macodes sanderianum **Pl. 4B**
Malaxeae (Malaxideae) 5, 6
Masdevallia pelecaniceps,
    pollen **35**
Maxillarieae 63
Mexipedium 106, 112, 114,
    161–4
    M. xerophyticum 109, 112,
        114, 161, **162**,
        163–4, **Pl. 16D**
    chromosomes **76**
Microtis 57
Miltonia 82, 83
Monandrae 3, 4, 5, 7, 8, 81
Monopodiales 5, 7, 9, 23

Neotinea 82–3, 88
Neottianthae 6, 10
Neottieae 4, 5, 6, 9, 11, 20,
    23, 34, 35, 37, 52–3, 57
Neottioideae 5, 10, 11, 33, 37
Nervilia 54
Neuwiedia 6, 19, 21, 25, 26,
    27, 31, 33, 34, 37, 38,
    40, 48–9, 57, 64, 94,
    95, 96–100
    N. annamensis 98
    N. borneensis 98, 100
    N. inae 98
    N. siamensis 98
    N. veratrifolia **97**, 100,
        Pl. 9A
    embryo **49**
    N. zollingeri var.
        singapureana 100
Nigritella 88

Odontoglossum 82, 83
Oncidiinae 66, 82, 87, 88
Oncidium 82, 83, 84, 88
    O. heteranthum 20
    O. maculatum, seed **62**
    O. reichenheimii, pollen **36**
    O. splendidum **Pl. 4D**
Ophidion cymbula, helical
    thickenings, leaf **29**
Ophryeae (Ophrydeae) 5, 9,
    23
Ophrys 20
    O. apifera 21
    protocorms **Pl. 8D**
Orchidaceae 6, 10, 39, 56,
    91–3
Orchideae 4, 5, 6, 9, 10, 19,
    24, 28, 29, 51, 61, 64
Orchidinae 28, 51
Orchidoideae 5, 6, 7, 9, 10,
    19, 20, 23, 25, 26, 27,
    31, 33, 37, 57, 65, 66,
    68, 81, 93
Orchis 20, 33, 82, 88
Ornithocephalinae 82

Palmorchis 59, 65
Paphiopedilum 25, 27, 34, 40,
    47, 106, 109–10, 111,
    112, 114, 137–53, 163,
    164
    P. adductum 152
    P. appletonianum 140
    P. armeniacum 145, 146,
        149, 151, **Pl. 13A**
    P. barbatum 141, 148
    P. barbigerum 138, 145,
        146, 152
    P. bellatulum 145, **Pl. 13C**

P. bougainvilleanum 140,
    152
P. bullenianum 140, 147
P. callosum 140, 147
P. charlesworthii 152
P. ciliolare 152
P. concolor 140, 145
P. dayanum 140, 147
P. delenatii 140, 146, 149
P. dianthum 138, 145
P. druryi 137, 140, 141,
    142, 143, 147
P. emersonii 138, 145, 146
P. exul 147
P. fairrieanum 141, 142,
    145, 146, 152
P. gigantifolium 146
P. glanduliferum 140, 141,
    145, 146
P. glaucophyllum 146
P. godefroyae **Pl. 5D**
    column of **Pl. 6D**
P. haynaldianum 140, 143
P. helenae 146, 152
P. hennisianum 148
P. henryanum 152
P. hirsutissimum 140, 141,
    145
    var. *esquirolei* 138, 145,
        146
P. hookerae 142
    var. *volonteanum* 145,
        147, 148
P. insigne 137, 141, 142,
    148, 152, **Pl. 14A**
P. javanicum 138, 140,
    143, 147
    var. *virens* 147, 148
P. liemianum **Pl. 13D**
P. lowii 140, 141, 142,
    145, 147, **Pl. 14D**
P. malipoense 138, **138**,
    145, 146
P. mastersianum 140
P. micranthum 138, 145,
    146, 149, 151, **Pl.
    13B**
P. niveum 142, 145
P. papuanum 140, 152
P. parishii 140, 141, 145
P. philippinense 140, 145,
    147, 148, 150
P. purpuratum 138, 140,
    147, 148
P. randsii 152
P. rothschildianum 140,
    141, 146, 147, 148,
    149, **Pl. 14B**
P. sanderianum 145, 146,
    Pl. 14C

P. spicerianum 141, 142,
    146
P. stonei 140, 141, 145,
    146, 150
P. sukhakulii 141
P. supardii 146
P. superbiens 147
P. tigrinum 145
P. tonsum 141
P. venustum 141, 142
P. victoria-regina 141
P. villosum 140, 145, 147,
    148, 152
P. violascens 140, 152,
    Pl. 15A
P. wentworthianum 140,
    147, 152
sect. *Barbata* 141, 142,
    144, 149, 152
sect. *Cochlopetalum* 140,
    142, 144
sect. *Coryopedilum* 140,
    142, 144, 145, 151,
    152
sect. *Paphiopedilum* 140,
    142, 143, 144, 145,
    151, 152
sect. *Pardalopetalum* 140,
    142, 144, 145
subg. *Brachypetalum* 140,
    142, 143, 144, 145,
    149, 150, 151
subg. *Paphiopedilum* 140,
    144, 145, 152
subg. *Parvisepalum* 140,
    142, 143, 144, 149,
    151
Phalaenopsis, root **Pl. 2D**
Phragmipedium 26, 27, 34,
    40, 106, 110, 111, 112,
    114, 153–61, 163
    P. besseae 157, 159, 160,
        Pl. 15B
    P. boissierianum 156, 157,
        159, 160
    P. caricinum 156, 157, 159,
        160
    P. caudatum 153, 155, 156,
        157, 159, 160
    P. czerwiakowianum 157,
        160
    P. dalessandroi 160
    P. ecuadorense 157
    P. exstaminodium 159
    P. hirtzii 159
    P. kaiteurum 157
    P. klotzchianum 157, 159
    P. lindenii 153, 155, 156,
        157, 160, **Pl. 16C**
    P. lindleyanum 157, **Pl. 15C**

P. longifolium **154**, 155, 156, 159, **Pl. 15D**
embryo **48**
P. pearcei 156, 157, 159, 160, **Pl. 16A**
P. richteri 160
P. sargentianum 156, 157, 159
P. schlimii 156, 157, 160
P. × sedenii 156
P. vittatum 159
P. wallisii 157, **Pl. 16B**
P. warscewiczianum 157
sect. *Lorifolia* 155, 156, 157
sect. *Micropetalum* 155, 156, 157
sect. *Phragmipedium* 155, 156, 157
sect. *Platypetalum* 155, 156, 157
Platythelys querceticola, pollen **35**
Pleonandrae 7, 8
Pleuranthae 5, 7, 9, 23
Pleurothallidinae 19, 25, 28, 88
Pleurothallis 55, 88
P. circumplexa, leaf idioblasts **29**

P. mystax, pollen **35**
Podochileae 55
Pogonia 65
Polychondreae 5, 8, 9, 10, 23
Prasophyllinae 20
Prasophyllum parvifolium, tubers **24**
Prescottinae 65
Prosthechea (=Encyclia) cochleata **Pl. 1B**
Pterostylis 57
P. × ingens, embryo **51**

Renanthera imschootiana, velamen **30**
Restrepiella ophiocephala pollen **35**
root transection **30**
Rhizanthella 22
R. gardneri **Pl. 2C**
Robiquetia mooreana, chromosomes **79**
Rodriguezia venusta, pollen **36**
Selenipedium 22, 27, 34, 59, 64, 105, 106, 108, 110, 112, 114, 118, 120, 123, 132–7
S. aequinoctiale 136

S. chica **133**, 134, 135, 136, **Pl. 12C,D**
S. palmifolium 134, 135
S. steyermarkii 136
Sinorchis 21
Sobralia decora, root sections **31**
Spiranthes
S. australis, embryo **52**
S. vernalis, seed **63**
Spiranthinae 8, 23, 51–2, 64
Spiranthoideae 6, 7, 10, 11, 19, 25, 26, 27, 28, 37, 57, 66, 81
Stanhopea 26
Stanhopeinae 87
Stenoglottis fimbriata, embryo **51**
Sunipia 20
Sympodiales 5, 7, 9, 23

Taeniophyllum trachypus **Pl. 3A**
Tangtsinia 21
Telipogoninae 82
Thelymitroideae 7, 12
Thraurosphaereae 5, 9
Tipularia discolor, seed **64**

Tolumnia variegata, seed **64**
Townsonia 50
Traunsteinera 83
Trichocentrum pfavii, seed **63**
Trichoglottis smithii **Pl. 3C**
Triphoreae 36
Tropidia 66
Tropidieae 20, 28, 57
Tropidioideae 7, 12

Vandeae 5, 6, 10, 23, 37, 55–6, 57
Vandoideae 6, 7, 10, 11, 31, 33
Vanilla 22, 53, 59, 64, 91
V. planifolia, embryo **53**
V. pompona **Pl. 2B**
V. sp., seed **65**
Vanillaceae 4
Vanilleae 11, 19, 57
Vanilloideae 7, 12, 22, 35, 36, 53, 61, 64, 81, 93

Zeuxine 52
Z. elongata, seed **62**
Zygopetalum intermedium var. *peruvianum*, velamen **30**

SUBJECT INDEX

Page numbers in boldface refer to relevant illustrations.

anatomy 24–32; *see also* root;
    stems; leaves; floral
  of *Apostasia* 27, 95
  of *Apostasioideae* 95
  of *Cypripedium* 27, 118–19,
    118
  of Diseae 27–8
  of Diurideae 28
  of Laeliinae 28
  of *Mexipedium* 163
  of *Neuwiedia* 27, 95
  of Orchideae 28
  of *Paphiopedilum* 27, 141–2
  of *Phragmipedium* 27,
    155–6
  of Pleurothallidinae 28, **29**
  of *Selenipedium* 27, 134–5
  of Spiranthoideae 27
  vegetative 24, 26–32
aneuploidy 69
anthers 19
  acrotonic 19
  basitonic 19
  development 19
  endothecium 26
    thickening types 26, **27**
autogamy 21

Bentham, G. 4, 6–7
Blume, C. L. von 3
Brieger, F. G. 10
Brown, R. 3, 24
Burns-Balogh, P. 6, 11

capsules 21–2, **Pl. 8C**
caudicles 20
chromosomes **71, 72, 76, 79**;
    see also cytogenetics
  B-chromosomes 69
  morphology 68
  number 68–80, 98, 102,
    108–12, 142–3, 163
cladistics 84–8; *see also*
    molecular systematics;
    phylogenetics
cladograms 85–6
classification, infrageneric
  Apostasia 102

Cypripedium 115–16
Paphiopedilum 140–1, 144–5
Phragmipedium 155, 158
column 19, **Pl. 6D, 7A–D**
corms 15
cultivation
  of *Apostasia* 104
  of *Cypripedium* 127–31
  of *Mexipedium* 164
  of *Neuwiedia* 100
  of *Paphiopedilum* 149–52
  of *Phragmipedium* 160
  of *Selenipedium* 136
cymes 17, **Pl. 5A**
cytogenetics 67–80
  of *Apostasia* 102
  of Cypripedioideae 106,
    108–12
  of *Cypripedium* 108, 111,
    112, 115, 119
  of *Mexipedium* **76**, 109,
    112, 163
  of *Neuwiedia* 98
  of *Paphiopedilum* 109, 110,
    111, 112, 142–3
  of *Phragmipedium* 108,
    111, 112, 156–7
  of *Selenipedium* 108, 135
  techniques 67–8

derivation of name
  Apostasia 100
  Cypripedium 115
  Mexipedium 161
  Neuwiedia 96
  Orchidaceae 91
  Paphiopedilum 137
  Phragmipedium 153
  Selenipedium 132
description
  of *Apostasia* 100, **101**
  of Apostasioideae 94
  of Cypripedioideae 105
  of *Cypripedium* 115, **116**
  of *Mexipedium* 161, **162**,
    163
  of *Neuwiedia* 96, **97**, 98
  of Orchidaceae 91

of *Paphiopedilum* 137, **138**
of *Phragmipedium* 153,
  154, 155
of *Selenipedium* 132, **133**,
  134
Dioscorides 3
distribution
  of *Apostasia* 102, **103**
  of Apostasioideae 94
  of Cypripedioideae 106, **107**
  of *Cypripedium* 115, **117**,
    131
  of *Mexipedium* **131**
  of *Neuwiedia* 98, **99**
  of Orchidaceae 91, **92**
  of *Paphiopedilum* **131**,
    137–8, **139**, 140
  of *Phragmipedium* **131**,
    155, **155**
  of *Selenipedium* **131**, 134,
    134
Dodson, C. H. 5, 9
Dressler, R. L. **ii**, 5, 6, 7, 9,
    10, 11, 13
dysploidy 69

ecology
  of *Apostasia* 102
  of *Cypripedium* 124–5
  of *Mexipedium* 164
  of *Neuwiedia* 98, 100
  of *Paphiopedilum* 144–7
  of *Phragmipedium* 159
  of *Selenipedium* 135–6
elaiophores 26
elastoviscin 19
embryo (sac) 39–58
  endosperm 44–5
  suspensor 42–4
  types 39–58, **48–54**
embryogenesis 38–58
  megagametogensis 39–40
  megasporogenesis 39
  microgametogenesis 38–9
  microsporogenesis 38–9
  phylogenetic significance
    56–8
endodermis 24

endothecium 26
  thickening types 26, **27**
exine, *see* pollen
exodermis 14, 24

floral anatomy 25–6
  elaiophores 26
  osmophores 26
  vasculature 26
flowers 18–21, **Pl. 6A–B**
  anthers 19
  caudicles 20
  column 19, **Pl. 6D, 7A–D**
  dimorphic 20–1, **Pl. 8B**
  hamulus 20
  labellum 18
  massulae 19
  nonresupinate 18
  ovary 18
    placentation 18, **Pl. 5D**
  petals 18
  pollen 19
  pollinarium (-a) 19,
    Pl. 7B–D, 8A
  pollinium (-a) 19, **Pl.**
    7B–D, 8A
  resupinate 18
  rostellum 20
  sepals 18
  spur 18–19, **Pl. 6C**
  stigma 20
  stipes 20, **Pl. 7B–D**
  tegula 20
  teratological 21
  viscidium 20, **Pl. 7B–D,**
    8A
fruits, *see* capsules
Funk, V. 6, 11

Garay, L. A. 5, 9, 10
growth
  monopodial 15, **Pl. 3C**
  sympodial 15, **Pl. 3B**
gynostemium, *see* column

Hennig, W. 10

idioblasts 28, **29**

inflorescences 17–21, **Pl. 4C–D, 5A–C**

Jussieu, A. L. de 3

key, artificial
　　to genera of Apostasioideae 94
　　to genera of Cypripedioideae 106
　　to subfamilies of Orchidaceae 93
Kränzlin, F. 8

labellum 18
　　spur 18–19
leaves 16–17
　　anatomy 26–9, 31, 95, 118–19, **118**, 134–5, 141–2, 155–6
　　arrangement 16
　　conduplicate 16, **Pl. 4A**
　　indumentum 16
　　plicate 16, **Pl. 3D**
　　shape 16
　　sheaths 17
　　size 16
　　terete 16
　　tessellations 16
　　venation 16
　　vernation 16
Leitgeb, H. 25
Lindley, J. 4–6
Link, H. F. 24
Linnaeus, C. von 3
lip, *see* labellum

Mansfeld, R. 5, 9
massulae 19, 39
mentum 18
　　spur 18–19
Möbius, M. 25
molecular systematics 81–8; *see also* phylogenetics; cladistics
　　of Apostasioideae 96
　　of Cypripedioideae 87, 112, 113
　　of *Cypripedium* 120, 123
　　DNA sequences 81, 83–8
　　of Oncidiinae 82, 83, 88
　　of Orchidinae 82–3, 88
　　and phylogenetics 83–4
　　of Pleurothallidinae 88
morphology 13–23

orchid, *see also* orchids; flowers
　　anatomy 24–32
　　　floral 25–6
　　　vegetative 24, 26–32

classification (*see also* phylogenetics)
　　history of 3–12, 81
flowers 18–21
fruits 21–22, **Pl. 8C**
inflorescence types 17–20, **Pl. 4C–D, 5A–C**
leaves 16–17
morphology 13–23
protocorms 22–3, **Pl. 8D**
roots 14–15
　　aerial 14, **Pl. 2D**
　　basket 14
　　function of 14–15
　　tubers 14–15
seedlings 22–3
seeds 22, **Pl. 8C**
stems 15
orchids
　　achlorophyllous 14
　　epiphytic 13, **Pl. 1A–B**
　　habit of 13–14
　　'jewel' 7, **Pl. 4B**
　　leafless 15, **Pl. 3A**
　　leaves of 16–17
　　lithophytic 13, **Pl. 1C**
　　monopodial 15
　　photosynthetic 15
　　saprophytic 14, 16, **Pl. 2C**
　　size of 14
　　sympodial 15
　　terrestrial 13, **Pl. 1D, 2A**
　　vining 13–14, **Pl. 2B**
osmophores 26
ovary 18, **Pl. 5D**

palynology 33–7; *see also* pollen
panicles 17, **Pl. 4D**
parsimony 84–6
peduncle 17
Pfitzer, E. 4, 7–8, 25
phylogenetics 56–8, 81–8
　　of *Apostasia* 96
　　of Apostasioideae 96
　　of Cypripedioideae 110–12, 114, **114**
　　of *Cypripedium* 110, 114, **114**, 120, 123, **123**
　　of *Mexipedium* 112, 114, **114**, 163–4
　　of *Neuwiedia* 96
　　of *Paphiopedilum* 112, 114, **114**, 143–4, **143**, 144, 145
　　of *Phragmipedium* 112, 114, **114**, 157, **157**, 159
　　of *Selenipedium* 110, 114, **114**

phytochemistry
　　of *Apostasia* 102
　　of *Cypripedium* 120
　　of *Mexipedium* 163
　　of *Neuwiedia* 98
　　of *Paphiopedilum* 143
　　of *Phragmipedium* 157
　　of *Selenipedium* 135
pollen 19, 33–7, **35, 36**, 38–9
　　aperture 37
　　exine 19, 33–7
　　　evolution of 36–7
　　　sculpture 33–4
　　　structure 34
　　of *Apostasia* 33–4, 38, 95–6
　　of Apostasioideae 95–6
　　of Arethuseae 35–6
　　of Caladeniinae 34
　　of Chloraeinae 34
　　of *Cypripedium* 34, 119
　　of Diseae 34
　　of Gastrodieae 35–6
　　of *Lecanorchis* 34–5
　　of *Mexipedium* 163
　　of Neottieae 34–5
　　of *Neuwiedia* 33–4, 38, 95–6
　　of *Paphiopedilum* 34, 142
　　of *Phragmipedium* 34, 156
　　of *Selenipedium* 34, 135
　　of Triphoreae 35–6
　　of Vanilloideae 36
　　unit 37
pollinarium (-a) 19, **Pl. 7B–D, 8A**
　　caudicles 20
　　hamulus 20
　　stipes 20, **Pl. 7B–D, 8A**
　　tegula 20
　　viscidium (-a) 20, **Pl. 7B–D, 8A**
pollination
　　of *Apostasia* 102, 104
　　of *Cypripedium* 125–7
　　of *Mexipedium* 164
　　of *Neuwiedia* 100
　　of *Paphiopedilum* 148–9
　　of *Phragmipedium* 159–160
　　of *Selenipedium* 136
pollinium (-a) 19, **Pl. 7B–D, 8A**
　　number 19–20
polyploidy 69
propagation, *see* cultivation
protocorm 22–23, **Pl. 8D**
　　morphology 22–3
　　and classification 23
　　size 22
pseudobulbs 15

racemes 17, **Pl. 4C**
Reichenbach, H. G. 6, 8
Reichenbach, H. G. L. 8
Rheede tot Draakestein, H. A. von 3
rhizoids 22
rhizomes 15
　　anatomy 119, 135, 142, 156
Richard, L. C. 3
Rolfe, R. A. 8
roots 14–15
　　anatomy 24, 25, 26–32, **30, 31**, 95, 119, 135, 142, 156
　　types, taxonomic 29
rostellum 20
Rumphius, G. E. 3

Schlechter, R. 5, 8–9
seed 22, **60–5, Pl. 8C**
　　coat 22
　　　characters 59–66
　　　morphology 22, 59–66
　　　size 22
seedlings 22–3
spikes 17
spiral thickenings 28, **29**
spiranthosomes **28**
spur 18–19
stems 15
　　anatomy 26–8, 31, 95, 119, 135, 142, 156
　　corms 15
　　pseudobulbs 15
　　rhizomes 15
stigma 20
　　rostellum 20
　　types 25–6
stipes 20, **Pl. 7B–D**
stomata 28–9
subsidiary cells 28–9
Swartz, O. 3, 4
Szlachetko, D. L. 7, 11–12

taxonomic literature
　　for *Apostasia* 104
　　for *Cypripedium* 131–2
　　for *Mexipedium* 164
　　for *Neuwiedia* 100
　　for *Paphiopedilum* 152–3
　　for *Phragmipedium* 160–1
　　for *Selenipedium* 136–7
taxonomy
　　of *Apostasia* 100, 102
　　of Apostasioideae 94, 96
　　of Cypripedioideae 105, 106, 110–14
　　of *Cypripedium* 114–16, 120, 121, 122, 123

of *Mexipedium*   161, 163–4
of *Neuwiedia*   96, 98
of *Paphiopedilum*   137,
      140–1, 143–4, 143,
      144, 145
of *Phragmipedium*   153,
      155, 157–9

of *Selenipedium*   132, 136
Thouars, A. A. du Petit   3
tilosomes   31, **31**
tubers   14–15, 24, **24**

uses
      of *Apostasia*   104

of *Cypripedium*   127
of *Neuwiedia*   100
of Orchidaceae   91
of *Paphiopedilum*   149
of *Phragmipedium*   160
of *Selenipedium*   136

vegetative anatomy   24, 26–32
velamen   14, 24–5, 29–31, **30,**
      31
      functions of   14
Vermeulen, P.   , 10
viscidium (-a)   19, **Pl. 7B–D**
Vogel, S.   26

DUE